Evolution of Bioenergetics from Elements to Life

David F. Stowe

Evolution of Bioenergetics from Elements to Life

Emergence of High Energy Mitochondria

Springer

David F. Stowe
Departments of Anesthesiology Research, Physiology
and Biomedical Engineering
Medical College of Wisconsin
Milwaukee, WI, USA

ISBN 978-3-031-76667-1 ISBN 978-3-031-76668-8 (eBook)
https://doi.org/10.1007/978-3-031-76668-8

This Springer imprint is published by the registered company Springer Nature Switzerland AG
The registered company address is: Gewerbestrasse 11, 6330 Cham, Switzerland

Acknowledgments

This book is dedicated to all the scientists and students I have had the pleasure to work with over many years, for the most part at the Medical College of Wisconsin. There are tough times and good times in research. But the eureka moments bring out the best in us. I am especially grateful for my interactions over many years with all my colleagues at MCW, including three research-supportive chairs of the Department of Anesthesiology, Drs. John P. Kampine, David C. Warltier, and Cindy Lien. The decades long generosity of the Research Service, Veterans Administration, has been absolutely essential for my research endeavors. I offer my heartfelt thanks to my long-term Laboratory Manager and Senior Technologist, James S. Heisner, who has not only acquired the vast technical skills required for bench research but also taken on a huge role in educating and training fellows and students in the use and function of research equipment. I thank my collaborators in research who make research fun and highly rewarding. These include my close colleagues for decades: Drs. Amadou K.S. Camara and Wai-Meng Kwok. Many other collaborators over time include faculty at MCW and other institutions: Drs. M.L. Riess, S.H. Audi, D.H. Harder, M.T. Jiang, T.E. Ebert, P. Pagel, J.P. Kampine, J.T. Ells, R.K. Dash, D.A. Beard. J.J. Smith, M. Bienengraeber, G.F. Gross, D.L. Roerig, K. Madsen, L.A.Turner, B.W. Palmisano, A. Geurts, N. Hogg, S.C. Smart, D. Lathrop, Z. Iqbal, J.S. Bernstein, H.J. Woehick, W.E. Antholine, M.W. Lawlor, J.C. Hong, J. Kim, K.M. Ropella, S. deHert, E.J. Lesnefsky, C.L. Hopple, S.A. Glantz, K. Chatterjee, W.W. Parmley, M.S. Jahania, and P.M. Heerdt. Funding is impossible without our present and former Research Associates: J. Mishra, B.T. Wakim, M.Y. Yang, G.K. Natarajan, M. Aldakkak, and A. Gadicheria. Post-doctoral trainees making discoveries include: J. Haumann, A.D. Bolens, C.A. Blomeyer, B. Agarwal, J.N. Bazil, J. Sprung, J. Marijic, L.G. Kevin, Q. Chen, A. Heinen, S.S. Rhodes, J.Z. An, R. M. Jordan, E. Martin, B. Fischer, W.C. O'Brien, S.G. Tewari, A. Abdoli, L.A. Gallenberg, R.T. Rooney, N. Buljubasic, M. Boban, M.B. Graf, H. Habazettl, E. Novalija, S. Fujita, and S.G. Varadarajan. I am also highly fortunate to date to have the opportunity to take part in the research training of over 50 students that are always so eager to learn about science and thrilled to undertake original studies. I must include gratitude to my Ph.D. mentors Drs. Jerry B. Scott, Joseph Dabney, and Francis

J. Haddy and post-doctoral mentor Dr. John V. Tyberg. I also thank the following scientists for reading and critiquing the draft manuscripts of this book: Professor Amadou K.S. Camara, Ph.D., Medical College of Wisconsin; Professor Cameron B. Gundersen, Ph.D., University of California, Los Angeles; and Professor David R. Harder, Ph.D., University of Maine.

David F. Stowe
January 2025

Introduction

The long-term interest in knowing "if we are alone" in the Universe and how we came to be is beguiling to humankind. Our curiosity to know about our Earth and all the single-cell species, animals, and plants is so strong that many thousands of biological scientists, geologists, cosmologists, astrophysicists, mathematicians, and philosophers work every day to enhance the physical and spiritual cognition of our Universe in its past, present, and future. While our knowledge gained about the cosmos, our solar system, and Earth is growing exponentially, the answers attained only beg for new answers to many more complex questions.

Life would not exist if early molecules were not able to harness the photon energy of the sun, heat, and nascent energetic chemical bonds to form new macromolecules as building blocks, which could perform photosynthesis and respiration to make higher energy molecules that catalyzed even more sophisticated macromolecules. This is bioenergetics, a branch of biochemistry. This book aims to trace its origins and to describe its function in energy transduction using modern day cardiac mitochondria as the cell's energy organelle.

It is becoming increasingly likely that life, as we know it, is abundant in the Universe. Supporting this idea is the longevity and changes in the Universe that we have observed. Another is that all the "star stuff" is the same throughout the cosmos as we know it. Indeed, there are differences in the abundance of ions, elements, and molecules, but they are everywhere, and the most abundant nuclei at the beginning, hydrogen and helium, remain today as progenitors of all the other elements.

As the elements were created in increasing atomic mass by the forces and energy of the beginning Universe, their atoms had increasingly different properties that allowed different binding possibilities with different bonding strengths. That meant that some atoms bind weaker or stronger with others as a universal feature of bonding chemistry. The molecules that resulted had properties different than their individual atoms. Using the same elements, lifeforms on other worlds might form useful compounds that are toxic to us, e.g., carbon tetrachloride, organophosphates, and heavy metals, while molecules and compounds utilized on Earth, e.g., oxygen molecules and nitrogen compounds, and lighter alkali and transition metals, might be toxic to extra-terrestrial life.

The premise of this book is to first build on information of the genesis of the elements, each with distinct numbers of protons, neutrons, and electrons, to form diverse molecules and to explain how certain atoms became useful in forming unique bonds and chains with other atoms that retained energy that could be released to make new compounds. This book is also about evolutionary biochemistry. Although we don't know yet if all these reactions could have occurred spontaneously on Earth with the materials at hand, or if some elements and compounds needed to arrive on the protoplanet Earth via its accretion disk (Fig. 1.1) [1] or later after Earth's formation from meteorites or comets, in some way the pre-life molecules were able to form semipermeable, membrane vesicles where confined atoms and molecules could attract or repel each other.

The one key molecule that cosmologists look for in the development of lifeforms is liquid water. Its unique property to allow mobile ions to move around H_2O molecules and to form a medium for other ions and molecules to exist is truly a unique property of hydrogen (H) bonding with oxygen (O). H as a cation (proton), H bonding with carbon (C), and O bonding with phosphorus (P) are essential for life; these atoms plus nitrogen (N), and sulfur (S), make up the abundance and backbone of the most important elements of lifeforms, i.e., H, C, O, N, P, S.

This book is divided into 3 parts, 9 chapters, and 35 sections. A given section has frequent references to other sections to better integrate a given topic; e.g., what is the role of magnesium in bioenergetics? Why, how, where, and when does it contribute to sustaining life? Part I presents the reader with the early nucleosynthesis of the lighter elements and their properties as atoms reformed into molecules and compounds that would become the building blocks of all future lifeforms. That began about 13.8 billion years ago (Bya) with the "Big Bang." Next, it discusses the role of heavier elements as cofactors with atoms having several possible valences (equal number of protons and electrons but unbalanced electron orbits), which make them versatile for forming different bonding configurations. An atom's bonding strength (electronegativity) and characteristics, e.g., whether polar, nonpolar, ionic, or covalent, are key to understanding how more complex molecules were created. Next, a description is given about how the early fatty acids, sugars, amino acids, and nucleobases were formed and then condensed and polymerized to form lipid, peptide, saccharide, and nucleotide chains.

Attention is given to the importance of protons, water, phosphate, and other organic factors to bind molecules of each of these classes together to make compounds such as nicotinamide adenine dinucleotide (NADH), ADP, and RNA that are some of the essential ingredients, with energy from our Sun, for life to develop. This prebiotic period ended about 3.7–3.9 Bya when the first protokaryotes appeared. This is evidenced by the advent of high ratios of C^{12} relative to its isotope C^{13} found in areas where life first formed. The last universal common ancestor (LUCA) is based theoretically on a common genetic code and amino acid chirality for all lifeforms to this day.

Part II jumps ahead to about 1.5 Bya to present how it was possible that a facultative aerobe was taken up by another microbe to be phagocytosed, but escaped into the host cell cytoplasm and began to participate with the host's metabolism. Over time this symbiont evolved selectively into an organelle, the mitochondrion. Other symbionts, like the nucleus, led to Eukaryotes with a rather fast and large expansion and diversity of species. These evolutionary events are followed by a detailed examination of a modern bioenergetic product of evolution, the high-energy mitochondrial machinery of cardiac myocytes.

Since mitochondria first came around during evolution, they have become much more complicated and are likely more efficient with systems that interact with other parts of the cell to regulate metabolism, repair, and cell demise. Muscle and nerve cells have a very high energy demand for locomotion and for information transfer. Thus, cardiac mitochondrial function is examined in detail. Other topics in Part II explain the roles of mitochondrial-induced release of reactive oxygen species, sequestration of Ca^{2+} from the ionized pool, mitochondrial ion channels and exchangers, and examples from experimental studies on how to evaluate mitochondrial function in living cells and in isolated mitochondria.

In Part III, the reader is first presented with a brief review of the origin of the heart and circulation in triploblastic animals, particularly, the mammalian heart because of its greater energy output. Next, our current understanding of the work of the mitochondrion in cardiac myocytes to provide the ever-changing energy requirements for cardiac muscle contraction and relaxation is discussed. This part includes action potential conduction to drive the heart as a pump, the pivotal role of Ca^{2+} in contraction/relaxation of the myofilament apparatus, mitochondrial buffering of Ca^{2+}, and protective versus anti-protective roles of reactive oxygen species.

The compilation of this book into these parts was inevitable because the author is primarily a cardiovascular physiologist with a keen interest in mitochondrial bioenergetics; thus, the material is resourced and written in terms of mammalian heart physiology and biochemistry with an emphasis on cardiac mitochondria as the essential organelle of bioenergetics. Derived from an ancient bacterium, mitochondria provide the abundant, readily available energy required for the work of the heart (and skeletal muscle) in vertebrates (and birds) to hunt for food and to escape prey or other danger.

The later sections include some relevant publications from the author's research group on how to study cardiac mitochondria, as well as how these mitochondria are altered and respond to cellular stress. But much of the discussion could be applied to mitochondria present in other tissues, as well as in other organisms including non-mammalian lifeforms. An apology to plant biologists that the genesis of the kingdom Plantae was not included and photosynthesis as an essential bioenergetic process is given short shrift. But hopefully, students of chloroplasts, as well as "mitochondriacs," will enjoy this treatise on evolutionary bioenergetics.

This text is intended to be evidence-based, that is, founded on scientific discovery alone, with the understanding that the "facts" will likely be modified, changed, or repudiated with increasing discoveries over time. Thus, the text intentionally does not portend to portray any "beliefs" or "thoughts." An attempt was made to resource the most pertinent discoveries on the chemical origins of early life and recent knowledge in animal biochemistry and bioenergetics, along with a few classic discoveries. The author acknowledges that many additional excellent reviews, historical accounts, and original research articles are readily available but were not utilized in developing this book. However, the reader is encouraged to read those references of interest and investigate their contents more vigorously. Most of the figures were sourced from experts in their respective fields; the author thanks them all for their contributions in helping others to observe the workings of Nature from a bioenergetics point of view.

Reference

Halliday A, Canup RM. The accretion of planet Earth. Nat Rev Earth Environ. 2023;4:19–36. https://doi.org/10.1038/s43017-022-00370-0.

Contents

About the Author

David Frank Stowe, MA, MD, PhD, Professor of Anesthesiology, Physiology, and Biomedical Engineering at the Medical College of Wisconsin, Milwaukee, attended the Free University of Berlin, Germany, and Indiana University, Bloomington, where he received BA and MA degrees in biology and cell biology. After lab bench experience working for a pharmaceutical firm, he went on to earn a PhD in physiology at Michigan State University, followed by a Cardiovascular Research Institute Fellowship at the University of California, San Francisco. He first joined the Medical College of Wisconsin as an Assistant Professor of Physiology; it is there where he later completed his MD degree and a residency in Anesthesiology. Dr. Stowe practiced clinically and trained anesthesiology residents for 35 years. He has received nearly 50 research grants and has mentored more than 200 graduate, medical, undergraduate, and post-doctoral fellows in basic science research projects. He has authored with collaborators collectively over 200 book chapters, review articles, and original papers, and he has served on numerous national and international research merit review panels for the funding of research projects and for review of journal articles. Based on a publication composite score (c = 3.3) that is based on a lifetime number of publications, views, and citations, Dr. Stowe is ranked within the top 2% of over 7 million scientists worldwide across 22 broad and 176 sub-scientific fields, including biomedical research. He also ranks within the top 1% for publications in the fields of anesthesiology and cardiovascular physiology (Statistics and Meta-Research Innovation Center, at Stanford University https://doi.org/10.1371/journal.pbio.3000384. subset: Table_1_Authors_career_2022_pubs_since_1788_wopp_extracted_202310.xlsx). His recent work involves: identification and activation of mitochondrial SK_{Ca} and BK_{Ca} channels and their cardioprotective mechanisms; biophysical mechanisms and roles of mitochondrial ion exchange channels (e.g., NCE, CHE, KHE) in mitochondrial bioenergetic function; functions of mitochondrial Ca^{2+} and reactive oxygen and nitrogen species in cardiac damage/protection in isolated mitochondria and intact hearts; and transport of substances via mitochondrial channels and exchangers in health and disease. Dr. Stowe greatly appreciates the education he has received at each of his institutions of learning, and he dedicates this book to all his prior mentors and colleagues, but especially to all the eager students he has mentored over the years in basic scientific discovery.

Abbreviations[1]

Bya	billion years ago
DM	dark matter
PBHs	primordial black holes
EM	electromagnetic
MJ	megajoules
STP	standard temperature and pressure
ETS	electron transport system
OxPhos	oxidative phosphorylation
Mya	million years ago
CSE	cystathionine- γ-lyase
TRP	transient receptor potential
P38 MAPK	p38 mitogen-activated protein kinase
ERK	extracellular signal-regulated kinase
PKB	protein kinase B
NHE	Na^+-H^+ exchange or exchanger
NCE	Na^+-Ca^{2+} exchange or exchanger
KHE	K^+-H^+ exchange or exchanger
CHE	Ca^{2+}-H^+ exchange or exchanger
IMM	inner mitochondrial membrane
ICF	intracellular fluid
sGAG	sulfated glycosaminoglycans
NOS	nitric oxide synthases
SR	sarcoplasmic reticulum
COS	carbonyl sulfide
LUCA	last universal common ancestor
TCA	tricarboxylic acid
FA	fatty acid
αKG	α-ketoglutarate
SAM	S-adenosyl methionine
TIM	triose-phosphate-isomerase

[1] In order of appearance.

TIM	translocase inner membrane of mitochondria
IDP	intrinsically disordered peptides
PTC	peptidyl transferase center
SGC	standard genetic code
HGT	horizontal (lateral) gene transfer
FUCA	first universal common ancestor
pmf	proton motive force
OMM	outer mitochondrial membrane
ANT	adenine nucleotide translocase (of IMM)
ACC	ADP/ATP carrier protein
IM	inner membrane of cell
OM	outer membrane of cell
TOM	translocase outer membrane of mitochondria LUCA
VDAC	voltage-dependent anion channel (of OMM)
ROS	reactive oxygen species
FECA	first eukaryotic common ancestor
LECA	last eukaryote common ancestor
RT	retrotranscribing virus
NADH	nicotinamide adenine dehydrogenase (reduced)
NAD^+	nicotinamide adenine dehydrogenase (oxidized)
NMN	nicotinamide mononucleotide
NAM	nicotinamide
PARP	poly(ADP-ribosyl) polymerase
PPP	pentose phosphate pathway
FMN	flavin mononucleotide (oxidized)
FAD	flavin adenine dinucleotide (oxidized)
$FMNH_2$	flavin mononucleotide (reduced)
$FADH_2$	flavin adenine dinucleotide (reduced)
PMC	pyruvate mitochondrial carrier
IMS	inner membrane space
$\Delta\Psi_m$	mitochondrial membrane potential
$CoQH_2$	ubiquinol (reduced)
$CoQH^{\bullet -}$	semi-ubiquinone radical
CoQ	ubiquinone (oxidized)
COX	cytochrome *c* oxidase
FET	forward electron transport
UCPs	uncoupling proteins
PDHC	pyruvate dehydrogenase complex
PDKs	pyruvate dehydrogenase kinases
PDH	pyruvate dehydrogenase
CCL	citryl-CoA lyase
CS	citrate synthase
IDH	isocitrate dehydrogenase
OGDH	oxoglutarate dehydrogenase
OAA	oxaloacetate

HIFα	hypoxia-inducible factor
PGK	phosphoglycerate kinase
LDH	lactate dehydrogenase
SERCA	sarcoendoplasmic reticular Ca^{2+} ATPase
SF	selectivity filter
MCU	mitochondrial calcium uniporter
mRyR	mitochondrial ryanodine receptor (channel)
MAM	mitochondrial-associated membranes
CATR	carboxyatractyloside
OMN	oligomycin
mBK_{Ca}	mitochondrial big potassium Ca^{2+}-dependent channel
mSK_{Ca}	mitochondrial small potassium Ca^{2+}-dependent channel
MnSOD	manganese superoxide dismutase
RET	reverse electron transfer
SS	sub-sarcolemmal
PN	peri-nuclear
IM	intra-myofibrillar
MQC	mitochondrial quality control
PGC-1α	proliferator-activated -γ cofactor 1α
TFAM	mitochondrial transcription factor A
PPAR-γ	peroxisome proliferator-activated receptor γ
MPTP	membrane permeability transition pore
MAC	mitochondrial apoptosis channel
mTOR	target of rapamycin
MICOS	mitochondrial contact site and cristae organizing system
IF1	ATP-ase inhibitory factor 1
MERCs	mitochondria ER/SR contact sites
CICR	calcium-induced calcium release
E-C	excitation contraction
$m\beta_{Ca}$	mitochondrial calcium buffering capacity
CRC	calcium retention capacity
Cyp D	cyclophilin D
CSA	cyclosporin A
RuR	ruthenium red
DHE, ETH	dihydroethidium
TBAP	manganese tetrakis (4-benzoic acid) porphyrin
PUFAs	polyunsaturated fatty acids
EDRF	endothelium-dependent relaxing factor
L-NAME	N^9 L-arginine methyl ester
MEF2	myocyte enhancer factor 2
CNS	central nervous system
ANS	autonomic nervous system
SNS	sympathetic nervous system
NTS	nucleus tractus solitarius

VLM	ventral lateral medulla
HRV	heart rate variability
NOS	nitric oxide synthase
GC	guanylyl cyclase
EDHF	endothelium-dependent hyperpolarizing factor
EDCFs	endothelium-dependent constriction factors
GPCRs	gene protein-coupled receptors
PLC-β	phospholipase C-β
PIP_2	phosphatidylinositol 4,5-bisphosphate
DAG	diacylglycerol
PKC	protein kinase C
IP_3	inositol 1,4,5-triphosphate
ER	endoplasmic reticulum
MLC	myosin light chain
MLCP	myosin light chain phosphatase
CPO	cardiac power output
MAP	mean aortic pressure
RAP	right atrial pressure
VSW	ventricular stroke work

Part I
Elements and Molecules in Biochemical Evolution

Overview of Part I: From the creation of early crucial elements came atoms that became bounded to make inorganic and organic molecules, that with energy mostly from the Sun, made up the early groups of sugars, fatty acids, amino acids, and nucleic acid bases that began to condense and polymerize into macromolecules. Under prebiotic conditions, metal catalysts and later organic cofactors facilitated the first expansion of biosynthetic ingredients that could eventually retain and replicate information allowing stability and growth. The utilization of light and heat, with formation of small, precursor chemical substrates for energy, is the basis of chemiosmosis and the evolution of bioenergetics that drives the biosynthesis of lipid membranes, proteins, nucleotides, and polysaccharides in all lifeforms. That life arose on Earth at all is an astonishing feat that likely required just the most distinctive elements and conditions at a given time from the accretion of the Earth and the heat of the solar system to the cooling and the condensation of water vapor and steam to liquid water. Part I describes how these circustancers, and many more, were essential for life to arise and evolve on Earth.

Chapter 1
Elements of the Cosmos as Precursors for Lifeforms

Abstract Some of the photon energy of the Big Bang was converted to quark particles and gluon plasma that condensed to form protons and neutrons. The first element to form was likely the first isotope of hydrogen, deuterium, composed of one proton and one neutron, but no electron. Atomic fusion of hydrogen to form helium, and of helium to form beryllium, required very high energy to overcome the forces that repel protons. Subsequently, stellar nucleosynthesis fused hydrogen, helium, or beryllium nuclei to create nuclei with increasing atomic numbers (protons) and isotopes (variable neutrons). Later when these positively charged protons attracted negatively charged electrons, atoms of these elements were generated. The bonding or sharing of unbalanced orbital electrons among atoms led to the development of molecules and compounds. Of the 94 naturally occurring elements, the binding properties of hydrogen, carbon, oxygen, nitrogen, phosphorous, and sulfur make them the primary ingredients that formed into the small molecular compounds that provided for the structure and function of macromolecules that led to lifeforms. Twenty other elements in small amounts are also essential for life.

Keywords Accretion disk · Acid-base chemistry · Archaean Eon · Atomic forces · Atomic fusion · Atomic nuclei · Big Bang · Carbon element · Cations and anions · Chemical solubility · Coulomb barrier · Covalent bonding · Dark energy · Dark matter · Deuterium nucleus · Electromagnetism · Electron orbitals · Gibbs free energy · Gravity · H^+ bonding · Hadean Eon · Helium element · Hydrogen element · Hydrocarbons · Ionic bonding · Isotopes · Nitrogen element · Nucleosynthesis · Oxygen element · O_2 pressure · Pauling electronegativity scale · Phosphorus element · Positrons · Subatomic particles · Strong nuclear interactive force · Sulfur element · Up and down quarks · Valence chemical elements · Weak nuclear interactive force

D. F. Stowe, *Evolution of Bioenergetics from Elements to Life*,
https://doi.org/10.1007/978-3-031-76668-8_1

1.1 The Creation of Atomic Nuclei, Atoms, Elements, and Ions

The First Particles and Nuclei Earth only began to coalesce from the Sun's accretion disk approximately 4.6 billion years ago (Bya) (Fig. 1.1) [1], while models indicate that the Universe began approximately 13.8 Bya. According to the "Big Bang" theory, in the first microseconds of the expansion photons were released with such high energy, one estimate is 10^{68} joules (J), that they combined to make pairs of positrons ("+" charged anti-particles) and electrons ("–" charged particles) that as quickly collided to reform photons [2, 3]. By 15 s the equilibrium among positrons, electrons, and photons was lost so that electrons exceeded positrons by one billion to 1. Even now the Universe remains composed of matter with very little existence of anti-matter.

Small particles called quarks were also formed with variable masses that were a bit larger than electrons. Less than a fifth of the matter created by the Big Bang can be measured today; the rest is called dark matter (DM). Theoretical calculations

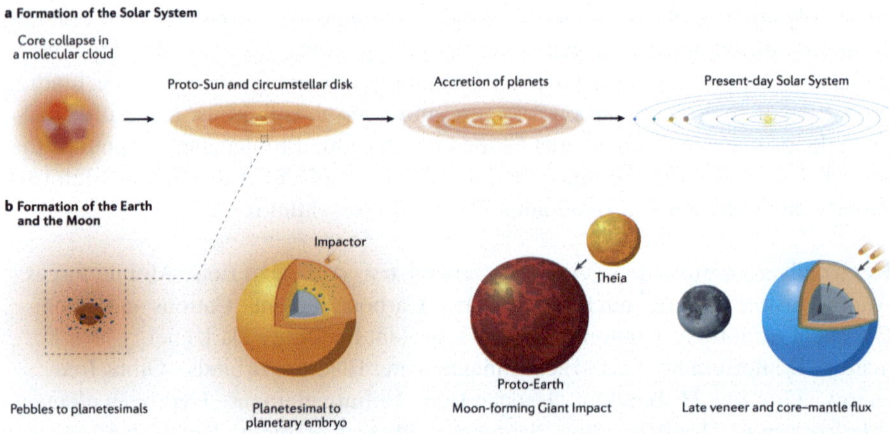

Fig. 1.1 Formation of the Solar System, Earth, and Moon: (**a**) The Solar System formed from the gravitational collapse of a molecular nebula cloud, rich in H_2 and He gases, into a circumstellar disk with the proto-Sun at its center. Within the disk, dust particles accreted to eventually form the planets. (**b**) Collisions between initially μm-sized dust grains are thought to have led to the growth of mm-to-dm-sized pebbles, which in turn accreted into ~100-km-sized planetesimals. Planetesimals further accreted into ~1000-km-sized planetary embryos, some with segregated cores. As the nebula dissipated, gravitational interactions between planetary embryos caused their orbits to cross, leading to protoplanet collisions. It is strongly held that a Mars-sized protoplanet called Theia collided with Earth (The Giant Impact), and the resulting impact ejecta formed the Moon. After the Giant Impact, a small portion (<0.5%) of Earth's mass has been argued to be delivered by bombardment of much smaller impactors, called the late veneer. However, the evidence for the late veneer is unclear, and the isotopic evidence could be explained partially by a flux of deep mantle and/or core components into the mantle over geological time. (Cartoon (Fig. 1) and its legend resourced with permission from copyright owner of Halliday and Canup [1] (review))

based on quantum chromodynamics (QCD) suggest that dark matter remains confined in minuscule (asteroid-size), primordial black holes (PBHs) that might have arisen immediately (<1 μs) during the Big Bang when unbound (free) quark particles and gluon plasma, both derived from photon energy, were absorbed to form huge populations of minuscule PBHs [4, 5]. PBHs are proposed to form whenever a very hot quark gluon plasma collapses under gravity. Like massive black holes, the merger of PBHs may produce a detectable gravitational wave.

Within minutes and after cooling, the plasma composed of quarks and massless gluons condensed to form protons and neutrons. These quark particles rapidly reacted with each other, in a process called nucleosynthesis, to form the first isotope of hydrogen, deuterium ($_1H^2$), a nucleus with one proton and one neutron, but no electrons.

An isotope has a distinct atomic number that defines the individual element by the number of protons but has a different mass due to more, or fewer, neutrons that defines the isotope. By 5 min after the Big Bang, it is postulated that there were isotopes of the nuclei hydrogen, $_1H^2$ (deuterium nucleus), helium, for example, $_2He^3$, $_2He^4$ (dominant isotope) and much less so lithium, $_3Li^7$, where the subscript is the atomic number Z, that is, number of protons (one proton = 1 atomic mass unit) and the superscript is the number of protons plus neutrons. Without electrons (e-), the isotopes were simply nuclei; atoms, which by definition contain electrons in orbit around the nuclei, did not exist until about 0.3–0.4 million years (Myr) later [2]. Figure 1.2 [2] Although He is the second most abundant element in the Universe, on Earth it is nonreactive with 2 e- in its only orbital shell; with no free e- to exchange or share, it has 0 valence, the main characteristic of the noble gases. He^0, that is, $_2He^4$ and especially $_2He^3$, are rare in the Earth's crust, and are products of radioactive decay; when neutral He^0 is released, it gradually escapes from the atmosphere because its mass is so small.

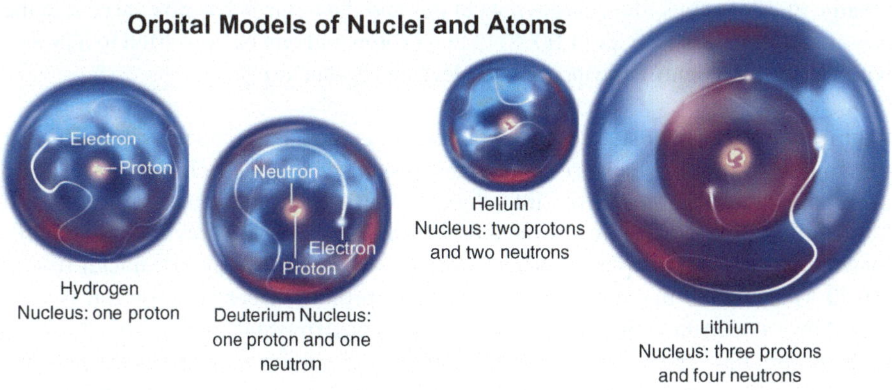

Fig. 1.2 Early nuclei and atoms: Atoms are shown in orbital models to reveal their subatomic constituents. (Models credited to Elena Hartley and legend resourced with permission from copyright owner of Fortenberry [2])

Name	Strength	Range	How Operates
strong nuclear force	1	short nuclear size of about 10^{-5}	on nucleons and heavier elementary particles
electromagnetic force	10^{-2}	infinite	on charged particles
weak nuclear force	10^{-13}	very short subnuclear size of about 10^{-16}	on some elementary particles, particularly electrons and neutrinos
Gravitational force	10^{-39}	infinite	on all objects in Universe

Fig. 1.3 Four fundamental forces of nature: The strong force is nominally set at 1. (Table assembled from various resources)

Four Fundamental Forces Of the four named fundamental forces of Nature, two are long-range gravity and electromagnetism that guide the fundamental structure and activity of the Universe; then there are the subatomic strong and weak nuclear forces (or interactions) that govern nuclear reactions within atoms (Fig. 1.3) [6, 7]. There is cosmic evidence for two additional forces, dark matter (DM) and dark energy (DE), but they remain to be identified and characterized [8]. The strong nuclear interactive force binds quarks to hold protons and neutrons tightly together by the subatomic particle carriers, gluons, to make isotopes of atoms. The weak interactive force utilizes Boson carriers to change the "color" of quarks so that protons (the original matter of nuclei) can be converted into neutrons, which is essential for new nuclei (and atoms) to form. In this way the color charge of quarks dictates the ratio of protons and neutrons that form. Protons are composed of 1 down quark (charge of −1/3) and 2 up quarks (charge of +2/3 each) for a net charge of +1. Neutrons are composed of 2 down quarks (charge of −1/3 each) and 1 up quark (charge of +2/3); thus, the charges cancel out and the neutron has no charge like the proton. A neutron is heavier (1.0008) than a proton and can be converted to a proton with release of an anti-neutrino and an electron (β^--decay).

Atomic Fusion The fusion of smaller nuclei to make a larger nucleus requires that the Coulomb barrier, which arises from two protons repelling each another, must be overcome by forcing the protons into close proximity with each other by gravity and heat (as in the core of the Sun) so that the strong nuclear force prevails and the protons merge. As an example of nucleosynthesis, two $_1H^1$ (proton, H^+) nuclei fuse at 10–20 M kelvin to make one $_1H^2$ (unstable di-proton), where one proton is converted into a neutron with release of a positron, a neutrino and energy. This β^+-decay of the proton into a neutron occurs through the weak force where one of the proton's up quarks is converted into a down quark with emission of a W Boson. In this way a proton decays into a neutron, a positron, and an electron neutrino such that $_2He^3$ gains one neutron. Simultaneously, another $_1H^1$ nucleus collides with $_1H^2$ to form

${}_2He^3$ (2 protons, one neutron) and 2 ${}_2He^3$ collide to form beryllium, ${}_4Be^6$, and with release of energy, to form ${}_2He^4$ (helium nucleus) plus two ${}_1H^1$ nuclei.

Alternatively, two ${}_2He^3$ nuclei fuse to form ${}_2He^4$ plus two ${}_1H^1$ nuclei, or a ${}_2H^2$ nucleus fuses with a ${}_1H^1$ nucleus to form ${}_2He^3$ plus ${}_1H^1$ [9]. Whichever the pathway, the mass of the product ${}_2He^4$ (2 protons + 2 neutrons), also called an alpha particle (${}_2He^{4\,2+}$), that is, minus 2 *e*-, is 0.645% less than the mass of ${}_1H^2$ fusing with ${}_1H^3$ (2 protons + 3 neutrons) with release of one neutron. Thus, in a star like our Sun the lost neutron mass is converted to energy ($E = mc^2$) in the form of alpha-particles and low-energy photon radiation, which is sunlight with an electromagnetic (EM) photon-dispersion spectrum of approximately 54% infrared (heat), 45% visible light, and 1% ultraviolet (waves). Heat is sensed mostly from invisible infrared wave particles (photons) being absorbed onto surfaces. For each 3-part nuclear reaction where one helium nucleus (${}_2He^4$) is formed, the energy output, calculated from $\Delta E = \Delta mc^2$, is about 26.7 MeV, or $4.3e^{-12}$ joules (J) per reaction; this is equivalent to $7.3 \bullet 10^{-9}$ watt-hours or $6.3 \bullet 10^{-9}$ Kcal per reaction. Thus $1 \bullet 10^9$ such reactions would yield 6.3 Kcal.

As protons are "+" charged, each proton in a nucleus eventually attracted an equivalent number of "–" charged electrons (*e*-) (called "recombination") that orbit the nucleus in shells representing the energy within electron clouds. Again, this happened about 0.3–0.4 By after the Big Bang. The first shell contains 2 *e*- of opposite spin, the second shell up to 2 + 6 = 8 electrons, the third shell up to 2 + 6 + 10 = 18 electrons, and so on. Although *e*- have mass, this is so small that it is generally not considered. As noted above, the first atom to form was not ${}_1H^1$, but helium (${}_2He^4$).

The First Atoms and Elements With stellar cooling after the Big Bang, the helium nucleus (alpha particle ${}_2He^{4\,2+}$) readily attracted 2 *e*- in the first complete quantum shell to orbit around the 2 proton, 2 neutron nucleus to form the charge neutral helium atom (${}_2He^4$ or He^0). Eventually, helium shared its 2 *e*- with the ${}_1H^1$ nucleus (proton without *e*-) to form helonium, HeH^+, the first and most abundant molecule and ion in the early Universe [2]. As the Universe expanded and cooled, it is proposed that hydrogen nuclei, ${}_1H^1$, collided with HeH^+ to first form the H^+ ion, that is, ${}_1H^1$. But with no electron, the H^+ ion is only a proton nucleus, not an atom, and so has a radius much smaller than a hydrogen atom (1 proton, 1 *e*-). As discussed extensively throughout this text, the proton plays an essential role in energy transfer within the watery environment of all lifeforms.

The element hydrogen comprises 90% of the atoms in the Universe today and accounts for two-third of the atoms in lifeforms [3]. But atomic hydrogen (1 proton + 1 *e*-) is a free radical with an unpaired *e*-, so it is highly reactive and rarely exists. H_2 is a stable gas molecule that covalently shares the 2 *e*-, *one* from each H. H_2 gas has a higher energy density of 120 megajoules/kg (MJ/kg H_2) than any carbon fuel per weight [10]. H_2 gas can be made by breaking its covalent bond with another atom, for example, from CH_4 (methane) or by reacting an active metal, for example, solid sodium (Na) with H_2O. Previously not detected, it is now known that there are vast underground reserves of natural (white) H_2 gas formed continuously deep in the

Earth's crust; it is proposed to be formed by the endothermic (energy requiring) reaction of a metal with H_2O steam (e.g., $3Fe^o + 4H_2O \rightarrow 4H_2 + Fe_3O_4$) at a high temperature and pressure below Earth's surface [11, 12]. H_2 gas liquifies at −253 °C, but if not contained at standard temperature and pressure (STP), H_2 gas readily escapes Earth's atmosphere as it is 14 times lighter than air with its heavier 80% N_2 content.

Most of the heavier elements up to copper ($Z \geq 6$ to 26) were formed by intense heat (>4 M kelvin), pressure, and containment (including nuclear and electromagnetic forces, and gravity) from the fusion of the protons and neutrons of hydrogen, helium, beryllium, and lithium in stars by stellar nucleosynthesis [13, 14]. For example, after H_2 came H_3^+ and from that came CH^+ and CH_2^+ [2]. Figure 1.4 depicts several early atoms and molecules as orbital models, as *e*- clouds orbiting nuclei, and as nuclei (balls) of the same or different atoms shared by *e*- bridges (sticks) [2]. All the natural elements are found throughout the Universe, and some molecules are

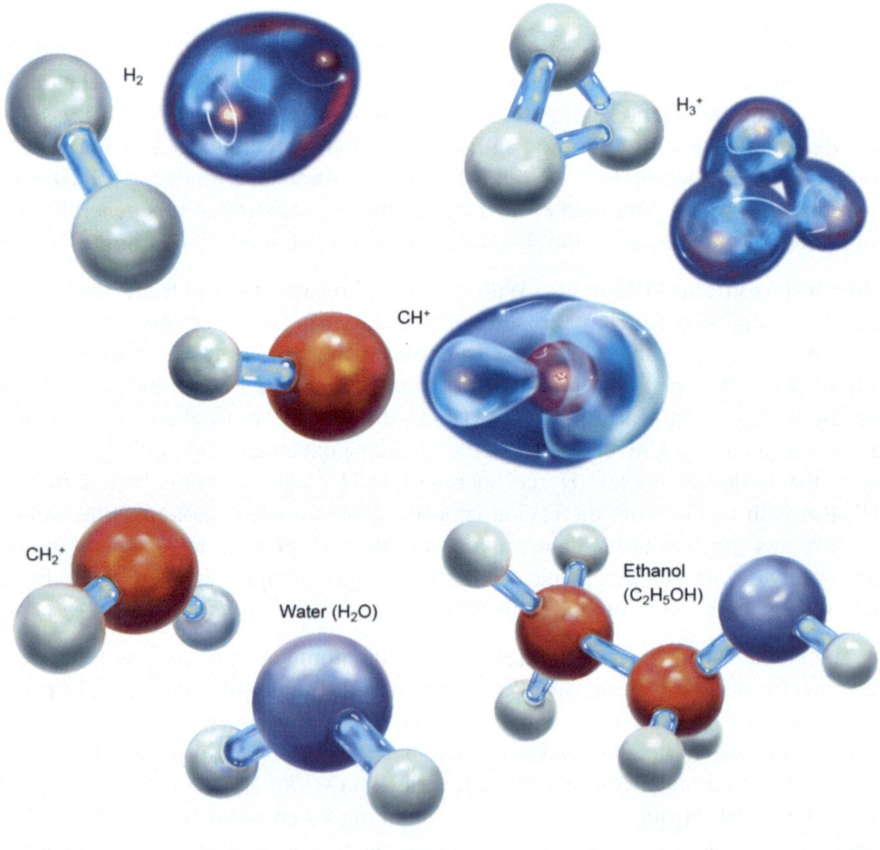

Fig. 1.4 Earliest molecules: Bonds of two or more atoms represented as balls for atoms and sticks as electrons shared by the atoms. (Models credited to Elena Hartley and resourced with permission from copyright owner of Fortenberry [2])

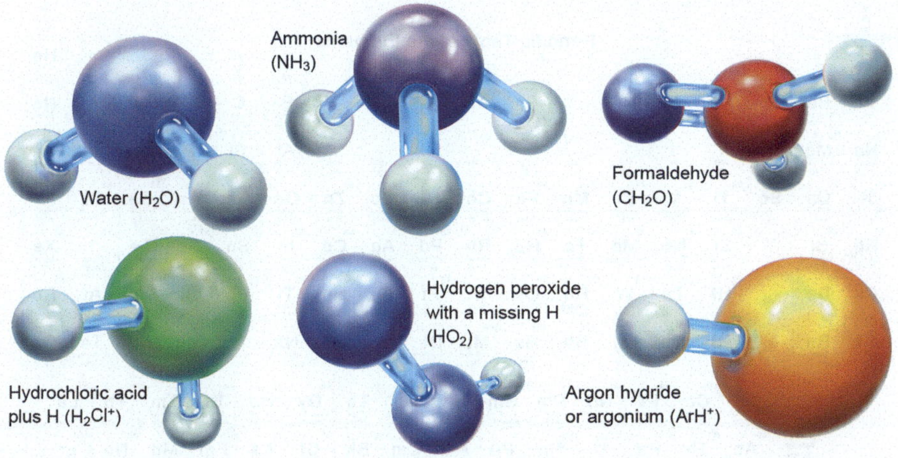

Fig. 1.5 Common molecules found throughout the Universe: Bonds of two or more atoms represented as balls for atoms and sticks as electrons shared by the color-coded atoms. (Models credited to Elena Hartley and resourced with permission from copyright owner of Fortenberry [2])

universally found, for example, H_2O, ammonia (NH_3), methane (CH_4), and formaldehyde (CH_2O) as depicted in Fig. 1.5 [2], however, these molecules may not necessarily exist in the structural forms (allotropes) found on Earth. Many other molecules that have been identified in space are non-Earth-like molecules that were formed selectively within the specific environments of various systems or regions of the Universe.

1.2 Elements Utilized to Make Ions, Organic Molecules, and Biochemical Compounds

Elements of Life Lifeforms on Earth incorporate about 25 of the lightest 92 naturally occurring elements of the Universe and are assigned into chemical groups in the periodic table as shown in Fig. 1.6 [15]. Fourteen elements ranked here by *Z* are the most important: ${}_{1}H^{1}$, ${}_{6}C^{12}$, ${}_{7}N^{14}$, ${}_{8}O^{16}$, ${}_{11}Na^{23}$, ${}_{12}Mg^{24}$, ${}_{15}P^{31}$, ${}_{16}S^{32}$, ${}_{17}Cl^{35}$, ${}_{19}K^{39}$, ${}_{20}Ca^{40}$, ${}_{25}Mn^{56}$, ${}_{26}Fe^{56}$, and ${}_{29}Cu^{64}$. The elements H, C, O, N, and P make up to 97–99% of the mass of most organisms [16]. For cell and mitochondrial functions, H, C, O, N, Na, Mg, P, S, Cl, K, Ca, Mn, Fe, and Cu stand out as the most important elements for life to develop and exist, so their roles are discussed in detail in this book. Other elements essential to life [15, 17] are fluorine (${}_{9}F^{19}$), vanadium (${}_{23}V^{51}$), chromium (${}_{24}Cr^{52}$), cobalt (${}_{27}Co^{59}$), nickel (${}_{28}Ni^{59}$), zinc (${}_{30}Zn^{65}$), selenium (${}_{34}Se^{79}$), molybdenum (${}_{40}Mo^{96}$), tin (${}_{50}Sn^{119}$), and iodine (${}_{53}I^{127}$). They are utilized in very small quantities but are important in mediating vital biochemical reactions by functioning as cofactors for many enzymes, as well as acting as centers for stabilizing structures of enzymes and other proteins. Their roles are not further considered here.

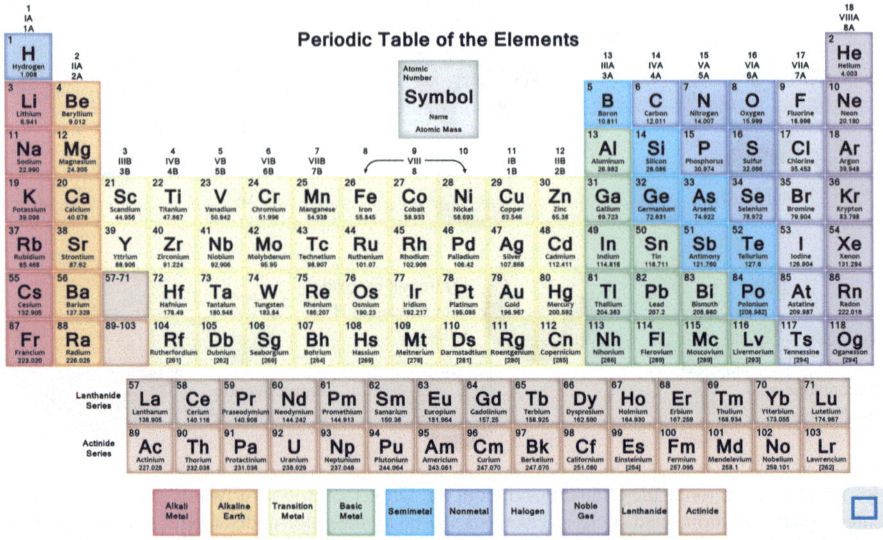

Fig. 1.6 Periodic table of the elements: Chemical group numbers are given above columns; chemical group descriptions are given by the color coded boxes below the table. Above the symbol is the atomic number *Z*. Below the symbol is the element name and atomic mass. (Table available from https://pubchem.ncbi.nlm.nih.gov/ptable/)

As discussed above, neutral He^0, and beryllium ($_4Be^6$) with 2 valence *e*-, are essential for the stellar formation of succeeding elements of higher atomic number; but these two elements are not utilized in lifeforms. Indeed, Be does not form ionic bonds and its covalent bonds with other elements make compounds that are very toxic to body tissues. It is interesting that silicon ($_{14}Si^{28}$) with 4 valence electrons, as in SiO_2 (quartz) is nontoxic but also not utilized in lifeforms. Silicon can bond well with other elements like fluorine, and organosilanes are quite stable. However, the low utilization of silicon in biological systems is primarily due to its natural forms (SiO_2 and silicates) being largely insoluble and chemically inert. Most other nonessential elements are toxic to lifeforms [18], especially metals as grouped in the periodic table (Fig. 1.6) [19]. Potentially toxic elements, when entered into an organism, include the less toxic, light metals beryllium ($_4Be^6$), lithium ($_3Li^7$), and aluminum ($_{13}Al^{27}$), and the more toxic, heavy metals arsenic ($_{30}As^{75}$), cadmium ($_{48}Cd^{112}$), barium ($_{56}Ba^{137}$), mercury ($_{80}Hg^{201}$), and lead ($_{82}Pb^{207}$) [18–20]. Of these, only the role of Li is discussed further.

For the homonuclear, diatomic molecular gases H_2, O_2, and N_2, and the solid free elements such as S_8, and C_n (i.e., graphite, diamond), each atom shares its available *e*- by covalent bonding with the same element, so they are all electron neutral molecules. Many other elements can be found naturally as stable pure molecules, for example, iron ($_{26}Fe^{56}$), copper ($_{29}Cu^{64}$), silver ($_{47}Ag^{108}$), platinum ($_{76}Pt^{195}$), and gold ($_{79}Au^{197}$), whereas others can be extracted from ores to their elemental forms, for example, silicon ($_{14}Si^{28}$), titanium ($_{22}Ti^{48}$), nickel ($_{28}Ni^{59}$), and zirconium ($_{40}Zr^{91}$)

(Fig. 1.6). All these pure elements have no net charge and so are described as Fe^o, Au^o, Pt^o, and so on. The octet rule of atoms is that the stability of an element occurs when eight valence *e-* occupy the valence shell; the rule applies to the s^2p^6 *e-* orbitals. The noble gases have filled valence shells and so are inert. The number of protons in an element is always equal to the number of *e-*. The valence shell is the outermost orbital shell of an atom where *e-* bind or share *e-* with other atoms. After Neon, the number of protons in an element exceeds the number of *valence* shell *e-*. They are unstable ions and to make their valence shells complete they must either bind to any other atom either by a) donating *e-* (ionic bonding) to another element, or b) by sharing their *e-* (covalent bonding) with another element to form molecules or compounds.

Molecules from Atomic Bonds As stated above, most of the elements after Neon have atomic instability because they have fewer valence *e-* than protons (atomic number) and therefore must seek out bonding partners to make stable molecules (bonding of same elements, e.g., H_2, Cu_n^0), and compounds (bonding of different elements to form, for example, H_2O, CH_2O, and ammonia (NH_3), each of which is commonly found throughout the Universe (Fig. 1.5). The presence, orbital location, and number of valence electrons (incomplete orbital shells) determine the chemical bonding properties of the element, that is, whether it can bond, how weakly or strongly it bonds, and with how many other elements it can readily bond [21]. Therefore, a given element's reactivity with other elements and the properties of the products are highly dependent on the electronic configuration of each reactant. Note that all compounds are molecules, but molecules of the same elements are not compounds, for example, ozone (O_3), N_2, and Cu_n^0 are molecules. In contrast to covalent bonding, atoms with unfilled orbital shells often combine their nuclei by ionic bonding to make stable compounds.

Chemical Bonds and Solubility The sharing or transfer of electrons is how most molecules are formed. A simple example of ionic bonding is the dehydration of salt water to solid NaCl. The Cl^- ion, which has one *e-* less in its valence shell for stability, accepts an *e-* from Na^+, which has one *e-* more than it needs for stability; thus, NaCl remains stable but Na and Cl are still ions as a solid. When the salt ions are surrounded by water (solvation), they are often described as cations and anions, for example, Na^+ and Cl^- as the bonds break due to electrostatic (Coulombic) intervention (H-bonding, van der Waals forces of H_2O). Only ionic bonded molecules (e.g., NaCl, $CaCl_2$) dissociate (solvate) into ions in water. So NaCl is very soluble in water.

Solubility of a substance in a solvent is dependent on temperature, pressure, and pH, but at the atomic level, the balance of intermolecular forces between the solvent and solute and the entropy change also determine solubility. Basically, "like dissolves like.". A polar bond is a bond in which atoms unequally share their *e-* so that the electronegativity difference between atoms is greater than zero. Thus, it is notable that covalently bonded molecules that are nonpolar (e.g., methane (CH_4) generally will not dissolve in polar H_2O, whereas polar NH_3 and glucose ($C_6H_{12}O_6$) will dissolve in H_2O. However, other factors such as molecular size of the substance and hydrophobicity of side chains in H_2O solvents or hydrophilicity of side chains in nonpolar solvents are also important determinants of solubility.

As an example of chemical bonding suitable for lifeforms to develop, the existence and linking of molecules with polar and nonpolar ends led to the formation of the monolayer lipid membrane over 4 Bya, and later the amphiphilic bilayered lipid membrane [22]. When shaped into a spheroid structure, the bilayer lipid membrane, today composed largely of glycerophospholipids, was formed, which isolated the budding new molecules from their aquatic environment [23]. The coexistence of polar and nonpolar bonds to form selectively permeable membranes to isolate ions, atoms, and compounds, was a requirement for abiogenesis, the evidence for the origin of life. Indeed, life was thought to begin as microbes that had amphiphile-based plasma membranes where chemiosmotic (ionic) gradients could form [19, 23]. Membranes are discussed in more detail in Sects. 2.1, 3.1, 3.2 and 6.3.

Ion Concentrations In multicellular eukaryotes today, based on the selective permeability of their membranes to ions, ion pumps, ion channels and exchangers, some ion gradients are very large while others are very small. Starting in extracellular fluid, cation, anion, and polyanion concentrations are approximately: Na^+ 150 mM, K^+ 5 mM, Ca^{2+} 1.2 mM, H^+ 40 nM (pH 7.4), Cl^- 100 mM, Mg^{2+} 1 mM, and PO_4^- 1 mM. Note that extracellular (e) free e[H^+] is only about $2.7 \bullet 10^{-7}$ to $4 \bullet 10^{-5}$ that of the other ions. The ratios of cytosolic to extracellular ion concentrations are approximately: Na^+ 0.07, K^+ 30, Ca^{2+} 0.00025, H^+ 1.5, Cl^- 0.6, Mg^{2+} 1, and PO_4^- 1. The ratios of mitochondrial to cytosolic ion concentrations are approximately: Na^+ 0.5, K^+ 0.2, Ca^{2+} 0.3, H^+ 0.4, Cl^- 0.3, Mg^{2+} 0.5, and PO_4^- 15 [24–26]. In general, these ratios indicate that in the resting cell, Na^+, Cl^-, and especially Ca^{2+}, are basically kept out of the cytosol, while K^+ and H^+ are retained in the cytosol. The ratios of ions in the mitochondrial matrix to that of the cytosol indicate that all the ions in the matrix are kept lower than in the cytosol except for PO_4^-. Note that within the cytosol and organelles of living cells, H^+ and Ca^{2+} concentrations are maintained at nM to low μM concentrations, unlike the other ions that are at mM concentrations. It is the movement of these ions against their resting concentration gradients that drives metabolism, bio-electricity, sensation, and movement in all organisms.

1.3 Molecules from Hydrogen, Oxygen, and Carbon, as Precursors for Earth's Lifeforms

H, C, O, N, P These elements [16, 27] form the inorganic (e.g., O_2, H_2O, CO_2) and organic (carbon + hydrogen = hydrocarbons, C-H) molecules like glucose ($C_6H_{12}O_6$) that form the basic biochemical structures for biosynthesis, metabolism, replication, and repair. Because of the specific bonding characteristics of these first five essential elements, simple organic molecules were formed that were likely the precursors of more complicated molecules like ATP and RNA that could lead to development of the earliest lifeforms. The history and great importance of H, particularly the H^+

ion, and H-bonding in molecules containing O or N, as essential for life on Earth are discussed first.

Hydrogen ($_1H^1$), the most abundant element (Fig. 1.2) in the Universe, lists in the periodic table with the alkali metals (Fig. 1.6). As introduced in Sect. 1.1, it can take the form of a nucleus, that is, a proton ($_1H^{1+}$, 1 proton nucleus with 0 *e*-), an unstable atom ($_1H^1$, 1 proton + 1 *e*-), a hydride ion ($_1H_1{}^{1-}$, 1 proton + 2 *e*-), and a molecule (H_2, 2 protons + 2 shared *e*-). Only H_2 is stable by itself; but its combustion in air produces abundant energy with an enthalpy of −286 KJ/mol. The H^+ ion in the form of the hydronium ion (H_3O^-) can exist in H_2O or it must be covalently bonded to other elements, as its atomic form is a free radical. It is important to remember that chemical reduction is the gain of *e*-, while oxidation is the loss of *e*-. So when an *e*- carrier molecule becomes reduced (*e*- gain), a proton neutralizes the negative charge so both the *e*- and H^+ move together [28]. But in a bilayer lipid membrane impermeable to H^+, when the *e*- moves to the next carrier, the energy released can be used to discharge the proton across the membrane, that is called proton pumping, in a process called chemiosmosis.

Most ionic compounds readily dissociate into ions in the presence of H_2O forming "electrolytes" like $Na^+ + Cl^-$. The H^+ (proton) does not actually exist in aqueous solutions. Protons in H_2O are highly mobile, transiting from one H_2O molecule to its neighbor forming H_3O^+ (hydronium ion) [28, 29]. Basically, $2H_2O \rightarrow [H_3O]^+ + [OH]^-$, where one H_2O self-ionizes to OH^- and the released H^+ attaches to the other H_2O to make $[H_3O]^+$; like H^+, $[H_3O]^+$ has one less valence *e*- than the number of protons. Thus, as will be detailed for mitochondrial respiration, ultimately $O_2 + 4[H_3O]^+ + 4e\text{-} \rightarrow 6H_2O + 4e\text{-}$, so O_2, as the final *e*- acceptor, is reduced to H_2O at complex IV (Sects. 5.2 and 5.3) [30].

Acid-Base Scale The number of protons as $[H_3O]^+$ in a volume of pure H_2O is small and forms the basis of the acid-base scale, where both $[OH]^-$ and $[H_3O]^+ = 1 \bullet 10^{-7}$ mol/L, that is, 100 nM (pH = 7). pH 7 is considered neutral, and lower and higher proton concentrations are considered basic and acidic, respectively. In isolated mitochondria bathed at 25 °C in buffered, external (e) pH solutions of 7.6–6.9, i.e., 25–126 nM e[H^+], the matrix (m) (inside) m[H^+] can also vary between approximately 25–115 nM, reflecting H^+ (proton) pumping and inward H^+ leaking under different experimental conditions [26, 31]. Proton pumping and proton leak are discussed in detail in Sects. 5.2, 5.3, 6.2, and 6.4.

Hydrogen binds covalently with all nonmetal, natural elements, for example, as in the most simple hydrocarbon, CH_4, where up to 4 H atoms can share their *e*- with one C atom. H can also form covalent bonds with transition metals, called metal hydrides, in which hydrogen is attached to a less electronegative element like sodium, as in sodium hydride (NaH). H-C bonds are polar because carbon is more electronegative than hydrogen, although they are not as polar as N-H and O-H bonds. For instance, in methane (CH_4), the oxidation number of carbon is −4. If the attraction for *e*- is unbalanced (different electronegativities), the bonding is polar (asymmetrical) as the bonding *e-* are shared unequally. In general, polar covalent molecules dissolve in polar solvents and nonpolar molecules dissolve in nonpolar solvents. Covalent bonds between H and O include H_2O, H_2O_2, alcohols (-O-H) (Figs. 1.4 and 1.5), and carboxyls (HO-R = O).

H-Bonds "H-bonding" is very different from ionic or covalent bonding. It is an electrostatic attraction between parts of the same or different molecules. As a non-ionic, noncovalent bond it exerts a weak attractive force in which a H atom that is covalently bound to O or N (or F) atoms can also form weak H-bonds with lone *e-* pairs on other nearby O or N atoms that are not already covalently bonded [32]. This interaction brings molecules weakly together that have highly polar, covalent bonds between H to either O or N (e.g., like H_2O to H_2O (Fig. 1.7), or H_2O to ammonia (NH_3), or NH_3 to NH_3 [32]. For example, one H_2O molecule can be weakly attracted to a maximum of 4 other H_2O molecules by H-bonding. As illustrated in Fig. 1.7, all H-bonding is broken in water vapor, variable in steam and liquid water, and unbroken in solid ice, where it forms a relatively strong crystal lattice.

H-bonding is essential for life because it brings molecules loosely together (stickiness), which gives stability for protein confirmation and enzyme function, and furnishes the scaffolding for 3-D molecular structures like the DNA backbone and phospholipid membranes [32–34]. As an important example, double stranded DNA is two chains of DNA molecules held together by H-bonds (Sects. 1.3 and 3.2). There is some evidence for a weaker force for H-bonding when another H is covalently bound to C [32, 35].

H_2O Although not an element, this important molecular compound elicits special mention here. H_2O is a polar molecule with one δ "−" O bond and two δ "+"

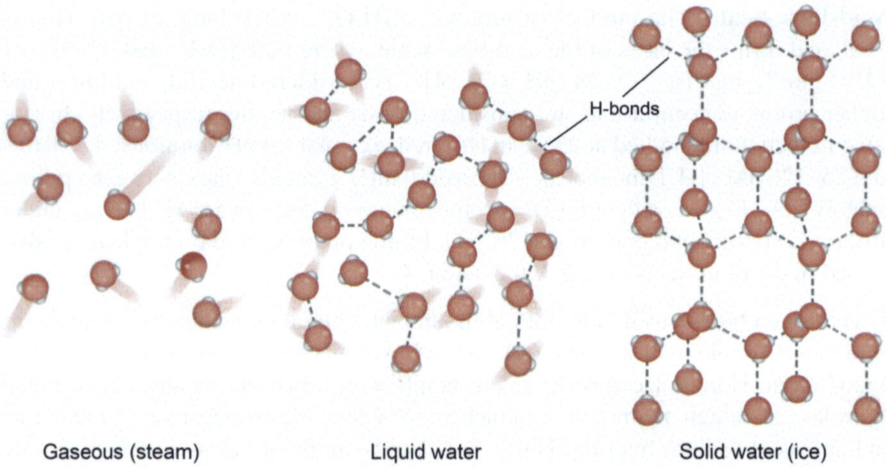

Fig. 1.7 Hydrogen bonds form in water and ice, but not in water vapor: Hydrogen atom = white; oxygen atom = red; H-bond = dashed blue line. (Models available from https://www.coursehero.com/sg/cell-biology/hydrogen-bonding-between-water-molecules/)

H-bonds as shown and discussed above (Fig. 1.7). Liquid water, with its variable and weak H-bonding, is essential for all of Earth's lifeforms to function because life requires the formation, movement, and interaction of molecules, and the ionization of compounds like salts. Polar H-bonding allows dissociation of ionic bonds in liquid H_2O to form different molecules [36]. Liquid H_2O provides the "stickiness" of organic molecules associated via H-bonds in H_2O. H_2O, frozen as ice (persistent H-bonds) in the stellar cold (average temperature of 2.7 kelvin or −270°C), is now known to be very abundant in the Universe; but because many reactions can produce H_2O as a by-product, it has not been clear how H_2O originally formed in the Universe.

Adding molecular H_2 and O_2 together with a spark will make H_2O with an explosion of energy. But molecular O_2 is very scarce in the Universe. Unlike O_2, H_2O is present and abundant as amorphous water ice on dark, cold interstellar dust grains that serve as catalysts [37]. Experiments simulating cosmic conditions at a temperature of 10 kelvin show that H^+ and O^+ ions that are trapped in dust form and multiply as water ice [38].

Source of Water The H_2O present on Earth today originated largely from H_2O trapped as water ice in dust grains within the accretion disk around the Sun (Fig. 1.1) that formed the four terrestrial planets; some H_2O ice was also likely carried in by asteroids colliding with Earth, and less likely by comets [39]. During the Heavy Bombardment Hadean eon between 4.6 and 3.9 Bya [40], H_2O in the form of steam, remained in the thick, hot atmosphere (27 x STP atmosphere (atm), 230 °C) [41], also containing CH_4, CO_2, NH_3, N_2, and other gas molecules. The lightest gases, He and H_2, mostly escaped from the forming terrestrial planets. Because of the high atmospheric pressure, H_2O eventually fell from clouds to the Earth as rain once its surface had cooled sufficiently [39, 42]. Newer evidence indicates this began as early as 4.3 Bya [43] and with continuous rain, the oceans formed within millions of years. The heavy continuous rain occurred during the later Archaean eon between 3.9 and 2.8 Bya [40]. H_2O vapor was also released by the very high volcanic activity at the time [44]. All lifeforms require liquid H_2O and although H_2O turns to solid ice below 0 °C, at 1 atm, lifeforms exist even at 15,750 psi (1000 atm) at a H_2O ocean depth of almost 11,000 m in the Mariana Trench [45]. Here, H_2O density is 5% greater than at the surface but H_2O remains liquid at the coldest temperatures of 1–3 °C, so lifeforms can exist there.

During Earth's atmospheric and water cycles, warm air at the surface carries H_2O vapor upward from surface liquid H_2O; vapor saturation and cooling of the air releases the vapor back to liquid water (fog, rain). In general, water evaporates as water temperature rises, and or the dew point temperature of the air is higher than the surface temperature. The energy needed to convert one g of liquid H_2O to H_2O vapor is 41 kilojoules (KJ)/mol. At this point H-bonds break up and invisible H_2O molecules are released from the liquid. Conversely, H_2O gas condenses when water temperature falls, the surface temperature is cooler than the dew point temperature, or the water vapor in air is saturated. Condensation releases energy (−41 KJ/mol) and H-bonds reform converting H_2O gas to visible fog, steam, or liquid H_2O.

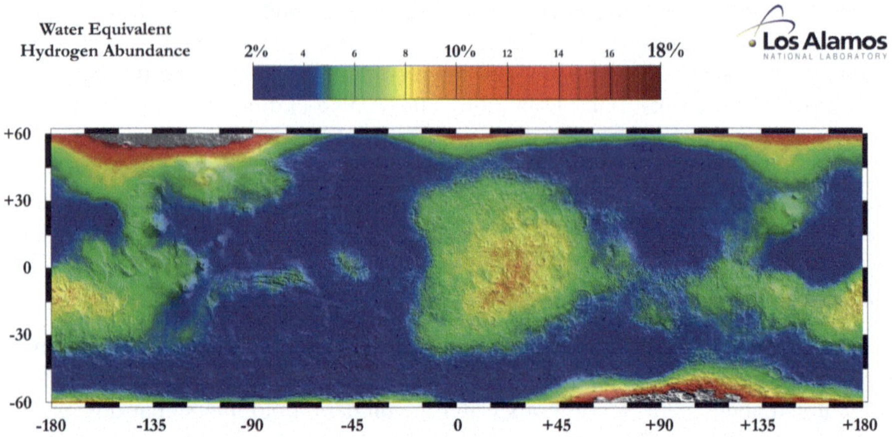

Fig. 1.8 Global distribution of near-surface hydrogen (H_2O) on Mars: Robinson projection of the water-equivalent hydrogen content of the semi-infinite layer of water-bearing soils derived from epithermal-neutron counting rates (mass % of water) using the Neutron Spectrometer aboard Mars Odyssey probe between February 2002 and April 2003. Overlay of water equivalent hydrogen abundances and a shaded relief were derived from Mars Orbiter Laser Altimeter (MOLA) topography. (Color plot (Fig. 4) and modified legend provided by NASA/Los Alamos National Lab and reused in Feldman et al. [47])

Geological evidence for surface liquid H_2O on Mars many Bya [46], the evidence of massive H_2O ice sheets at the poles of Mars as mapped in Fig. 1.8 [39, 47, 48], evidence of exposed dusty H_2O ice in gullies on Mars [49], and new, clear evidence of evaporated liquid H_2O lakes on Mars [34], all present the possibility that simple aqueous lifeforms may have once existed on Mars. Many planetary scientists expect lifeforms have or will emerge on exoplanets that have the conditions that permit liquid H_2O [50, 51], that is, an appropriate, and stable, distance from its star(s) (liquid H_2O habitable zone), both a dry and wet rocky planet, and sufficient pressure and gravity to hold an atmosphere that can retain H_2O vapor.

Other likely factors required for evolution of lifeforms are that a neoplanet forms from a protoplanetary disk containing C, N, S and metal ions, the presence of greenhouse gases (e.g. CH_4, CO_2, N_2O) in the atmosphere to retain the planet's heat, and a magnetic shield in its atmosphere to protect electrons from being stripped from atoms that results in production of damaging ions (cosmic ionizing radiation). One interesting estimate for a chance for other life in the Universe is based on the number of stars with an abiogenesis event (prior to life) vs. the length of an RNA strand (Fig. 1.9) [52].

Carbon (${}_6C^{12}$), the fourth most abundant element in the Universe, was formed much later after the Big Bang when exploding supernova stars, at 1–2 • 10^8 kelvin, caused the fusion of three ${}_2He^4$ (alpha particles) nuclei to make atomic ${}_6C^{12}$. Alternatively, the intrastellar fusion of + ${}_4Be^8$ + ${}_2He^4$ to make ${}_6C^{12}$ (Fig. 1.6) is more favorable [53]. C has a valence of 4, meaning it has 4 *e*- to share in covalent binding. C atoms can form single, double, or triple bonds with H atoms often filling in for the last *e*- to share. CH_4 is the lightest stable hydrocarbon. The capacity of C atoms to

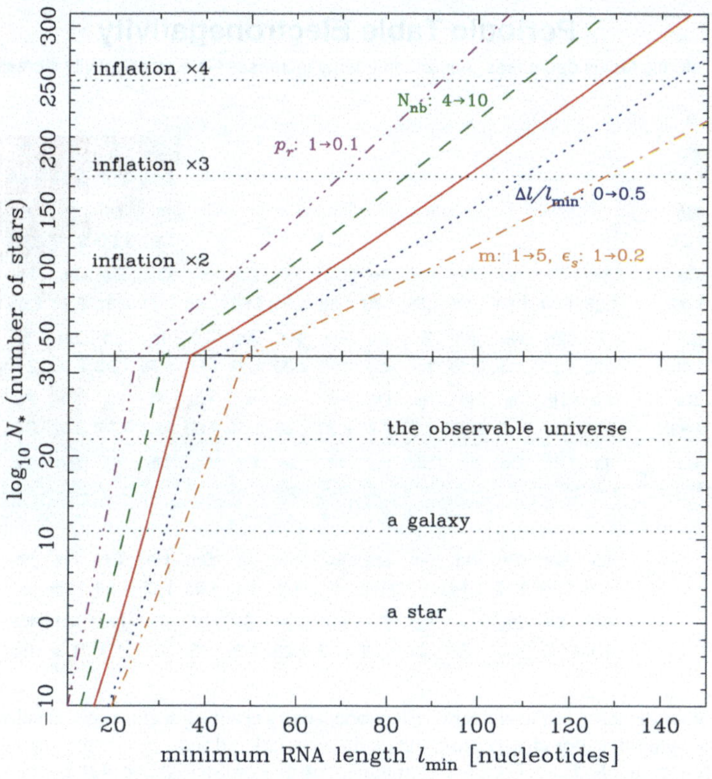

Fig. 1.9 Chance of life in Universe: Logarithm of the number (N) of stars necessary to expect at least one abiogenesis event ($\log_{10}N^*$) versus the minimal RNA length (l) required to show a biological activity leading to abiogenesis (l_{min}): The difference between the top and bottom panels is just the scale of the vertical axis. Important values of lgN* are indicated by horizontal dotted lines; "inflation Å ~ 2" means the universe size when the inflation lasted twice as long as that required to make the observable Universe homogeneous. The red solid curve is the relation using the baseline model parameter values, and other curves are when some of the model parameters are changed from the baseline values, as shown. (Data plot (Fig. 1) and its legend resourced with license (https://creativecommons.org/licenses/by/4.0/) from Totani et al. [52])

bond with one another in various chain, ring structures, and three-dimensional conformations (called "catenation") as pure molecules, for example, graphite and diamond, as well as to link with other atoms, notably H and O, accounts for much of the complexity and diversity of organic forms of which there are millions known.

C-C covalent bonds are very strong (around −345 KJ/mol) where one *e-* is shared from each C. Also, bonds of C≡C (acetylene, C_2H_2) are stronger than C=C (ethylene (C_2H_4) bonds that are also stronger than C-C (ethane, C_2H_6) bonds. C=C bonds are found in unsaturated fatty acids (FAs); C≡C bonds are not found in Nature. The bond enthalpies for C atoms binding with O atoms are not as strong as C-C bonds because of the greater electronegativity of O (Pauling scale 3.40) over C (Pauling score 2.55) as laid out in this version of the periodic table (Fig. 1.10). But bond strength is determined by other factors, including the size of each atom involved in

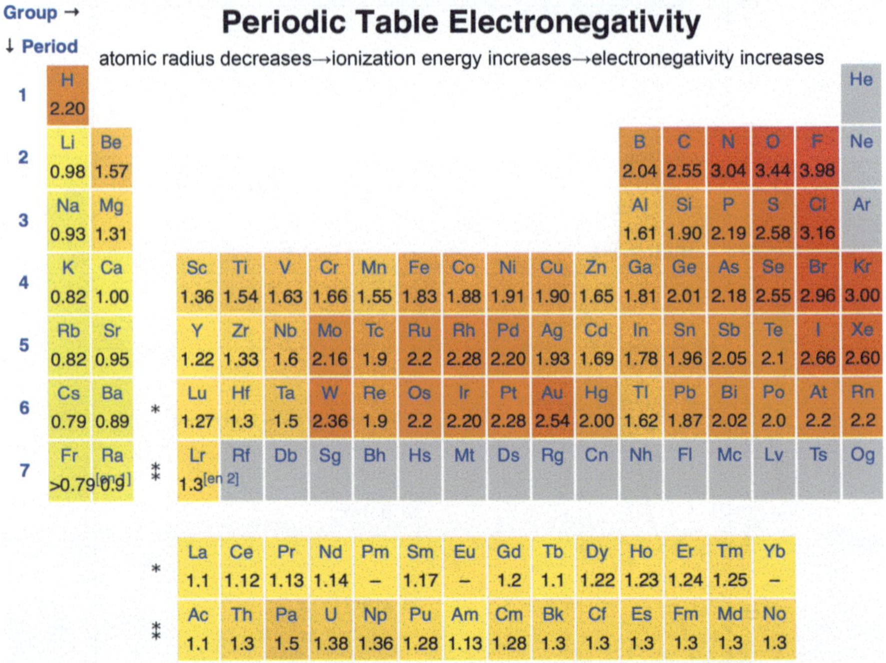

Period																			
1	H 2.20																		He
2	Li 0.98	Be 1.57												B 2.04	C 2.55	N 3.04	O 3.44	F 3.98	Ne
3	Na 0.93	Mg 1.31												Al 1.61	Si 1.90	P 2.19	S 2.58	Cl 3.16	Ar
4	K 0.82	Ca 1.00		Sc 1.36	Ti 1.54	V 1.63	Cr 1.66	Mn 1.55	Fe 1.83	Co 1.88	Ni 1.91	Cu 1.90	Zn 1.65	Ga 1.81	Ge 2.01	As 2.18	Se 2.55	Br 2.96	Kr 3.00
5	Rb 0.82	Sr 0.95		Y 1.22	Zr 1.33	Nb 1.6	Mo 2.16	Tc 1.9	Ru 2.2	Rh 2.28	Pd 2.20	Ag 1.93	Cd 1.69	In 1.78	Sn 1.96	Sb 2.05	Te 2.1	I 2.66	Xe 2.60
6	Cs 0.79	Ba 0.89	*	Lu 1.27	Hf 1.3	Ta 1.5	W 2.36	Re 1.9	Os 2.2	Ir 2.20	Pt 2.28	Au 2.54	Hg 2.00	Tl 1.62	Pb 1.87	Bi 2.02	Po 2.0	At 2.2	Rn 2.2
7	Fr >0.79[en 1]	Ra 0.9	**	Lr 1.3[en 2]	Rf	Db	Sg	Bh	Hs	Mt	Ds	Rg	Cn	Nh	Fl	Mc	Lv	Ts	Og

*	La 1.1	Ce 1.12	Pr 1.13	Nd 1.14	Pm –	Sm 1.17	Eu –	Gd 1.2	Tb 1.1	Dy 1.22	Ho 1.23	Er 1.24	Tm 1.25	Yb –
**	Ac 1.1	Th 1.3	Pa 1.5	U 1.38	Np 1.36	Pu 1.28	Am 1.13	Cm 1.28	Bk 1.3	Cf 1.3	Es 1.3	Fm 1.3	Md 1.3	No 1.3

Fig. 1.10 Periodic table of the elements: Electronegativity (Pauling scale) is given below name of element and symbol. Color change from light yellow to dark red indicates increasing electronegativity. (Table available from Creative Commons Attribution-ShareAlike 4.0 License; en.wikipedia.org)

the bond and the relative energies of the atomic orbitals forming the bond. Thus, a single C-C bond is actually slightly weaker than a C-O bond [54]. C-O bonds include mainly covalently single bonded alcohols (C-O-H), ethers (C-O-C), and double bonded carbonyls (C=O), i.e., carbonate esters (R–O–C(=O)–O–R′) and inorganic CO_2 (O=C=O) and carbonates like HCO_3^-, as in the solvation of $Na_2CO_3 + H_2O \rightarrow HCO_3^- + OH^- + 2Na^+$, where the O-C(=O)-O is protonated. C-N bonds are discussed below. In chemical formulas, "R" represents any sized, residual, variable C-H groups, that depending on the nature of the atoms in a group, determine the characteristic of the overall molecule, for example, as in the different R groups of amino acids.

Oxygen (${}_8O^{16}$) lies behind H_2 and ${}_2He^4$ as the third most abundant element in the Universe. It was created in collapsing stars depleted of H by fusion of ${}_2He^4$ and ${}_6C^{12}$ to form ${}_8O^{16}$ [53]. O as an atom is too reactive and so does not exist naturally. Unlike H_2 gas and the noble gas He, diatomic O_2 gas, liquid, or solid is extremely rare in the Universe (except on Earth) because atomic O is most often bound to dust particles [37] blocking them from sharing *e-* to form molecules, or it is bound to H^+ as water ice [38]. A possible source of molecular O_2 in the cosmos is an incomplete reaction:

$2CO_{(gas)} + 4H_{2(gas)} \rightarrow 2CH_{4(gas)} + O_{2(gas)}$; another possible reaction is: $2CO_{2(gas)} + 4H_2O_{(ice)}$ (+ photon energy) $\rightarrow 2CH_3OH + 3O_2$ [55].

On Earth, oxygen arrived mostly as metal oxides from asteroids and meteors. Molecular O_2 first formed in the seas of Earth about 3.5 Bya because of photosynthesis in cyanobacteria. With the uptake of abundant CO_2, photosynthesis in H_2O led to the byproducts H_2, which escaped if it reached the surface, and heavier O_2, which remained in the seas to react with metals like iron to form ferric oxides (see also Sect. 2.3) [30, 56]. Once underwater metal oxygenation was exhausted, O_2 escaped from the oceans into the atmosphere during the Great Oxygenation Event, beginning 2.6 Bya [57, 58]. The high O_2 level was toxic to many anaerobic species but would be beneficial to newer species that evolved to use O_2 as an *e*- acceptor during the stepwise chemical evolution of the electron transfer system (ETS) and oxidative phosphorylation (OxPhos) somewhere around 2–3 Bya [59].

Figure 1.11 shows three models of Earth's atmospheric O_2 pressure that peaked much later during the Phanerozoic Eon, Carboniferous Era, 350 million years ago (Mya) [57]. This event, lasting approximately 100 My, eventually caused the partial pressure of O_2 in the atmosphere to rise above 20 kPa to about 27 kPa, which was

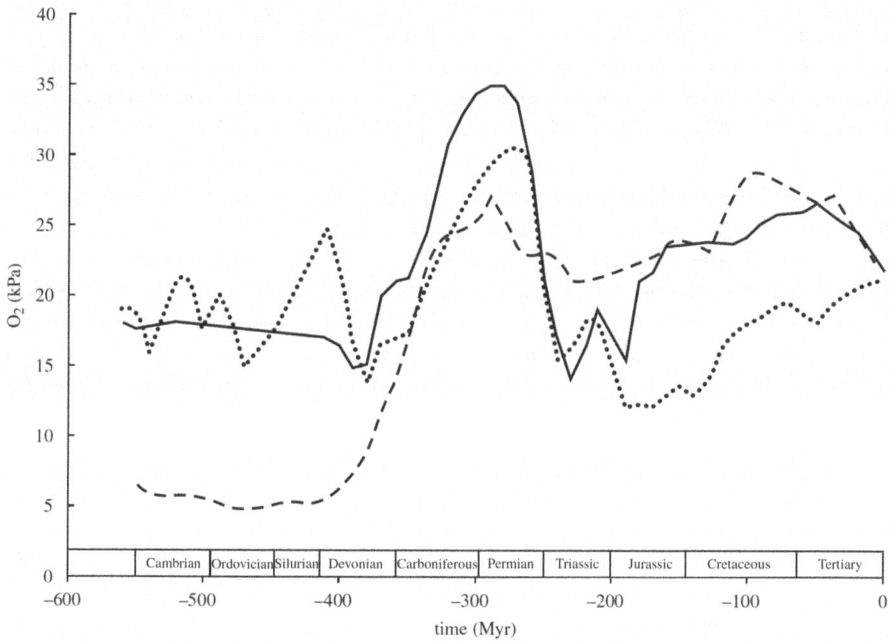

Fig. 1.11 Oxygen pressure on Earth's surface: Various patterns of modelled atmospheric O_2 pressure at the Earth's surface over the Phanerozoic Eon covering the Periods listed from about 540 Mya to 200 Mya. See citation for origins of original data. (Data plot (Fig. 1) and modified legend resourced with permission from copyright owner of Harrison et al. [57])

higher than the sea level pressure of 20 kPa (150 mmHg) that exists on Earth now [56].

Di-atomic O_2 (O=O) is now the most common gas molecule at the Earth's surface. As much as 70% of the O_2 today is made in the oceans by phytoplankton such as prochlorococcus [56]. In addition to the 2% O_2 made by cyanobacteria, plants (mostly in rain forests) make up about 28% of the remaining O_2 by photosynthesis. Ozone (O_3) is formed in the stratosphere by UV radiation, which spits molecular O_2 into two atoms of O and recombines three atoms of O to form O_3. As noted above, O, is a very strong electronegative (Pauling scale 3.4) atom (Fig. 1.10) that attracts *e*- from other atoms to form strong covalent bonds with H, C, N, O, and P as well as weak electrostatic H^+ bonds between molecules containing O or N. The important impact of O in the position 2 OH group of the sugar ribose is brought up in Sect. 3.1. The importance of the superoxide anion ($O_2^{\bullet-}$), which results from a 1 *e*- reduction of O_2, on mitochondrial and cell function is discussed in more detail in Sects. 2.4, 5.3, 6.1 and 8.4.

1.4 Molecules from Nitrogen, Phosphorous, and Sulfur as Precursors for Earth's Lifeforms

Nitrogen ($_7N^{14}$), the fifth or sixth most abundant element in the Universe, was formed in stars from protons and neutrons of He by stellar nucleosynthesis when temperature and pressure were sufficient to cause atomic fusion reactions [13]. Atomic N does not exist naturally. N_2 is a triple bonded gas molecule (N≡N) and as such is nonreactive (inert). But in the presence of sunlight and energetic particles from Saturn's magnetosphere, N_2 reacts with CH_4 in Saturn's moon Titan to form negatively charged, nitrile and C chain anion pairs such as CN^-/C_2H, C_3N^-/C_4H^-, and C_5N^-/C_6H^- that are the precursors of cyclic rings, for example, C_6H_6 (benzene) [60]. N_2 is the most abundant gas in Earth's atmosphere and like O_2 is too heavy to escape the Earth's atmosphere. The arrival of N on Earth is thought to have come from the inner and the outer disk of our solar system (Fig. 1.1) with N locked in iron meteorites that fell to Earth, as well as from inside dust clouds that came from far beyond Jupiter [61].

The simplest N compound is ammonia, NH_3. N is a little less electronegative (3.04) than O (Fig. 1.10), so N's *e*- are attracted and shared by O to make covalent bonds that are polar. Molecular examples are the nitrogen free radical, nitric oxide ($NO^{\bullet}$), nitrous oxide (N_2O), nitrogen dioxide (NO_2), and the amino acid arginine ($C_6H_{14}N_4O_2$). N also participates in H^+ bonding with other adjacent molecules containing O or N. N is abundant (15–20% by weight) in amino acids, nucleic acids, and nucleobases. N bonds are very stable and require energy to break them. C-N bonds are polar, covalent, and the C's are strongly polarized to N due to the higher electronegativity of N (3.04) than C (2.55) (Fig. 1.10). Common organic N compounds are amines (-C-NH_2), nitros (-C-NO_2), nitriles (-CN), and amides (-$CONH_2$).

The role of N in nicotinamide, a component of NAD^+, as an *e*-receptor, is discussed in Sect. 5.1. The regulatory role of the free radical nitric oxide, $NO^•$, on mitochondrial and cell function is discussed in Sects. 7.2, 8.4 and 9.3.

Phosphorus ($_{15}P^{31}$) is a highly reactive element and is essentially nonexistent except in molecules; it too was created in stars by stellar nucleosynthesis and disseminated on Earth in comets and meteorites [13, 14]. In the cosmos, P is difficult to measure but it is not nearly as abundant as H, C, O, and N. There is evidence that P was present on Earth earlier than 3.5 Bya. Meteorites containing a phosphorus-bearing mineral called schreibersite that crashed onto Earth is likely to have provided the phosphates essential for terrestrial life [62, 63]. Schreibersite is a phosphide, that is, a phosphorus ion bonded to a metal, for example, $[(Fe, Ni)_3P]$. When synthetic schreibersite is exposed to H_2O the metal nickel ($_{28}Ni^{59}$) reacts with it; as the H_2O evaporates, it creates non-polar P-O bonds on the surface of the schreibersite [63], making the P bioavailable to life as PO_4^{3-} [62, 64]. Figure 1.12 provides an overview of possible pathways for the prebiotic origin of phosphates [64]. During cycles of evaporation and rehydration, when the surface of the schreibersite dries, it allows PO_4^{3-} molecules to join into longer chains, or polymers, with release of H_2O [62].

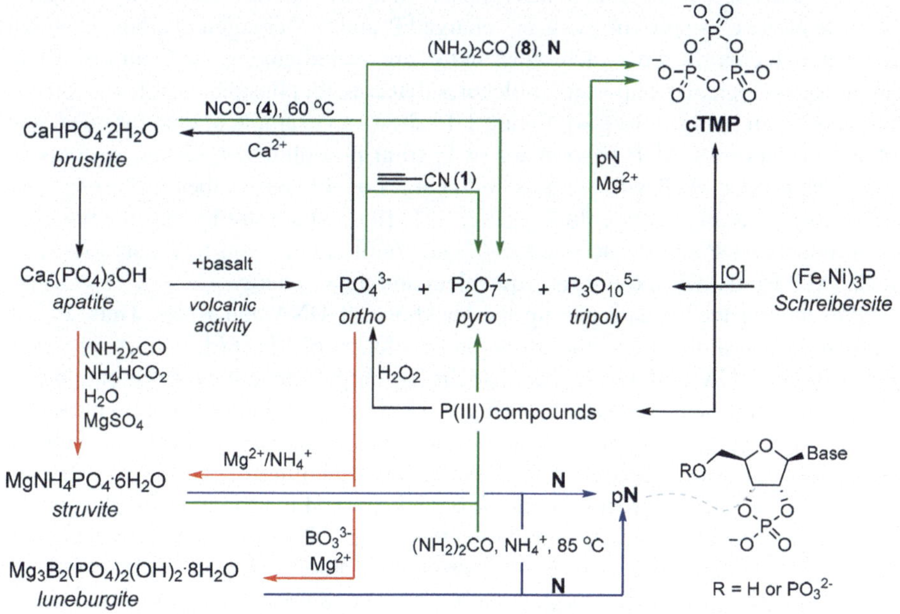

Fig. 1.12 Origin of phosphates: An overview of suggested prebiotic routes to orthophosphate and condensed phosphates production (black arrows), phosphate interconversion (green arrows), phosphate solubilization (red arrows), and urea-mediated phosphorylation (blue arrows). *N* nucleoside, *pN* nucleotide, *cTMP* cyclotriphosphate. (Diagram (Scheme 1) and legend resourced with license (https://creativecommons.org/licenses/by/4.0/) from Fernandez et al. [64] (review))

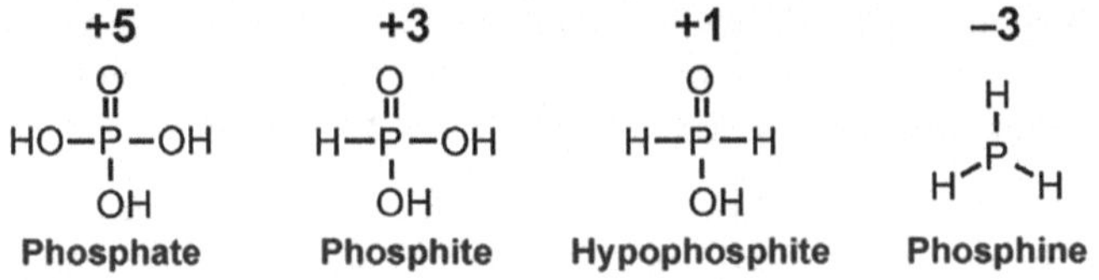

Fig. 1.13 Structures of inorganic phosphate compounds: Each phosphorous compound has different redox states (valences with charges of +5, +3, +1, and −3). (Drawing (Fig. 3) resourced with license (https://creativecommons.org/licenses/by-nc-nd/4.0/) from Kitadi and Maruyama [16] (review))

P-O Bonds Figure 1.13 shows structures of some inorganic phosphate compounds with different valences [16]. Due to its usual valence of 5, P covalently binds to more electronegative O which readily attracts P's electrons to form many kinds of phosphate groups and in nature, mostly orthophosphates with 3 P-O bonds and 1 $P{=}O^-$ bond [16], usually abbreviated as inorganic phosphate (P_i), With unshared *e*-, an orthophosphate is a polyanion, PO_4^{3-}. With P binding to O, examples are Na_3PO_4 and $Ca_3(PO_4)^2$. Recall that compounds like Na_3PO_4 can dissociate in H_2O to $3Na^+$ + PO_4^{3-} but the P-O covalent bonds are not broken. Phosphates are additionally discussed in Sects. 3.1, 3.2 and 6.2.

The bonds between P_i molecules (phosphoanhydride bonds) are highly energetic because of the close proximity of "+" charged P_i and "−" charged O atoms in which the negative charges repel each other and *e*- are shared among the O atoms. There are many biologically important molecules that contain phosphates; a few common ones are given in Fig. 1.14 [64]. Figure 1.15 shows an example of enzyme-mediated phosphorylation of ADP by transfer of P_i from phosphoenopyruvate to pyruvate [65]. The product ATP, which has 4 "−" charges on 4 O atoms then prefers to eject a P_i group to attain greater enthalpy (Fig. 1.15) [65], which strengthens the bonds in the products ADP and P_i and releases 7.3 kcal/mol of energy at STP conditions. To make ATP from ADP requires an equivalent amount of energy.

Phosphodiester bonds make up the backbone of DNA and RNA. Thus, P_i is a very important molecule in the origin and evolution of life [66], with its presence not only in RNA and DNA, but also in the important cofactor, co-enzyme A (Fig. 1.16) [67], and in ADP, ATP (Fig. 1.17) [65], creatine phosphate, phosphoenolpyruvate (Fig. 1.14) and phospholipids, and for its role in post translational phosphorylation mechanisms [65]. A better understanding of prebiotic phosphate chemistry, i.e., the mechanisms for incorporation and exploitation of phosphates with other molecules is key in elucidating the origins of life on Earth as indicated by several probable, prebiotic routes to incorporate orthophosphates (Fig. 1.14) [64].

P-C bonds can be made synthetically, as in organophosphates, also known as phosphate esters, $O{=}P(OR)_3$, but they are very rare in nature [65, 68], and many of them are toxic to lifeforms because they inhibit the enzyme acetylcholinesterase, which deactivates the neurotransmitter acetylcholine, causing dangerous levels of acetycholine [69, 70]. Until recently, the P atom, in the form of P_i, had never been

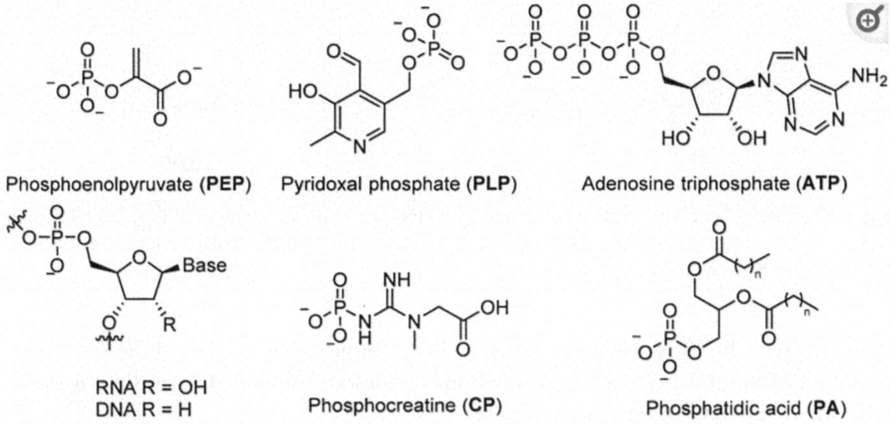

Fig. 1.14 Common biologically important phosphate containing molecules: These examples exemplify the central role phosphates play in biological information transfer, structure, and energy metabolism. (Structures (Fig. 1) and legend resourced with license (https://creativecommons.org/licenses/by/4.0/) from Fernandez-Garcia et al. [64] (review))

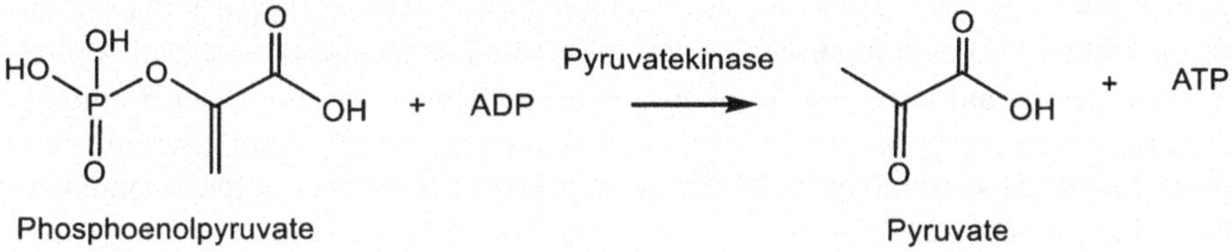

Fig. 1.15 An example of substrate level phosphorylation. (Structures (Fig. 3) resourced with license (https://creativecommons.org/licenses/by/4.0/) from Kolodiazhnyl [65] (review))

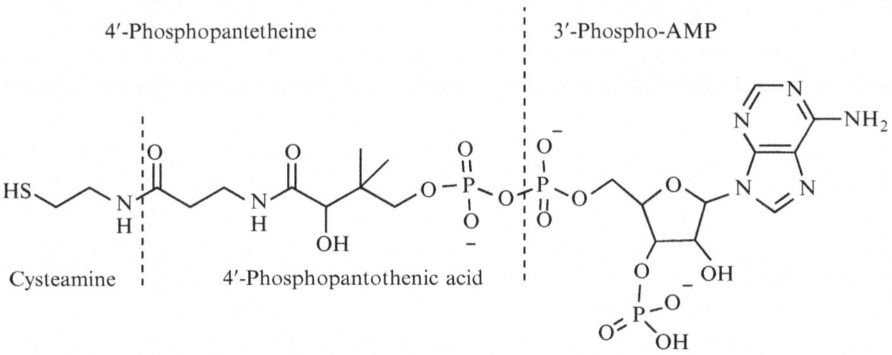

Fig. 1.16 Coenzyme A: Structure of CoA and its components. Note sulfur (S) atom. (Structure (Fig. 15) resourced with license (https://creativecommons.org/licenses/by/4.0/) from Francioso et al. [67])

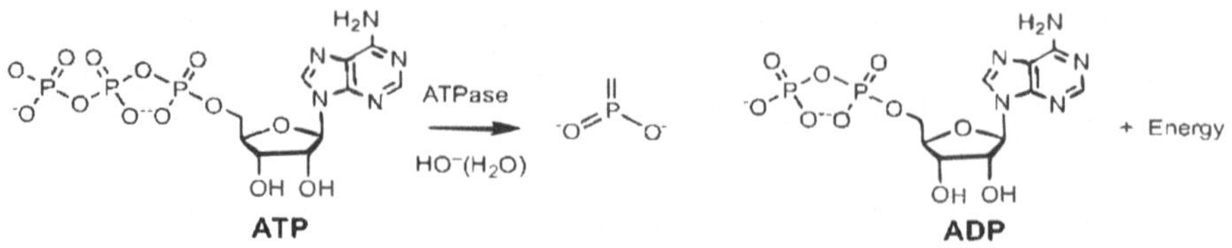

Fig. 1.17 Catalysis of ATP: ATP is converted by ATPase to ADP with release of PO_4^- and 7.3 kcal/mol of energy. (Structures (Fig. 14) resourced with license (https://creativecommons.org/licenses/by/4.0/) from Kolodiazhnyl [65] (review))

found in space but this polyanion has now been found in the oceans of the sixth largest moon of Saturn, Enceladus, which hints at another important ingredient for possible life elsewhere in our Universe [71].

Sulfur ($_{16}S^{32}$) is in the oxygen group 16 location of the periodic table (Fig. 1.6) and is estimated to be the tenth most abundant element in the Universe. Sulfur (S) is created by stellar nucleosynthesis with fusion (at $2.9 \bullet 10^9$ kelvin) of one nucleus of helium ($_2He^4$) and one nucleus of silicon ($_{14}Si^{28}$). S is much less electronegative (2.6) than oxygen (3.4). The innermost of the four Galilean moons of the planet Jupiter, Io, has more than 400 active volcanoes which deposit lava so rich in S that its surface is yellow [72]. S is the fifth most common element (by mass) on Earth. S binds to most metals and is present in many minerals. It is an essential, but not highly abundant, element in all lifeforms and is incorporated only from environmental organic sulfurous compounds [73]. S in the form of S^0 or H_2S is the *e*- donor for anaerobic photosynthesis in purple-sulfur bacteria [67]. Evidence for this process 2.9 Bya, before cyanobacteria initiated the Great Oxidation Event, comes from the age of highly organic black shale [74]; this suggests that one of the earliest lifeforms utilized sunlight, sulfur, and CO_2 to make glucose. In anaerobic bacteria, S^0 and sulfates can also be reduced to function as final *e*- acceptors with sulfides, for example H_2S gas, as a product [75].

There are several unique and important properties of S for lifeforms: (1) S can exist in several oxidation states such as hydrogen sulfide (H_2S, −2), sulfur dioxide (SO_2, +4), and sulfuric acid (H_2SO_4, +6) that gives it versatility to react with many other elements and molecules [67, 73]. (2) S is a component of the essential amino acids cysteine (thio group, R-S-H) and methionine (thioether, R-C-S-R') where the former can form a di-sulfide (S-S) that makes a strong bond and reduces solubility [76]. (3) The structure of co-enzyme A, as given in Fig. 1.16, as in acetyl-CoA, is an essential molecule for moving carbon atoms through the food chain; it contains a thioester (R-C(=O)-SR') high energy bond that is particularly reactive [67, 77]. Hydrolysis of this thioester bond releases energy (exergonic reaction, −31.5 KJ/mol). (4) In H_2O, sulfate salts form a polyanion SO_4^{2-}, which like HCO^{3-} and PO_4^{3-}, and Cl^-, contributes to ionic homeostasis. (5) The large number of oxidation states that S can display makes it ideal for redox biological reactions and electron transfers (e.g., Fe-S clusters in the respiratory complexes (Sects. 2.3, 5.2 and 5.3). Glutathione, that is, GSH tripeptide, is a compound comprised of amino acids cysteine, glutamate, and glycine. GSH is the major intracellular thiol antioxidant. GSH has an unusual γ-glutamyl bond linking glutamate and cysteine. The cellular thiol groups are protected by the "thiol redox buffering system," of which the key components are GSH, either reduced (GSH) or oxidized (GSSG), and the families of antioxidant enzymes named glutaredoxins, thioredoxins, and peroxiredoxins [78–82].

H_2S is a gaseous mediator synthesized mostly from cysteine by cystathionine γ lyase (CSE) but also by other enzymes. H_2S has disparate effects that likely result from the well-known reducing activity of H_2S and/or its ability to promote sulfhydration of protein cysteine moieties within the cell [83]. H_2S inhibits cytochrome *c* oxidase (mitochondrial complex IV) to attenuate mitochondrial respiration and it activates transient receptor potential (TRP) channels (especially TRPA1) and K_{ATP} and other K^+ channels (Sects. 2.1, 6.2, 7.1, 7.2 and 8.1). H_2S may inhibit or activate NF-κB, a nuclear translocation factor, and it is known to alter the activity of numerous kinases including p38 mitogen-activated protein kinase (p38 MAPK), extracellular signal-regulated kinase (ERK), and protein kinase B (PKB) [83].

References

1. Halliday A, Canup RM. The accretion of planet Earth. Nat Rev Earth Environ. 2023;4:19–36. https://doi.org/10.1038/s43017-022-00370-0.
2. Fortenberry RC. First molecule in the universe. Sci Am. 2020;322(2):58–65. https://doi.org/10.1038/scientificamerican0220-58.
3. Grochala W. First there was hydrogen. Nat Chem. 2015;7(3):264. https://doi.org/10.1038/nchem.2186. PubMed PMID: 25698337.
4. Alonso-Monsalve E, Kaiser DI. Primordial black holes with QCD color charge. Phys Rev Lett. 2024;132(23–7):231402. https://doi.org/10.1103/PhysRevLett.132.231402.
5. Villanueva-Domingo P, Mena O, Palomares-Ruiz S. A brief review on primordial black holes as dark matter. Front Astron Space Sci. 2021;8. https://doi.org/10.3389/fspas.2021.681084.
6. Poole G. Energy electromagnetic force equivalence, E = Fe x r. J High Energy Phys Gravit Cosmol. 2019;5(4) https://doi.org/10.4236/jhepgc.2019.54058.
7. Javadi H. A conceptual model of how the physical universe works. General Science Journal [Internet]. 2023. Available from: https://www.researchgate.net/publication/371491022_A_conceptual_model_of_how_the_physical_universe_works.
8. Oks E. Brief review of recent advances in understanding dark matter and dark energy. New Astrom Rev. 2021;93:101632. https://doi.org/10.1016/j.newar.2021.101632.
9. Broggini C. LUNA and the sun. Prog Part Nucl Phys. 2010;64(2):429–31. https://doi.org/10.1016/j.ppnp.2009.12.067.
10. Molloy P. Run on less with hydrogen fuel cells. ACT News. 2019. Available from: https://www.act-news.com/news/fcevs-run-on-less/.
11. Hand E. HIdden hydrogen: does Earth hold a wealth of clean hydrogen? Science. 2023;379(6633):630–6.
12. Naujokaityte G. Hydrogen could be taken straight from the ground. Science Business. 2022. Available from: https://sciencebusiness.net/climate-news/news/hydrogen-could-be-taken-straight-ground.
13. Jones A. Stellar nucleosynthesis: how stars make all of the elements. 2023. In: ThoughtCo [Internet]. Available from: https://www.thoughtco.com/stellar-nucleosynthesis-2699311.

14. Croswell K. The Alchemy of the heavens: searching for meaning in the Milky Way. Anchor; 1996. 356 p. Available from: https://www.penguinrandomhouse.com/books/33879/the-alchemy-of-the-heavens-by-ken-croswell/.
15. Frieden E. The chemical elements of life. Sci Am. 1972;227(1):52–60. https://doi.org/10.1038/scientificamerican0772-52. PubMed PMID: 5044408.
16. Kitadi N, Maruyama S. Origins of building blocks of life: a review. Geosci Front. 2017;9(4):1117–53. https://doi.org/10.1016/j.gsf.2017.07.007.
17. Mertz W. The essential trace elements. Science. 1981;213(4514):1332–8. https://doi.org/10.1126/science.7022654. PubMed PMID: 7022654.
18. Cannas D, Loi E, Serra M, Firinu D, Valera P, Zavattari P. Relevance of essential trace elements in nutrition and drinking water for human health and autoimmune disease risk. Nutrients. 2020;12(7). Epub 20200713. https://doi.org/10.3390/nu12072074. PubMed PMID: 32668647; PMCID: PMC7400883.
19. Balali-Mood M, Naseri K, Tahergorabi Z, Khazdair MR, Sadeghi M. Toxic mechanisms of five heavy metals: mercury, lead, chromium, cadmium, and arsenic. Front Pharmacol. 2021;12:643972. Epub 20210413. https://doi.org/10.3389/fphar.2021.643972. PubMed PMID: 33927623; PMCID: PMC8078867.
20. Martinez-Finley EJ, Chakraborty S, Fretham SJ, Aschner M. Cellular transport and homeostasis of essential and nonessential metals. Metallomics. 2012;4(7):593–605. Epub 20120215. https://doi.org/10.1039/c2mt00185c. PubMed PMID: 22337135; PMCID: PMC4936191.
21. LibreTexts. Introduction to chemical bonding. 12 ed. Raven BIology; 2023. Available from: https://chem.libretexts.org/Bookshelves/Inorganic_Chemistry/Supplemental_Modules_and_Websites_(Inorganic_Chemistry)/Chemical_Compounds/Introduction_to_Chemical_Bonding.
22. Alberts B, Johnson A, Lewis J, Raff M, Roberts K, Walter P. The lipid bilayer. In: Molecular biology of the cell. 4th ed. New York: Garland Science; 2002. Available from: https://www.ncbi.nlm.nih.gov/books/NBK26871/.
23. Robertson JL. The lipid bilayer membrane and its protein constituents. J Gen Physiol. 2018;150(11):1472–83. Epub 20180925. https://doi.org/10.1085/jgp.201812153. PubMed PMID: 30254021; PMCID: PMC6219687.
24. Singh H. Mitochondrial ion channels in cardiac function. Am J Physiol Cell Physiol. 2021;321(5):C812–C25. Epub 20210922. https://doi.org/10.1152/ajpcell.00246.2021. PubMed PMID: 34550794; PMCID: PMC8616592.
25. An J, Varadarajan SG, Camara A, Chen Q, Novalija E, Gross GJ, Stowe DF. Blocking Na^+/H^+ exchange reduces $[Na^+]_i$ and $[Ca^{2+}]_i$ load after ischemia and improves function in intact hearts. Am J Physiol Heart Circ Physiol. 2001;281(6):H2398–409. https://doi.org/10.1152/ajpheart.2001.281.6.H2398. PubMed PMID: 11709405.
26. Malas KM, Lambert DS, Heisner JS, Camara AKS, Stowe DF. Time and charge/pH-dependent activation of K^+ channel-mediated K^+ influx and K^+/H^+ exchange in guinea pig heart isolated mitochondria; role in bioenergetic stability. Biochim Biophys Acta Bioenerg. 2022;1863(8):148908. Epub 20220809. https://doi.org/10.1016/j.bbabio.2022.148908. PubMed PMID: 35961396.
27. Pasek M, Harnmeijer JP, Buick R, Gull M, Atlas Z. Evidence for reactive reduced phosphorus species in the early Archean ocean. Proc Natl Acad Sci USA. 2013; https://doi.org/10.1073/pnas.1303904110.
28. Alberts B, Johnson A, Lewis J, Raff M, Roberts K, Walter P. Electron-transport chains and their proton pumps. In: Molecular biology of the cell. 4th ed. New York: Garland Science; 2002. Available from: https://www.ncbi.nlm.nih.gov/books/NBK26904/.
29. Silverstein TP. The proton in biochemistry: impacts on bioenergetics, biophysical chemistry, and bioorganic chemistry. Front Mol Biosci. 2021;8:764099. Epub 20211126. https://doi.org/10.3389/fmolb.2021.764099. PubMed PMID: 34901158; PMCID: PMC8661011.
30. Babcock GT. How oxygen is activated and reduced in respiration. Proc Natl Acad Sci USA. 1999;96(23):12971–3. https://doi.org/10.1073/pnas.96.23.12971. PubMed PMID: 10557256; PMCID: PMC33932.

31. Haumann J, Camara AKS, Gadicherla AK, Navarro CD, Boelens AD, Blomeyer CA, Dash RK, Boswell MR, Kwok WM, Stowe DF. Slow Ca^{2+} efflux by Ca^{2+}/H^{+} exchange in cardiac mitochondria Is modulated by Ca^{2+} re-uptake via MCU, extra-mitochondrial pH, and H^{+} pumping by F_OF_1-ATPase. Front Physiol. 2018;9:1914. https://doi.org/10.3389/fphys.2018.01914. Epub 20190204. PubMed PMID: 30804812; PMCID: PMC6378946.
32. Weiss MS, Brandl M, Suhnel J, Pal D, Hilgenfeld R. More hydrogen bonds for the (structural) biologist. Trends Biochem Sci. 2001;26(9):521–3. https://doi.org/10.1016/s0968-0004(01)01935-1. PubMed PMID: 11551776.
33. Deepak RN, Sankararamakrishnan R. Unconventional N-H…N hydrogen bonds involving proline backbone nitrogen in protein structures. Biophys J. 2016;110(9):1967–79. https://doi.org/10.1016/j.bpj.2016.03.034. PubMed PMID: 27166805; PMCID: PMC4939747.
34. Watties J. NASA rover finds 'clearest evidence yet' of an ancient lake on Mars. In: CNN, editor. Atlanta: CNN; 2023. Available from: https://www.cnn.com/2023/02/10/world/mars-nasa-curiosity-rover-ancient-waters-scn/index.html.
35. Taylor R, Kennard O. Crystallographic evidence for the existence of C–H…O, C–H…N, and C–H…Cl hydrogen bonds. J Am Chem Soc. 1982;104:5063–70. https://doi.org/10.1021/ja00383a012.
36. Nielsen FH. Evolutionary events culminating in specific minerals becoming essential for life. Eur J Nutr. 2000;39(2):62–6. https://doi.org/10.1007/s003940050003. PubMed PMID: 10918986.
37. Tielens A. The physics and chemistry of the interstellar medium. Cambridge: Cambridge University Press; 2005. ISBN 9780511819056 (online 2012). Available from: https://www.cambridge.org.
38. Dulieu F, Amiaud L, Congui E, Fillion JH, Matar E, Momeni A, Pirronello V, Lemaire JL. Experimental evidence for water formation on interstellar dust grains by hydrogen and oxygen atoms. Astron Astrophys. 2010;512:A30. https://doi.org/10.1051/0004-6361/200912079.
39. Drake M. Origin of water in the terrestrial planets. Meteorit Planet Sci. 2005;40(4):519–27. https://doi.org/10.1111/j.1945-5100.2005.tb00960.x.
40. Sleep NH. The Hadean-Archaean environment. Cold Spring Harb Perspect Biol. 2010;2(6):a002527. https://doi.org/10.1101/cshperspect.a002527. Epub 20100505. PubMed PMID: 20516134; PMCID: PMC2869525.
41. Sleep NH, Zahnle K, Neuhoff PS. Initiation of clement surface conditions on the earliest Earth. Proc Natl Acad Sci USA. 2001;98(7):3666–72. https://doi.org/10.1073/pnas.071045698. Epub 20010320. PubMed PMID: 11259665; PMCID: PMC31109.
42. Korenaga J, Planavsky NJ, Evans DAD. Global water cycle and the coevolution of the Earth's interior and surface environment. Philos Trans A Math Phys Eng Sci. 2017;375(2094):20150393. https://doi.org/10.1098/rsta.2015.0393. PubMed PMID: 28416728; PMCID: PMC5394256.
43. Mojzsis SJ, Harrison TM, Pidgeon RT. Oxygen-isotope evidence from ancient zircons for liquid water at the Earth's surface 4,300 Myr ago. Nature. 2001;409(6817):178–81. https://doi.org/10.1038/35051557. PubMed PMID: 11196638.
44. Pearson DG, Brenker FE, Nestola F, McNeill J, Nasdala L, Hutchison MT, Matveev S, Mather K, Silversmit G, Schmitz S, Vekemans B, Vincze L. Hydrous mantle transition zone indicated by ringwoodite included within diamond. Nature. 2014;507(7491):221–4. https://doi.org/10.1038/nature13080. PubMed PMID: 24622201.
45. Woods MW, MB. Seven natural wonders of the Arctic, Antarctica, and the oceans. Minneapolis: Twenty-First Century Books, Lerner Publishing Group; 2009. 978-0-8225-9075-0.
46. Baker V, Strom RG, Gulick VC, Kargel JS, Komatsu G, Kale VS. Ancient oceans, ice sheets and the hydrological cycle on Mars. Nature. 1991;352(6348):589–94. https://doi.org/10.1038/352589a0.
47. Feldman W, Prettyman TH, Maurice S, Plaut JJ, Bish DL, Vaniman DT, Mellon MT, Metzger AE, Squyres SW, Karunatillake S, Boynton WV, Elphic RC, Funsten HO, Lawrence J, Tokar RL. Global distribution of near-surface hydrogen on Mars. J Geophys Res. 2004;109(E9):CiteID E09006. https://doi.org/10.1029/2003JE002160.

48. Christensen PR. Water at the poles and in permafrost regions of Mars. Elements. 2006;3(2):151–5. https://doi.org/10.2113/gselements.2.3.151.
49. Khuller A, Christensen P. Evidence of exposed dusty water ice within Martian gullies. J Geophys Res Planets. 2021;126(2):e2020JE006539. https://doi.org/10.1029/2020JE006539.
50. Noack L, Snellen I, Rauer H. Water in extrasolar planets and implications for habitability. Space Sci Rev. 2017;212:877–98. https://doi.org/10.1007/s11214-017-0413-1.
51. Kite E, Ford EB. Habitability of exoplanet waterworlds. Astrophys J. 2018;864:ApJ 864 75. https://doi.org/10.3847/1538-4357/aad6e0.
52. Totani T. Emergence of life in an inflationary universe. Sci Rep. 2020;10(1):1671. https://doi.org/10.1038/s41598-020-58060-0. Epub 20200203. PubMed PMID: 32015390; PMCID: PMC6997386.
53. Carroll B, Ostlie DA. An introduction to modern stellar astrophysics. San Francisco: Addison Wesley; 2007. 978-0-8053-0348-3
54. Blanksby SJ, Ellison GB. Bond dissociation energies of organic molecules. Acc Chem Res. 2003;36(4):255–63. https://doi.org/10.1021/ar020230d. PubMed PMID: 12693923.
55. Monteiro-Carvalho A, Sigaud L, Montenegro EC. O^{+}_2 production coming from CO_2 single-event electron impact. Phys Rev Lett. 2024;132(15):3002. https://doi.org/10.1103/PhysRevLett.132.153002.
56. West JB. The strange history of atmospheric oxygen. Phys Rep. 2022;10(6):e15214. https://doi.org/10.14814/phy2.15214. PubMed PMID: 35347882; PMCID: PMC8960603.
57. Harrison J, Kaiser A, VandenBrookes JM. Atmospheric oxygen level and the evolution of insect body size. Proc R Soc. 2010;277:1937–46. https://doi.org/10.1098/rspb.2010.0001.
58. Luo G, Ono S, Beukes NJ, Wang DT, Xie S, Summons RE. Rapid oxygenation of Earth's atmosphere 2.33 billion years ago. Sci Adv. 2016;2(5):e1600134. https://doi.org/10.1126/sciadv.1600134. Epub 20160513. PubMed PMID: 27386544; PMCID: PMC4928975.
59. Alberts B, Johnson A, Lewis J, Raff M, Roberts K, Walter P. The evolution of electron-transport chains. In: Molecular biology of the cell. 4th ed. New York: Garland Science; 2002. Available from: https://www.ncbi.nlm.nih.gov/books/NBK26849/.
60. Desai R, Coates AJ, Wellbrock A, Vuitton V, Crary FJ, González-Caniulef D, Shebanits O, Jones GH, Lewis GR, Waite JH, Cordiner M, Taylor SA, Kataria DO, Wahlund JE, Edberg NJT, Sittler EC. Carbon chain anions and the growth of complex organic molecules in Titan's ionosphere. Astrophys J Lett. 2017;844(2):L18. https://doi.org/10.3847/2041-8213/aa7851.
61. Grewal D, Dasgupta R, Marty B. A very early origin of isotopically distinct nitrogen in inner Solar System protoplanets. Nat Astron. 2021;5:356–64. https://doi.org/10.1038/s41550-020-01283-y.
62. Gull M, Mojica MA, Fernandez FM, Gaul DA, Orlando TM, Liotta CL, Pasek MA. Nucleoside phosphorylation by the mineral schreibersite. Sci Rep. 2015;5:17198. https://doi.org/10.1038/srep17198. Epub 20151126. PubMed PMID: 26606901; PMCID: PMC4660433.
63. La Cruz NL, Qasim D, Abbott-Lyon H, Pirim C, McKee AD, Orlando T, Gull M, Lindsay D, Pasek MA. The evolution of the surface of the mineral schreibersite in prebiotic chemistry. Phys Chem Chem Phys. 2016;18(30):20160–7. https://doi.org/10.1039/c6cp00836d. PubMed PMID: 27157087.
64. Fernandez-Garcia C, Coggins AJ, Powner MW. A chemist's perspective on the role of phosphorus at the origins of life. Life (Basel). 2017;7(3):31. https://doi.org/10.3390/life7030031. Epub 20170713. PubMed PMID: 28703763; PMCID: PMC5617956.
65. Kolodiazhnyi OI. Phosphorus compounds of natural origin: prebiotic, stereochemistry, application. Symmetry. 2021;13(5):889. https://doi.org/10.3390/sym13050889.
66. Butusov M, Jernelöv A. The role of phosphorus in the origin of life and in evolution. New York: Springer; 2013. 978-1-4614-6802-8. https://doi.org/10.1007/978-1-4614-6803-5_1.
67. Francioso A, Baseggio Conrado A, Mosca L, Fontana M. Chemistry and biochemistry of sulfur natural compounds: key intermediates of metabolism and redox biology. Oxidative Med Cell Longev. 2020;2020:8294158. https://doi.org/10.1155/2020/8294158. Epub 20200929. PubMed PMID: 33062147; PMCID: PMC7545470.

68. Kittredge JS, Roberts E. A carbon-phosphorus bond in nature. Science. 1969;164(3875):37–42. https://doi.org/10.1126/science.164.3875.37. PubMed PMID: 4388111.
69. Cadez T, Kolic D, Sinko G, Kovarik Z. Assessment of four organophosphorus pesticides as inhibitors of human acetylcholinesterase and butyrylcholinesterase. Sci Rep. 2021;11(1):21486. https://doi.org/10.1038/s41598-021-00953-9. Epub 20211102. PubMed PMID: 34728713; PMCID: PMC8563940.
70. Peter JV, Sudarsan TI, Moran JL. Clinical features of organophosphate poisoning: a review of different classification systems and approaches. Indian J Crit Care Med. 2014;18(11):735–45. https://doi.org/10.4103/0972-5229.144017. PubMed PMID: 25425841; PMCID: PMC4238091.
71. Postberg F, Sekine Y, Klenner F, Glein CR, Zou Z, Abel B, Furuya K, Hillier JK, Khawaja N, Kempf S, Noelle L, Saito T, Schmidt J, Shibuya T, Srama R, Tan S. Detection of phosphates originating from Enceladus's ocean. Nature. 2023;618(7965):489–93. https://doi.org/10.1038/s41586-023-05987-9. Epub 20230614. PubMed PMID: 37316718; PMCID: PMC10266972.
72. Wild S. Violent volcanoes have wracked Jupiter's moon Io for billions of years. Nature. 2024; https://doi.org/10.1038/d41586-024-01138-w. Epub 20240418. PubMed PMID: 38637709.
73. Chatterjee S, Hausinger RP. Sulfur incorporation into biomolecules: recent advances. Crit Rev Biochem Mol Biol. 2022;57(5–6):461–76. https://doi.org/10.1080/10409238.2022.2141678. Epub 20221120. PubMed PMID: 36403141; PMCID: PMC10192010.
74. Jin Z, Wang X, Wang H, Ye Y, Zhang S. Organic carbon cycling and black shale deposition: an Earth System Science perspective. Natl Sci Rev. 2023;10(11):nwad243. https://doi.org/10.1093/nsr/nwad243. Epub 20230915. PubMed PMID: 37900193; PMCID: PMC10612131.
75. Kushkevych I, Cejnar J, Treml J, Dordevic D, Kollar P, Vitezova M. Recent advances in metabolic pathways of sulfate reduction in intestinal bacteria. Cells. 2020;9(3):698. https://doi.org/10.3390/cells9030698. Epub 20200312. PubMed PMID: 32178484; PMCID: PMC7140700.
76. Brosnan JT, Brosnan ME. The sulfur-containing amino acids: an overview. J Nutr. 2006;136(6 Suppl):1636S–40S. https://doi.org/10.1093/jn/136.6.1636S. PubMed PMID: 16702333.
77. Pietrocola F, Galluzzi L, Bravo-San Pedro JM, Madeo F, Kroemer G. Acetyl coenzyme A: a central metabolite and second messenger. Cell Metab. 2015;21(6):805–21. https://doi.org/10.1016/j.cmet.2015.05.014. PubMed PMID: 26039447.
78. Stowe DF, Camara AK. Mitochondrial reactive oxygen species production in excitable cells: modulators of mitochondrial and cell function. Antioxid Redox Signal. 2009;11(6):1373–414. https://doi.org/10.1089/ars.2008.2331. PubMed PMID: 19187004; PMCID: PMC2842133.
79. Pai HV, Starke DW, Lesnefsky EJ, Hoppel CL, Mieyal JJ. What is the functional significance of the unique location of glutaredoxin 1 (GRx1) in the intermembrane space of mitochondria? Antioxid Redox Signal. 2007;9(11):2027–33. https://doi.org/10.1089/ars.2007.1642. PubMed PMID: 17845131.
80. Inarrea P, Moini H, Han D, Rettori D, Aguilo I, Alava MA, Iturralde M, Cadenas E. Mitochondrial respiratory chain and thioredoxin reductase regulate intermembrane Cu,Zn-superoxide dismutase activity: implications for mitochondrial energy metabolism and apoptosis. Biochem J. 2007;405(1):173–9. https://doi.org/10.1042/BJ20061809. PubMed PMID: 17394422; PMCID: PMC1925252.
81. Buettner GR, Ng CF, Wang M, Rodgers VG, Schafer FQ. A new paradigm: manganese superoxide dismutase influences the production of H_2O_2 in cells and thereby their biological state. Free Radic Biol Med. 2006;41(8):1338–50. https://doi.org/10.1016/j.freeradbiomed.2006.07.015. Epub 20060721. PubMed PMID: 17015180; PMCID: PMC2443724.
82. Zhu W, Richards NGJ. Biological functions controlled by manganese redox changes in mononuclear Mn-dependent enzymes. Essays Biochem. 2017;61(2):259–70. https://doi.org/10.1042/EBC20160070. Epub 20170509. PubMed PMID: 28487402.
83. Li L, Rose P, Moore PK. Hydrogen sulfide and cell signaling. Annu Rev Pharmacol Toxicol. 2011;51:169–87. https://doi.org/10.1146/annurev-pharmtox-010510-100505. PubMed PMID: 21210746.

Chapter 2
Light Metal Ions and Oxyanions as Current Carriers for Early Lifeforms

Abstract The cations sodium and potassium have a valence of +1 and form ionic bonds with anions. Because of their low charge density and valency, these cations are easily separated from their anions and solvate in water. Trapped by ancient seas into nascent spheroid lipid membranes, these ions are essential for cell homeostasis. Their separation across membranes establishes an electrochemical gradient that with ion pumps and exchangers function as a battery to store energy that is released and regenerated. The anion chloride balances out most of the cell sodium and potassium in ionic binding. Oxygen containing carbonates, nitrates, phosphates, and sulfates also play important roles as polyanions in combination with sodium, potassium, and divalent cations. Important divalent and trivalent cations are magnesium, calcium, iron, and copper. Variable valence atoms are important for transferring electrons in the electron transport system, for mediating catalytic activity, as cofactors in redox reactions, for carriage of O_2 in muscle and blood, for intracellular signaling, and for many other functions.

Keywords Acid-base balance · Alkali metals · Antioxidants · Archaean Eon · Calcium element · Carbonates · Catalysts · Chlorine element · Copper element · Electron transport system · Electrochemical gradient · High-energy phosphates · Ion homeostasis · Iron element · Last universal common ancestor · Lithium element · Magnesium element · Manganese element · Metal cofactors · Na^+K^+-ATPase · Na^+-H^+ exchanger · Na^+-Ca^{2+} exchanger · Nitrates · Oxyanions · Oxypolyanions · pH scale · Potassium element · Phosphates · Oxidation-reduction reactions · Sodium element · Sulfates · Superoxide ion · Superoxide dismutases · Transmembrane gradients · Transition metals

2.1 Na^+ and K^+ as Batteries and Chemi-osmoregulators

Sodium and Potassium ($_{11}Na^{23}$ and $_{19}K^{39}$) are Group 1A alkali metal atoms (i.e., group H, Li, Na, K, Rb, Cs, and Fr), which form ionic bonds (Fig. 1.6). Only the H atom forms covalent bonds (e.g., CH_4) as well as ionic bonds (e.g., $H^+ + Cl^-$). The

D. F. Stowe, *Evolution of Bioenergetics from Elements to Life*,
https://doi.org/10.1007/978-3-031-76668-8_2

universal importance of the H atom in forming the cosmos and its indispensable roles in nature are discussed elsewhere in Sects. 1.1, 1.2, 1.3, 3.1, 5.1, 5.2, 5.3, 6.2, 6.4 and 7.1. Each of the remaining metal elements with a valency of +1 form ionic bonds like Na^+ with Cl^-, or Cs^+ with SO_4^-, by donating *e-* to anions to form salts, hydroxides, and oxyanions like carbonates, phosphates, nitrates, and sulfates. Salts of Group 1A alkali metals, especially Na^+ and K^+, are very soluble in water because their charge densities are low with the valency of +1; thus, they are easily separated from their anions or polyanions and solvated by polar solvents like water [1]. Of these Group 1A metal cations, only the abundant Na^+ and K^+ cations, dissolved in the watery fluids of early lifeforms, are used naturally as electrolytes with H^+ to establish membrane potentials and for biochemical reactions and cell and organelle communication.

Abundant in the universe as dusty water ice, and on Earth as atmospheric H_2O vapor and steam, once liquid H_2O rained down on Earth, Na^+ and K^+ were likely the cations selected to control ionic chemiosmotic balance because of their unique chemical properties. Early life existed in an excess of K^+ over Na^+ in hydrothermal fields so nascent enveloping membranes contained more K^+, as do archaea, bacteria, and eukaryotes today [2–4]. Mechanisms were needed to balance the increasingly Na^+-rich watery environment with the K^+-rich cytoplasmic compartment encased by a nascent lipid membrane (see Sects. 3.2 and 7.1). To counteract the Na^+ leaking into these primordial cells and to maintain cell $K^+_{in} > Na^+_{out}$, they needed to be able to extrude Na^+ ions [3].

Membranes Figure 2.1 [3] depicts one proposed scenario for evolutionary transition from an ion leaky, primitive membrane to a selectively permeable membrane that facilitated the development of ion pumps and ion exchangers that supported transmembrane ion gradients and ion homeostasis. In this way ions and molecules, differentially sequestered by a semipermeable membrane, provide the driving force behind the evolution of diverse redox, light, chemical or osmotic-dependent Na^+ export pumps, like the life-essential Na^+/K^+-ATPase pump we know today (Sects. 7.1 and 8.1). This pump essentially increases membrane impermeability and effectively establishes a resting electrochemical gradient across the cell membrane [3].

So it is well established that cell membranes and electrochemical gradients established by ion pumps and their associated ion exchangers are essential for life [5]. To carry out the complex regulation of cells and organelles, there are now known to be many families of Na^+ and K^+ channels that generally allow fast and selective movement of either cation according to their electrochemical gradient [6]. Many are post-transitionally modified after protein synthesis to regulate cation flux that performs a particular function. To maintain osmolar balance, there are slower exchangers for Na^+ and K^+, typically coupled to H^+ efflux, for example, a Na^+-H^+ exchanger (NHE) as well as a Ca^{2+} exchanger for Na^+ (NCE). The roles for a K^+-H^+ exchanger (KHE) and/or a Ca^{2+}-H^+ (CHE) exchanger in cell membranes are less clear. If present, they may be normally nonfunctional.

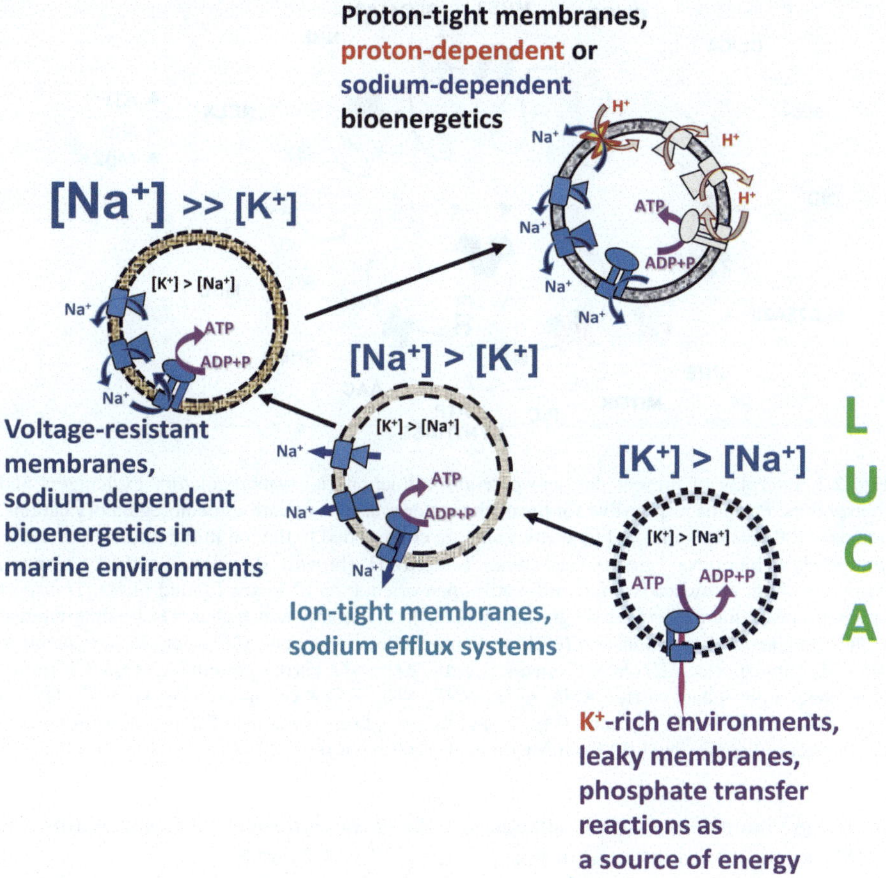

Fig. 2.1 Evolution of Na^+/K^+ homeostasis and membrane bioenergetics: The proposed scenario involves a transition from primitive membranes that were leaky to both Na^+ and H^+ (dotted lines), to membranes that could support Na^+ efflux pumps, and to membranes that were tight enough to support Na^+ dependent membrane bioenergetics (dashed lines). Transition occurred to the modern type membranes that are impermeable to H^+, K^+, and Na^+ ions (solid lines). The modern membrane is likely to have been fully functional well before the possible first separation of the protocell into bacterial and archaeal domains, called the Last Universal Common Ancestor (LUCA). (Cartoon (Fig. 7) and modified legend resourced with permission from copyright owner of Dibrova et al. [3])

The chemical and voltage gradients across the inner mitochondrial membrane (IMM) of mitochondria residing within cells are differently maintained than in their cell membranes. Likely because of the proton pumping complexes in mitochondria, but no longer in the cell membranes, there are no known Na^+ channels or Na^+ pumps in mitochondria. But there are many types of ligand-specific mitochondrial K^+ channels along with all four of the ion exchangers for Na^+, K^+, Ca^{2+}, and H^+. Figure 2.2 diagrams several known and putative mitochondrial ion channels and ion exchangers embedded within the inner mitochondrial membrane (IMM) [4]. The functions

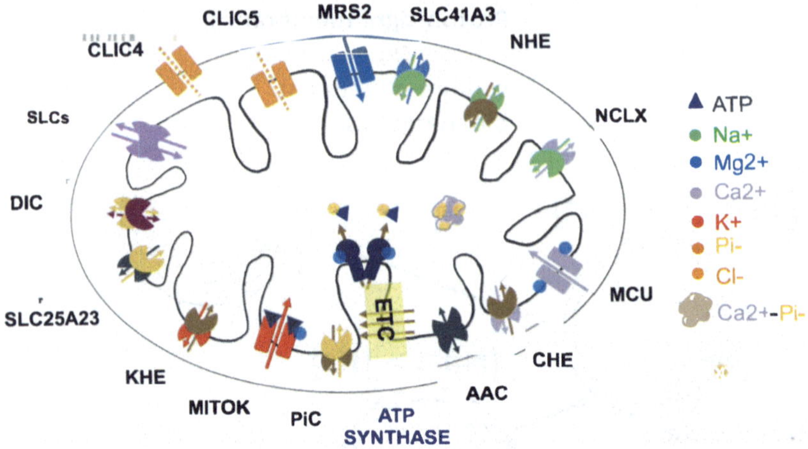

Fig. 2.2 Interplay of present day and potential mitochondrial osmoregulatory exchangers and transporters: Proteins responsible for the exchange or transport of various osmoregulatory cations, anions and fluxes are indicated by arrows and are color coded to the ion in the mitochondria diagrams: Mg^{2+} (blue), Na^+ (green), Ca^{2+} (lilac), K^+ (red), H^+ (brown), P_i (yellow), and Cl^- (orange). Arrows without arrowheads refer to the still open orientation of fluxes (dotted lines). Transport processes involving carriers are diagramed in semicircular components; channels are diagrammed with rectangles. Abbreviations: *CLIC* chloride intracellular channels, *SLC* solute carrier channels, *MRS* Mg^{2+} transporter, *MITOK* mK channels, *AAC, ADP, ATP* carrier protein (ANT), *MCU* mCa^{2+} uniporter, *Pi* phosphate carrier, *KHE, CHE, NHE, NCE, NCLX* cation exchangers, *ETC* electron transport chain (system). (Cartoon (Fig. 1) and its legend resourced with license (https://creativecommons.org/licenses/by/4.0/) from Austin et al. [4] (review))

of cation channels and ion exchangers, both in the cell membrane and within the IMM, are discussed in detail in Sects. 7.1, 7.2, 8.1, 8.2 and 8.3.

2.2 Cl^- and Oxyanions as Regulators of Ionic and Acid-Base Balance, and as Molecular Structural Components

Chlorine ($_{17}Cl^{35}$) as in Cl_2 is a gas but the element exists in nature mostly as a −1 charged anion with Na as sodium salt, $Na^+ + Cl^-$ (chloride). As an electrolyte, Cl^- comprises about two-thirds of all the anions in biological fluids, the rest being polyanionic carbonates, sulfates, nitrates, and phosphates. Cl^- has fundamental biological functions in all cells and organelles [7]. Cl^- homeostasis is a critical aspect of animal cell physiology that is often considered passive. But Cl^- has an important role in maintaining the charge balance across cell membranes, including the IMM. Cl^-, principally with the cations Na^+ and K^+, are responsible for cell osmotic pressure, ionic homeostasis, and indirectly, acid-base balance. Cl^- is the major contributor to the osmotic pressure gradient between the intracellular fluid (ICF) and extracellular fluid (ECF). Chloride maintains proper hydration and functions to balance the cations in the ECF to keep the electrical neutrality of this fluid. Cl^- is not in passive electrochemical equilibrium in cells, so its regulation involves the coordination of Cl^- channels (CLC), cotransporters, and exchangers that utilize electrochemical potential gradients to move two or more protons and ions in the same or opposite directions across the cell membrane. An overview of these mechanisms for transport of Cl^- and other anions is given in Fig. 2.3 [7].

Cl^- can ionize with organic molecules, as in the amino acid lysine prepared as lysine hydrochloride. Covalently bonded, chlorinated organic compounds are found in various plant biomolecules including alkaloids (N-containing), flavonoids, steroids, and fatty acids, but they are not naturally found in animal lifeforms [8]. Chlorinated hydrocarbons like chloroform, that is, trichloromethane ($CHCl_3$), and carbon tetrachloride (CCl_4) are very toxic to animal cells as they interact with natural cellular molecules to form oxidative free radical species [9]. The perchlorate (ClO_4^-) and chlorate (ClO_3^-) polyanions are also very toxic and not considered further here [10].

In addition to the Cl^- anion, the simple oxyanionic (oxides of nonmetals) compounds CO_3^{2-}, SO_4^{2-}, NO^{3-}, and PO_4^{3-} (Sects. 1.3, 1.4, 2.2, 6.2 and 8.3) also participate in cell and mitochondrial function. They are now mostly obtained from the diet. The O bonds with C, S, N, and P are covalent, except for the variable number of O atoms (n = 2,3) that are lacking e- that permits it to form the ionic bond with the cation.

Carbonate (CO_3^{2-}) is the second most abundant polyanion in body fluids after Cl^-. Carbonates of Na, like $NaHCO_3$, mixed with the weak acid carbonic acid, H_2CO_3, is a classic buffer that keeps the pH of body fluids stable within a narrow pH range. Ninety percent of the CO_2 produced during mitochondrial respiration is

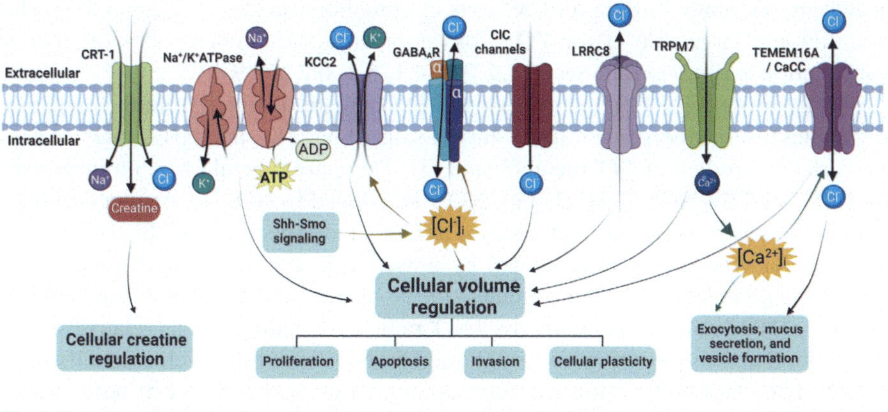

Fig. 2.3 Anion channels: Overview of chloride ion channels, ion cotransporters, and the creatine transporter in regulating cellular volume or creatine. CRT-1, creatine transporter 1; KCC2, K^+-Cl^- cotransporter 2; ClC channels, Cl^- channels; LRRC8, leucine-rich repeat-containing 8; TMEM16A, transmembrane protein 16A; TRPM7, transient receptor potential cation channel subfamily M member 7; CaCC, Ca^{2+}-activated Cl^- channel, and GABAAR, γ-aminobutyric acid (GABA) A receptor. (The diagram was created using BioRender.com. Cartoon (Fig. 1) and its legend resourced with license (https://creativecommons.org/licenses/by/4.0/) from Zhang et al. [7] (editorial))

converted to H_2CO_3 by carbonic anhydrase for transport around the body and is converted back to CO_2 only as it exits the blood in the lungs. Carbonates of Ca form the basic structures of skeletons and exoskeletons and so are essential for larger organisms to form shapes and for ambulation.

Sulfate (SO_4^{2-}) is the fourth most abundant polyanion in body fluids; most is obtained from two amino acids, methionine and cysteine, in food. As a hydrophilic anion, SO_4^{2-} like all oxyanions, cannot permeate membranes but there are known to be many cotransporters, often coupled with Na^+, to handle SO_4^{2-} flux [11]. SO_4^{2-} does not have a known catalytic or essential metabolic function in animals but it is involved in a variety of important biological processes including biosynthesis and detoxification via sulfation of many endogenous and exogenous compounds. In anaerobic bacteria, SO_4^{2-}, like S^0, can function as an *e*- acceptor in chemiosmosis [12]. Sulfate is required for cell matrix synthesis and for the maintenance of cell membranes. Sulfate conjugation also serves a role in the production of steroids, neurotransmitters, and bile acids [11]. SO_4^{2-} is essential for the biosynthesis of structural components of membranes and tissues, for example, cerebroside sulfate in myelin membranes of the brain, and in the formation of sulfated glycosaminoglycans (sGAG), a major component of cartilage [11].

Of the inorganic cations present in the mitochondrial matrix, only Na and K salts, for example, $NaHCO_3$, HK_2PO_4 (dipotassium hydrogen phosphate) readily dissociate, whereas Ca and Mg salts, for example, Ca_2SO_4 and $MgCO_3$, are relatively insoluble, and so with polyanions, Ca^{2+} and Mg^{2+} mostly form precipitates [13], as in tricalcium phosphate, $Ca_3(PO^4)_2$ discussed in Sect. 8.3. In precipitation, the atoms are bound with no remaining charge (0 valence) and so do not participate in homeostasis or function as chemical messengers. In mammals, bone and teeth bind up to 85% of the body's phosphate as part of calcium phosphate salts. Calcium phosphates are the major buffers for Ca^{2+} ions in mitochondria (Sects. 7.2 and 8.3). The essential and large role of the PO_4^{3-} oxyanion in adenine nucleotides on energy transduction is discussed in Sects. 1.4, 3.1, 4.1, 5.1, 6.2 and 9.4.

Mostly nitrates (NO^{3-}), but also nitrites (NO^{2-}), are obtained from the diet (raw vegetables); some inorganic nitrates are the salts of Na^+, K^+, and NH_4^+. NO^{2-} is also an oxidative product of $NO^•$ metabolism [14]. There are several known transporters for NO^{3-}, and for NO^{2-} [15]. Dietary NO^{3-} is reduced to NO^{2-} by commensal bacteria in the gastrointestinal tract because animals lack specific and effective nitrate reductase enzymes. $NO^•$ is generated by nitric oxide synthases (NOS) and is initially reduced to NO^{2-}. NO^{2-} is unique to the N-oxides in its redox position between oxidative ($NO_2^•$ radical) and reductive ($NO^•$ radical) signaling and its relative stability in blood and tissue. Of more recent interest, NO^{2-} can be recycled in cells back to $NO^•$ [14]. Once NO^{2-} is formed, there are numerous pathways for its reduction to $NO^•$ that utilize the Fe or Fe-S redox centers in hemoglobin, myoglobin, and xanthine oxidoreductase, among others [14]. Therefore, NO^{2-} can act as a storage pool for $NO^•$ when needed during cell stress. In the presence of superoxide ($O_2^{•-}$) and $NO^•$, and in the absence or dearth of superoxide dismutases, these two radicals combine to form the non-radical and charged, cell toxic, peroxynitrite ($O{=}N{-}O{-}O^-$) molecule [16]. The roles of generated $NO^•$ and its products are discussed in Sect. 8.4.

2.3 Mg^{2+}, Ca^{2+}, Fe^{2+}, and Cu^{2+} as Catalysts, Cell Modulators, and Messengers

Magnesium and Calcium ($_{12}Mg^{24}$ and $_{20}Ca^{40}$) are group 2A alkaline Earth metals that also include Be, Sr, Ba, and Ra. The elements Ca and Mg, in their form as divalent cations, are discussed here due to their functional importance in all lifeforms. Interestingly, the other four group 2A metals are not naturally present in lifeforms. During stellar nucleosynthesis Ca nuclei were formed by the fusion of argon ($_{18}Ar^{36}$) nuclei with He ($_2He^4$) nuclei; Mg nuclei were formed by fusion of Neon ($_{10}Ne^{20}$) nuclei with $_2He^4$ nuclei. Both elements are common in the Earth's crust in the form of precipitates of carbonates, sulfates, phosphates, and others [17]. Oxyanions of the Group 2A metals are much less soluble in water than those of the Group 1A alkali metals Na^+ and K^+, because of the higher charge densities at a valency of +2 versus +1 valence [1]. Most Ca^{2+} carbonates found today are derived from shells and skeletons of long dead lifeforms. Mg^{2+} is the third most abundant ion in seawater after Na^+ and Cl^-. The importance of both Ca and Mg atoms in life are as cations in electrolytic solutions. However, Ca^{2+} is also a very important regulator of many processes in the cytosol and in the mitochondrial matrix, and is discussed further in Sects. 7.2, 8.1, 8.2, 8.3 and 9.2. Recall that in cells, Mg^{2+} exists in mM range concentrations, whereas Ca^{2+} is restricted normally to high nM or low μM concentrations (Sect. 1.2).

Mg^{2+} as Catalyst Mg^{2+} holds a key position in the first chemical processes leading to the origin and early evolution of life because of its coordination properties [18]. Mg^{2+} can coordinate six O atoms efficiently in its first coordination shell and it can bind together two different phosphate (P_i) groups that are located at a distance from each other in a macromolecule as depicted in Fig. 2.4 [18], which allows for the folding of molecules like RNA. As for Ca^{2+}, cell Mg^{2+} levels are highly regulated via transport and buffering mechanisms to optimize the activity of enzymes and metabolic cycles [19]. Mg^{2+} acts as a catalyst for over 300 enzymes, for example, the large group of protein kinases. Mg^{2+} orients and stabilizes with the 2 P-O bonds of phosphates on ATP to function with ATPase as a coenzyme in the hydrolysis of ATP to ADP and P_i [20] as well as acting as a coenzyme in ATP synthesis (Fig. 2.4) [18, 21]. Mg^{2+} is incorporated into mitochondrial ATP-Mg^{2+}/P_i carriers that import adenine nucleotides into mitochondria and export P_i to the cytosol [22]. These carriers are regulated by Ca^{2+} to control the size of the adenine nucleotide pool in response to cellular energetic demands. Mg^{2+} is essential for nucleic acid synthesis in all organisms known [19]. The higher energy of formation for binding of Mg^{2+} over Ca^{2+} in polynucleotides and other phosphate-containing molecules may also explain the relative binding specificity of Mg^{2+} over Ca^{2+} in this role [23].

In almost any instance where the Ca^{2+} ion is an activator, excess Mg^{2+} as an ion can attenuate or inhibit the response. For example, in the presence of tropomyosin, Mg^{2+} competitively inhibits Ca^{2+} activation of troponin to initiate actin-myosin

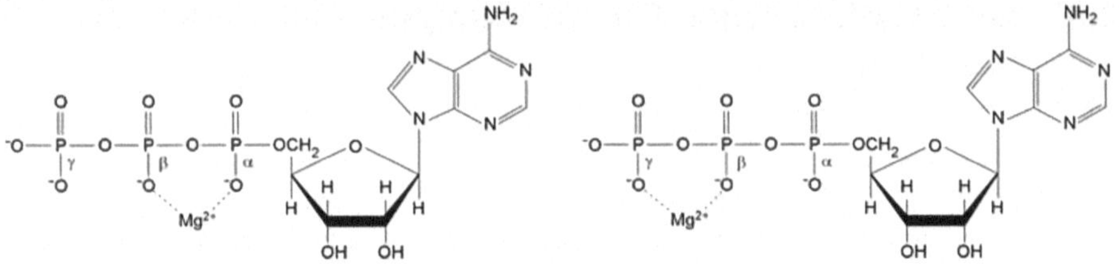

Fig. 2.4 Possible Mg^{2+} complexes with ATP: The most common complex is formed with oxygen of the β- and γ-P parts of the oxyanion (right figure). (Drawing (Fig. 1) and its legend resourced with license (https://creativecommons.org/licenses/by/4.0/) from Holm [18])

contraction [24]. Mg^{2+} also interferes with Ca^{2+}-ATPase dependent-sarcoplasmic reticulum (SR) Ca^{2+} reuptake during muscle relaxation [25]. The effects of mitochondrial Ca^{2+} influx via mitochondrial Ca^{2+} uptake (MCU) channels, Ca^{2+} buffering in the matrix, and Ca^{2+} efflux by Na^{+}-Ca^{2+} and Ca^{2+}-H^{+} exchangers (NCE and CHE) on mitochondrial functional bioenergetics, and the role of Mg^{2+} in modulating the numerous effects of Ca^{2+} are discussed in detail in Sects. 7.2, 7.4, 8.1 and 8.3.

Iron and Copper (${}_{26}Fe^{56}$ and ${}_{29}Cu^{64}$) are transition metals (Fig. 1.6) that are essential metal cofactors in redox reactions, notably in the electron transfer system (ETS) process of respiration. Both commonly have two oxidation states in biology [26, 27], Fe^{2+} (II, ferrous) and Fe^{3+} (III, ferric), and Cu^{1+} (I, cuprous) and Cu^{2+} (II, cupric). They individually oscillate between these two states during *e*- transfer in the ETS. Fe is a more reactive metal than Cu, so when pure Fe^0 is added to a $CuSO_4$ solution, the Fe^0 is oxidized to Fe^{2+}, and Cu^{2+} is reduced to pure Cu^0 (Eq. 2.1):

$$Fe^0 + 2e - + Cu^{2+} + SO_4^{\;2-} \rightarrow Cu^0 + 2e - + Fe^{2+} + SO_4^{\;2-} \tag{2.1}$$

Fe^{2+} the iron cation Fe^{2+} was abundant in early Earth seas and ferric oxides played a major role in the later release of O_2 into the atmosphere by early photosynthetic bacteria [28]. As O_2 was formed, almost all of it was quickly utilized to convert Fe^{2+} to Fe^{3+} in the formation of ferric oxides that now leave bands of iron ores in sedimentary rocks, thus defining the geologic period. When Fe^{2+} was exhausted, O_2 began accumulating in the atmosphere with gravity preventing its escape. In living forms, Fe cofactors are highly used in biochemical reactions, such as *e*- transfer by Fe-S clusters (Fig. 2.5) [26]. Fe is incorporated into proto-porphyrins where Fe is the center atom bonded to 4 N as seen in Fig. 2.5; these are called hemes, which vary functionally depending on the side groups attached to the porphyrins.

Heme Fe Function In addition to Fe-S clusters, a notable heme is the central component of the globular protein hemoglobin found encased in matured erythroblasts. Its primary function is O_2 binding and transport but it also has other functions [29]. Hemoglobin has four hemes, and consistent for all hemes, there is a Fe^{2+} ion held in the center of a porphyrin coordinated by the 4 N atoms of the porphyrin ring. Each

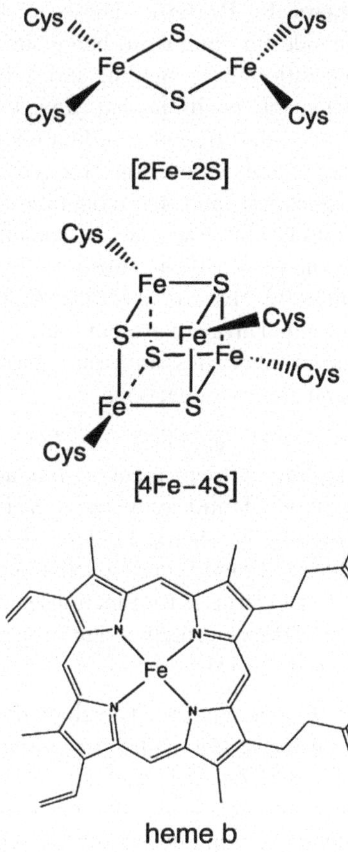

Fig. 2.5 Iron cofactors commonly utilized in biology: Top to bottom, [2Fe-2S] cluster, [4Fe-4S] cluster, and heme b. *Cys* cysteine. (Drawing resourced from Wikipedia Commons as shown (Fig. 1) in Outten and Theil [26] (review))

Fe^{2+} site of the hemoglobin tetramer can carry one O_2 molecule. Myoglobin, found mainly in muscle tissue, is monomeric, so each myoglobin molecule stores only one O_2 molecule.

Other hemes, like hemes *a-d*, are incorporated into metalloproteins such as the P-450 cytochromes and cytochromes *c, b* in the ETS [26]. The hemes are often linked to the parent protein by Cys amino acid residues (Fig. 2.5) and there may be more than one heme group within the cytochrome. Fe can also be directly coordinated by amino acids into polypeptide chains, forming mono- or di-nuclear Fe centers. The ligands for Fe vary, based on the Fe oxidation state. Ferroxidase is an antioxidant enzyme that oxidizes Fe^{2+} to Fe^{3+}. In addition to the relatively simple Fe-only centers, cells also can synthesize Fe-S cofactors that contain iron and sulfide organized into planar [2Fe-2S] or cuboidal [4Fe-4S] clusters (Fig. 2.5) that are functionally linked for *e*- transfer [26, 27]. This is discussed further in Sects. 5.1, 5.2 and 5.3.

Nonheme Fe Proteins These are families of enzymes that also use iron at the active side. In contrast to heme sites where the porphyrin Fe site leaves only one axial position for O_2 binding and activation, the non-heme Fe^{2+} sites have additional, exchangeable positions that can also allow substrate and or cofactor binding to the Fe^{2+} site with different coordination modes for O_2 activation [30–32]. Nonheme Fe enzymes catalyze reactions that typically involve O_2 and include monooxygenation, dioxygenation, and 4*e*- reduction of O_2 to H_2O. The mononuclear nonheme Fe enzymes fall into two classes: The intradiol dioxygenases and the lipoxygenases use Fe^{3+} centers to activate substrate for reaction with O_2. The other class of enzymes use an Fe^{2+} active site to activate O_2 in parallel to the heme systems. These enzymes use a limited array of protein folding and metal binding sites, but the oxidation reactions they catalyze span from complex rearrangements to uncommon bond formation and cleavage reactions.

Mono-Fe enzymes include α-ketoglutarate-dependent hydroxylation, α-ketoglutarate-dependent 4*e*- oxidation, hydroperoxidation, lipooxygenases, aconitase, ribonucleotide reductases, aromatic amino acid hydrolases, dioxygenases, and ferredoxins. Nonheme Fe-dependent enzymatic reactions play essential roles in building and modifying bioactive organic molecules in many pathways [30]. These enzymes are crucial for functions such as gene regulation (e.g., H3K36me3 demethylase (KDMA), protein maturation (e.g., prolyl-4 hydroxylases, hypoxia-inducible factor (HF1) [33], and metabolism (e.g., taurine dioxygenase).

Cu^{2+} (Copper cation). The element Cu is made by fusion of smaller elements plus $_2H^4$ in massive stars, much like other metals; Cu is often found with deposits of Pb, Ag, Au, and Zn. Pure Cu^0 appears in nature but most is found with oxypolyanion compounds such as acetates, carbonates, nitrates, and sulfates [34]. Cu is the heaviest element having important roles in biology. It is a component of metalloenzymes involved in hemoglobin formation, carbohydrate metabolism, catecholamine biosynthesis, and the cross-linking of collagen, elastin, and keratin, as well as in antioxidant defense mechanisms [35]. Moreover, Cu-dependent enzymes, such as cytochrome *c* oxidase (hemes containing Cu_B that traps O_2), superoxide dismutase, ferroxidases, monoamine oxidase, and dopamine *β*-monooxygenase, function to reduce reactive oxygen species (ROS) and hydrogen peroxide (H_2O_2) [35], In serum, Cu is found bound to ceruloplasmin, albumin, and amino acids, which may function to transport Cu [36]. The role of Cu in the electron transfer system (ETS) is discussed in Sects. 5.2 and 5.3.

2.4 Mn^{2+}/Mn^{3+} as an Enzyme Cofactor and $O_2^{\bullet-}$ Scavenger; Li^+ as a Cation Channel Blocker

Manganese ($_{25}Mn^{56}$) is a transition metal grouped just before Fe and Cu in the periodic table (Fig. 1.6). Mn is a rare but essential element for all lifeforms. Several enzymes are known to employ a mononuclear Mn cation that undergoes changes in redox state during catalysis [37]. Mn^{2+}/Mn^{3+} are the catalytically active oxidation

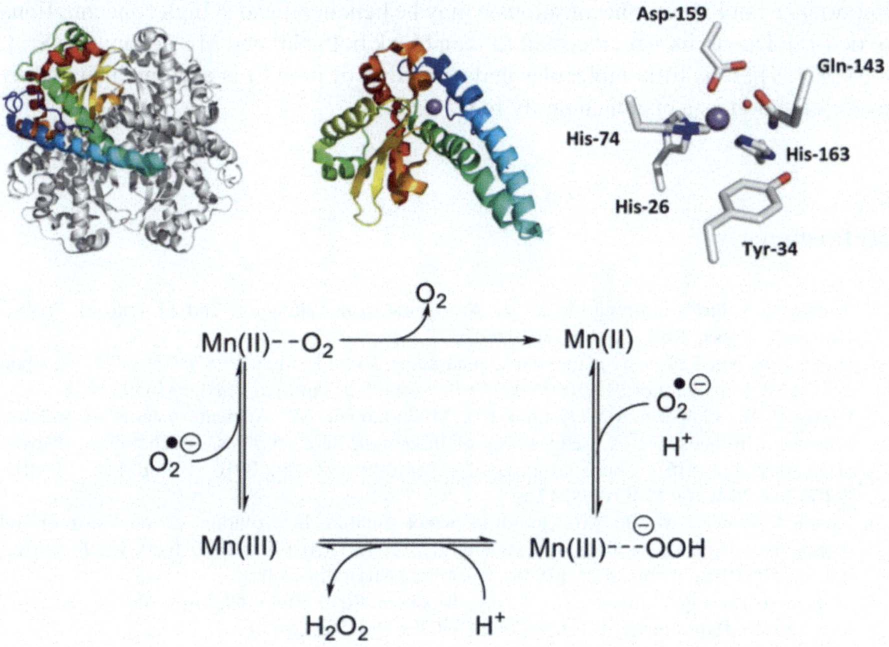

Fig. 2.6 Manganese superoxide dismutase (MnSOD): Biological assembly of the human mitochondrial MnSOD tetramer (left), the human MnSOD monomer (middle), and the Mn center with selected second-shell residues in the active site of human MnSOD. The Mn ion and water oxygen are rendered as purple and red spheres, respectively. Note the product of superoxide ($O_2^{\bullet-}$) is H_2O_2. (Proposed mechanism for MnSOD based on Ref. [36]. Structures (Fig. 1) and modified legend resourced with permission from copyright owner of Zhu and Richards [37] (review))

states of Mn in biology, and there is a 'second-shell' hydrogen bonding with the Mn ion that may underlie its unique properties [37]. The Mn ion is a cofactor in enzyme reactions that includes: oxidoreductases, transferases, hydrolases, lyases, isomerases, and arginases [38]. Other essentials are its role in biosynthesis by manganese ribonucleotide reductase and manganese lipoxygenase [39]. The ability of Mn-incorporated enzymes to modulate the redox properties of the Mn^{2+}/Mn^{3+} couple allows these enzymes to generate substrate-based radicals that neutralize oxygen free radicals ($O_2^{\bullet-}$). Mn cations have redox versatility to perform reactions involving O_2, and its derivatives, the $O_2^{\bullet-}$ ion, H_2O_2, and H_2O. During *e*- transfer, Mn-containing superoxide dismutase (MnSOD) is a very important mitochondrial enzyme in that it neutralizes excess $O_2^{\bullet-}$ released during *e*- transport in the ETS as shown in Fig. 2.6 [16, 37, 40]. This is discussed further in Sect. 6.1, 7.2 and 8.4.

Lithium ($_3Li^7$) is a Group 1A element (Fig. 1.6) that occurs naturally in soil and water, but mostly in low concentrations. Its importance in the early universe as the third lightest element is discussed in Sect. 1.1. Lithium is not one of the essential minerals for life, but it enters the food chain. Little is known about its biological importance but in low concentrations it may be beneficial and at high concentrations toxic [41]. Due to its size, ingested Li^+ can block both Na^+ and Mg^{2+} channels (Sect. 8.3) [42]. There is little molecular understanding of how Li is accumulated and its mode of action as a chemical entity [43].

References

1. Emsley J. Nature's building blocks: an A-Z guide to the elements. 2nd ed. Oxford: Oxford University Press; 2011. ISBN 9780199605637
2. Danchin A, Nikel PI. Why nature chose potassium. J Mol Evol. 2019;87(9–10):271–88. Epub 20191028. https://doi.org/10.1007/s00239-019-09915-2. PubMed PMID: 31659374.
3. Dibrova DV, Galperin MY, Koonin EV, Mulkidjanian AY. Ancient systems of sodium/potassium homeostasis as predecessors of membrane bioenergetics. Biochemistry (Mosc). 2015;80(5):495–516. https://doi.org/10.1134/S0006297915050016. PubMed PMID: 26071768; PMCID: PMC5898217.
4. Austin S, Nowikovsky K. Mitochondrial osmoregulation in evolution, cation transport and metabolism. Biochim Biophys Acta Bioenerg. 2021;1862(5):148368. Epub 20210108. https://doi.org/10.1016/j.bbabio.2021.148368. PubMed PMID: 33422486.
5. Watson H. Biological membranes. Essays Biochem. 2015;59:43–69. https://doi.org/10.1042/bse0590043. PubMed PMID: 26504250; PMCID: PMC4626904.
6. O'Rourke B. Mitochondrial ion channels. Annu Rev Physiol. 2007;69:19–49. https://doi.org/10.1146/annurev.physiol.69.031905.163804. PubMed PMID: 17059356; PMCID: PMC2712118.
7. Zhang J, Hartmann AM, Guo J. Editorial: chloride homeostasis in animal cell physiology. Front Physiol. 2023;14:1227565. Epub 20230605. https://doi.org/10.3389/fphys.2023.1227565. PubMed PMID: 37342798; PMCID: PMC10277798.
8. Wagner C, El Omari M, Konig GM. Biohalogenation: nature's way to synthesize halogenated metabolites. J Nat Prod. 2009;72(3):540–53. https://doi.org/10.1021/np800651m. PubMed PMID: 19245259.
9. Brennan RJ, Schiestl RH. Chloroform and carbon tetrachloride induce intrachromosomal recombination and oxidative free radicals in Saccharomyces cerevisiae. Mutat Res. 1998;397(2):271–8. https://doi.org/10.1016/s0027-5107(97)00225-x. PubMed PMID: 9541652.
10. Youngblut MD, Wang O, Barnum TP, Coates JD. (Per)chlorate in biology on Earth and beyond. Annu Rev Microbiol. 2016;70:435–57. Epub 20160725. https://doi.org/10.1146/annurev-micro-102215-095406. PubMed PMID: 27482739.
11. Markovich D. Physiological roles and regulation of mammalian sulfate transporters. Physiol Rev. 2001;81(4):1499–533. https://doi.org/10.1152/physrev.2001.81.4.1499. PubMed PMID: 11581495.
12. Kushkevych I, Cejnar J, Treml J, Dordevic D, Kollar P, Vitezova M. Recent advances in metabolic pathways of sulfate reduction in intestinal bacteria. Cells. 2020;9(3). Epub 20200312. https://doi.org/10.3390/cells9030698. PubMed PMID: 32178484; PMCID: PMC7140700.
13. Jasielec JJ, Filipek R, Dolowy K, Lewenstam A. Precipitation of inorganic salts in mitochondrial matrix. Membranes (Basel). 2020;10(5). Epub 20200427. https://doi.org/10.3390/membranes10050081. PubMed PMID: 32349446; PMCID: PMC7281443.
14. Lundberg JO, Weitzberg E, Gladwin MT. The nitrate-nitrite-nitric oxide pathway in physiology and therapeutics. Nat Rev Drug Discov. 2008;7(2):156–67. https://doi.org/10.1038/nrd2466. PubMed PMID: 18167491.

15. Galvan A, Fernandez E. Eukaryotic nitrate and nitrite transporters. Cell Mol Life Sci. 2001;58(2):225–33. https://doi.org/10.1007/PL00000850. PubMed PMID: 11289304.
16. Stowe DF, Camara AK. Mitochondrial reactive oxygen species production in excitable cells: modulators of mitochondrial and cell function. Antioxid Redox Signal. 2009;11(6):1373–414. https://doi.org/10.1089/ars.2008.2331. PubMed PMID: 19187004; PMCID: PMC2842133.
17. Gschneider KA. Physical properties and interrelationships of metallic and semimetallic elements. Solid State Phys. 1964;16 https://doi.org/10.1016/S0081-1947(08)60518-4.
18. Holm NG. The significance of Mg in prebiotic geochemistry. Geobiology. 2012;10(4):269–79. Epub 20120320. https://doi.org/10.1111/j.1472-4669.2012.00323.x. PubMed PMID: 22429303; PMCID: PMC3510310.
19. Romani AM. Magnesium in health and disease. Met Ions Life Sci. 2013;13:49–79. https://doi.org/10.1007/978-94-007-7500-8_3. PubMed PMID: 24470089.
20. Buelens FP, Leonov H, de Groot BL, Grubmuller H. ATP-magnesium coordination: protein structure-based force field evaluation and corrections. J Chem Theory Comput. 2021;17(3):1922–30. Epub 20210222. https://doi.org/10.1021/acs.jctc.0c01205. PubMed PMID: 33616388; PMCID: PMC8023659.
21. Ko YH, Hong S, Pedersen PL. Chemical mechanism of ATP synthase. Magnesium plays a pivotal role in formation of the transition state where ATP is synthesized from ADP and inorganic phosphate. J Biol Chem. 1999;274(41):28853–6. https://doi.org/10.1074/jbc.274.41.28853. PubMed PMID: 10506126.
22. Harborne SP, King MS, Crichton PG, Kunji ER. Calcium regulation of the human mitochondrial ATP-Mg/P_i carrier SLC25A24 uses a locking pin mechanism. Sci Rep. 2017;7:45383. Epub 20170328. https://doi.org/10.1038/srep45383. PubMed PMID: 28350015; PMCID: PMC5369052.
23. Marynick DS, Schaefer HF 3rd. Theoretical studies of metal-phosphate interactions: interaction of Li^+, Na^+, K^+, Be^{++}, Mg^{++}, and Ca^{++} with $H_2PO_4^-$ and $(CH_3O)_2^-$: implications for nucleic acid solvation. Proc Natl Acad Sci USA. 1975;72(10):3794–8. https://doi.org/10.1073/pnas.72.10.3794. PubMed PMID: 1060064; PMCID: PMC433081.
24. Rayani K, Seffernick J, Li AY, Davis JP, Spuches AM, Van Petegem F, Solaro RJ, Lindert S, Tibbits GF. Binding of calcium and magnesium to human cardiac troponin C. J Biol Chem. 2021;296:100350. Epub 20210203. https://doi.org/10.1016/j.jbc.2021.100350. PubMed PMID: 33548225; PMCID: PMC7961095.
25. Smith GA, Vandenberg JI, Freestone NS, Dixon HB. The effect of Mg^{2+} on cardiac muscle function: is CaATP the substrate for priming myofibril cross-bridge formation and Ca^{2+} reuptake by the sarcoplasmic reticulum? Biochem J. 2001;354(Pt 3):539–51. https://doi.org/10.1042/0264-6021:3540539. PubMed PMID: 11237858; PMCID: PMC1221685.
26. Outten FW, Theil EC. Iron-based redox switches in biology. Antioxid Redox Signal. 2009;11(5):1029–46. https://doi.org/10.1089/ars.2008.2296. PubMed PMID: 19021503; PMCID: PMC2842161.
27. Guengerich FP. Introduction to metals in biology 2018: copper homeostasis and utilization in redox enzymes. J Biol Chem. 2018;293(13):4603–5. Epub 20180207. https://doi.org/10.1074/jbc.TM118.002255. PubMed PMID: 29425098; PMCID: PMC5880126.
28. Alberts B, Johnson A, Lewis J, Raff M, Roberts K, Walter P. The evolution of electron-transport chains. In: Molecular biology of the cell. 4th ed. New York: Garland Science; 2002. Available from: https://www.ncbi.nlm.nih.gov/books/NBK26849/.
29. Ahmed MH, Ghatge MS, Safo MK. Hemoglobin: structure, function and allostery. Subcell Biochem. 2020;94:345–82. https://doi.org/10.1007/978-3-030-41769-7_14. PubMed PMID: 32189307; PMCID: PMC7370311.
30. Ushimaru R, Ikuro A. Unusual dioxygen-dependent reactions catalyzed by nonheme iron enzymes in natural product biosynthesis. ACS Catal. 2023;13(2):1045–76. https://doi.org/10.1021/acscatal.2c05247.
31. Solomon EI, Decker A, Lehnert N. Non-heme iron enzymes: contrasts to heme catalysis. Proc Natl Acad Sci U S A. 2003;100(7):3589–94. Epub 20030221. https://doi.org/10.1073/pnas.0336792100. PubMed PMID: 12598659; PMCID: PMC152966.

32. Song H, Naowarojna N, Cheng R, Lopez J, Liu P. Non-heme iron enzyme-catalyzed complex transformations: endoperoxidation, cyclopropanation, orthoester, oxidative C-C and C-S bond formation reactions in natural product biosynthesis. Adv Protein Chem Struct Biol. 2019;117:1–61. Epub 20190923. https://doi.org/10.1016/bs.apcsb.2019.06.002. PubMed PMID: 31564305.
33. Siddiq A, Aminova LR, Ratan RR. Hypoxia inducible factor prolyl 4-hydroxylase enzymes: center stage in the battle against hypoxia, metabolic compromise and oxidative stress. Neurochem Res. 2007;32(4–5):931–46. Epub 20070307. https://doi.org/10.1007/s11064-006-9268-7. PubMed PMID: 17342411; PMCID: PMC2576999.
34. Trammell R, Rajabimoghadam K, Garcia-Bosch I. Copper-promoted functionalization of organic molecules: from biologically relevant Cu/O_2 model systems to organometallic transformations. Chem Rev. 2019;119(4):2954–3031. Epub 20190130. https://doi.org/10.1021/acs.chemrev.8b00368. PubMed PMID: 30698952; PMCID: PMC6571019.
35. Iakovidis I, Delimaris I, Piperakis SM. Copper and its complexes in medicine: a biochemical approach. Mol Biol Int. 2011;2011:594529. Epub 20110615. https://doi.org/10.4061/2011/594529. PubMed PMID: 22091409; PMCID: PMC3195324.
36. Neumann PZ, Sass-Kortsak A. The state of copper in human serum: evidence for an amino acid-bound fraction. J Clin Invest. 1967;46(4):646–58. https://doi.org/10.1172/JCI105566. PubMed PMID: 6021211; PMCID: PMC442048.
37. Zhu W, Richards NGJ. Biological functions controlled by manganese redox changes in mononuclear Mn-dependent enzymes. Essays Biochem. 2017;61(2):259–70. Epub 20170509. https://doi.org/10.1042/EBC20160070. PubMed PMID: 28487402.
38. Li L, Yang X. The essential element manganese, oxidative stress, and metabolic diseases: links and interactions. Oxid Med Cell Longev. 2018;2018:7580707. Epub 20180405. https://doi.org/10.1155/2018/7580707. PubMed PMID: 29849912; PMCID: PMC5907490.
39. Rice DB, Massie AA, Jackson TA. Manganese-oxygen intermediates in O-O bond activation and hydrogen-atom tansfer reactions. Acc Chem Res. 2017;50(11):2706–17. Epub 20171024. https://doi.org/10.1021/acs.accounts.7b00343. PubMed PMID: 29064667.
40. Borgstahl GE, Parge HE, Hickey MJ, Beyer WF Jr, Hallewell RA, Tainer JA. The structure of human mitochondrial manganese superoxide dismutase reveals a novel tetrameric interface of two 4-helix bundles. Cell. 1992;71(1):107–18. https://doi.org/10.1016/0092-8674(92)90270-m. PubMed PMID: 1394426.
41. Szklarska D, Rzymski P. Is lithium a micronutrient? From biological activity and epidemiological observation to food fortification. Biol Trace Elem Res. 2019;189(1):18–27. Epub 20180731. https://doi.org/10.1007/s12011-018-1455-2. PubMed PMID: 30066063; PMCID: PMC6443601.
42. Jakobsson E, Arguello-Miranda O, Chiu SW, Fazal Z, Kruczek J, Nunez-Corrales S, Pandit S, Pritchet L. Towards a unified understanding of lithium action in basic biology and its significance for applied biology. J Membr Biol. 2017;250(6):587–604. Epub 20171110. https://doi.org/10.1007/s00232-017-9998-2. PubMed PMID: 29127487; PMCID: PMC5696506.
43. Shahzad B, Mughal MN, Tanveer M, Gupta D, Abbas G. Is lithium biologically an important or toxic element to living organisms? An overview. Environ Sci Pollut Res Int. 2017;24(1):103–15. Epub 20161027. https://doi.org/10.1007/s11356-016-7898-0. PubMed PMID: 27785724.

Chapter 3
Origin of Complex Biomolecules on the Dawn of Lifeforms

Abstract The details of the origin of life on Earth are unknown, but lifeforms must have originated and developed over a long period of chemical evolution from simple inorganic and organic compounds. This required consistent energy from our Sun and other local resources. Eventually, chemiosmosis took hold so that the energy of light, heat, and exothermic reactions could be transduced to endothermically form more complex molecules as in the carbon building of lipids. These molecules became lipid membrane structures to contain chemical reactions, carbohydrates for metabolic reactions, proteins to make catalytic enzymes, and nucleotides to provide high-energy phosphates and to form ATP, DNA, and RNA. Of all the 100 or so nucleobases found in Nature, only 5 were utilized and only 20 or so of the available amino acids were selected by the advent of the proposed last universal common ancestor. While this was likely a bottoms-up approach to prebiotic biosynthesis, parts of each of these groups paired up to make more complex, functioning macromolecules, as in ATP or DNA, so chemical evolution simplified the formation of compounds that would be needed for the onset of chemical replication, transcription of the genetic information, and translation of the genetic code to make polypeptides. In a broad view, some basic chemical compounds likely came from the cosmos and some molecules were likely formed in hydrothermal vents while some macromolecules were formed by condensation and polymerization, which probably occurred on land and sea in wet-dry cycles. Important events that led to life include development of semi-permeable lipid membranes, carbon building via acetate and coenzyme A, ascent of adenine as the preferred base for catalytic activity, generation of high-energy phosphate bonds to store energy, onset of a suitable chemiosmosis mechanism with oxidation and phosphorylation, dinucleotide cofactors to facilitate electron transfer and group transfer reactions, selection of a standard genetic code, genesis of the ribozyme with a peptidyl transfer center to match RNA with polypeptides, and naturally, a continuous supply of energy from our Sun.

Keywords Abiogenesis · Accretion disk · Ambiguity reduction · Amino acid translation · Amphiphilic molecules · Autotrophic hypothesis · Chemiosmosis hypothesis · Chemisorption · Coenzymes · Condensation · Disordered peptides · DNA replication · Electron transfer system · Enantiomers · Equilibrium fitness ·

D. F. Stowe, *Evolution of Bioenergetics from Elements to Life*,
https://doi.org/10.1007/978-3-031-76668-8_3

Group transfer cofactors · Hadean Eon · Heavy bombardment period · Heterotrophic hypothesis · Horizontal gene transfer · Hydrothermal vents · Last archaeal common ancestor · Last bacterial common ancestor · Last universal common ancestor · Mineral self-replicating structures · Molecular self-assembly · Non-coding RNA · Nucleobases · Nucleotides · Organic polymers · Oxidative phosphorylation · Patterned pathway evolution · Peptidyl transferase center · Permeability barrier · Photosynthesis · Phylogenomic analysis · Prebiotic pathways · Prokaryotes · Proto-ribosomes · Ribosomes · Ribozymes · RNA translation · RNA world concept · Standard genetic code · Surface hydrothermal fields

3.1 Abiogenesis: Origin of Simple Organic Biomolecules

Abiogenesis is a natural process from which the origin and development of Earth's lifeforms occurred during a long period of chemical evolution from simple inorganic and organic elements and compounds to complex structural and functional lipids, proteins, carbohydrates, and nucleotides [1–5]. It is well established that even simple lifeforms did not just spontaneously appear on Earth. The 100 -year old Oparin–Haldane hypothesis suggested that life arose gradually from simple inorganic and organic molecules, with "building blocks" like amino acids forming first and then combining to make complex polymers [6, 7]; this is now commonly called the heterotrophic hypothesis for life.

These complex molecules and polymers must have been derived from a long and gradual prelife process of chemical formation on Earth, perhaps with the help of some inorganic and organic molecules created extraterrestrially and transported to Earth that would be essential or helpful for life to develop. Some call this "pseudo-panspermia." This is different from the more than 160-year-old, unsubstantiated "panspermia hypothesis," which suggests that life has already developed in the Universe and that seeds or germs from elsewhere have arrived on Earth [8]. Although discoveries in chemical evolution have rapidly accumulated, there remain many complex scientific questions about the chemical actions and sequences that led to the beginnings of Earth's lifeforms.

Unknowns of Life's Origin Some important questions are as follows: (1) What was the Earth's physicochemical environment like when life first evolved?, (2) What kinds of chemical reactions could produce suitable organic molecules from the elements that were available and abundant in the early Earth's seas and fresh water-land surfaces?, (3) Were the precursor molecules de novo synthesized on Earth or extraterrestrially, or both?, (4) Which molecules were suitable for life to generate, and how were they set up to do so?, (5) How did early, simple biomolecules produce complex self-replicating organic macromolecules (condensation/polymerization) before enzymes?, (6) How were these molecules and macromolecules compartmentalized and separated from the general environment?, (7) How were templates made that led to the exact reproduction of macromolecules and then to genetic coding that allowed ready replication, transcription, and translation of molecules into chains for specific tasks?, (8) How was photon and thermal energy utilized to form just a few high energy chemical compounds that would be essential to drive metabolism, growth, and repair?, (9) Of the numerous amino acids and nucleoside bases found naturally, and capable of being synthesized, how was it that nearly only one base was selected for transduction of energy, and very few other bases and amino acids were incorporated into molecular chains to carry information?

These questions are hard to answer with firm evidence for now. It is evident that we will require a clearer knowledge about the primordial interactions of elements to form biomolecular precursors and a better understanding of the early and changing environmental conditions that were key for selection and development of bioorganic molecules and complex macromolecules leading to the emergence and diversity of early lifeforms. What follows here is a brief synopsis of what we know today.

Abiotic Environment and Chemical Characteristics Evidence for how life developed comes from the universal physicochemical characteristics of the lighter elements and the bonding characteristics of inorganic and organic molecules present on the early Earth's hot surfaces, as well as from the conditions of its atmosphere, and eventually its oceans, and new land masses formed from volcanoes and along the developing tectonic plates. Thus, specific ingredients, conditions, and timing were required before life could take form. The chemical ingredients and their bonding capabilities are discussed in detail in the sections of Chaps. 1 and 2. The delivery of these ingredients during the formation of our solar system (Fig. 1.1) is key to any development of prebiotic compounds that could lead to early life.

All of the solid planets formed individually from the bombardment of smaller bodies into larger ones with gravity pulling the smaller bodies together until there were only four rocky, inner terrestrial planets (Mercury, Venus, Earth, Mars) revolving around the Sun [9]. A timeline estimate for Earth's accretion is provided in Fig. 3.1. The crash by the planet Theia into Earth about 4.5 Bya added little to Earth's mass (8–10%) but tilted Earth's rotation (Fig. 3.1) giving it climate seasons and eventually creating a single moon [9], whose gravity provided for tidal basins to develop along the shores of seas of Earth causing wet and dry land cycles.

When the heavy rocky bombardments gradually receded approximately 3.5 Bya, the Earth had high volcanic activity with a hot mantle and surface; it had toxic (to life) atmospheric conditions, and high surface pressures [9]. The different geologic histories of the solid (especially Mars) and gaseous planets in the solar system positions Earth's accretion (Fig. 1.1) much later than Mars (Fig. 3.2) [9]. Examination of the geology of meteorites that hit Earth also yields clues about Earth's early formation. The water ice that arrived in the hot bombardment was all turned to steam so that liquid H_2O did not exist on Earth until some cooling occurred, perhaps as early at 4.3 Bya [10].

The early atmosphere likely contained large amounts of N_2, H_2O as vapor and steam, CO_2, and a smaller amount of H_2, as most H_2 escaped back into the cosmos. It also was composed of the gases H_2S, CO, CH_4, NH_3, HCN (hydrogen cyanide), and CH_2O (formaldehyde), but no O_2 or liquid H_2O at the surface. The

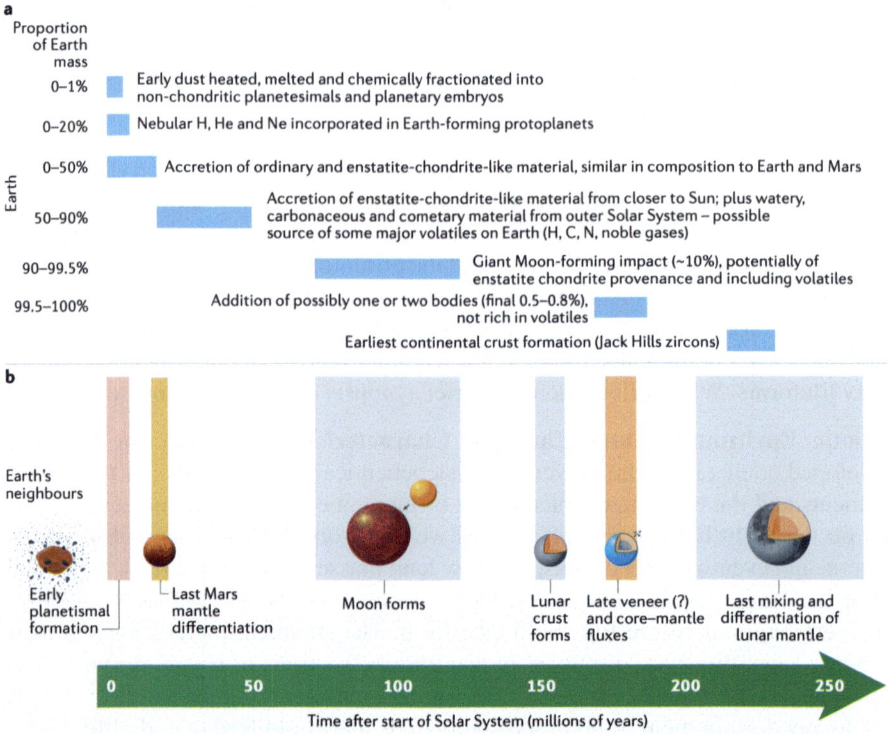

Fig. 3.1 Best estimates for chronology of Earth accretion: (**a**) Stages of Earth accretion by percentage of Earth's mass. (**b**) Comparison with stages of formation and differentiation of Mars and the Moon. Earth started to accrete early and fast, like the other planetary objects. However, it continued for longer, culminating in the Moon-forming Giant Impact. It is unclear if a substantial late veneer occurred; fingerprint evidence of highly siderophile elements in the silicate Earth could, in part, have originated from a flux from the core. (Cartoons with text (Fig. 4) and its legend resourced with permission from copyright owner of Halliday and Canup [9] (review))

elements, molecules, and biogenic compounds that arrived from the accretion disk of protoplanet Earth (Figs. 1.1 and 3.1) are first discussed in Sects. 1.2, 1.3. 1.4, and 3.1.

Genesis of Early Biomolecules

Early biochemical compounds almost certainly were generated by the spontaneous interaction of atoms and molecules present on the prebiotic Earth. It is inferred that the lifeforms on Earth today could not be recreated under the conditions on Earth today. The specific molecules that were formed were mostly dictated by the Earth's changing environmental conditions of slow cooling after high heat, a lowering of atmospheric pressure, cooling of H_2O vapor and steam to make liquid H_2O, reduced

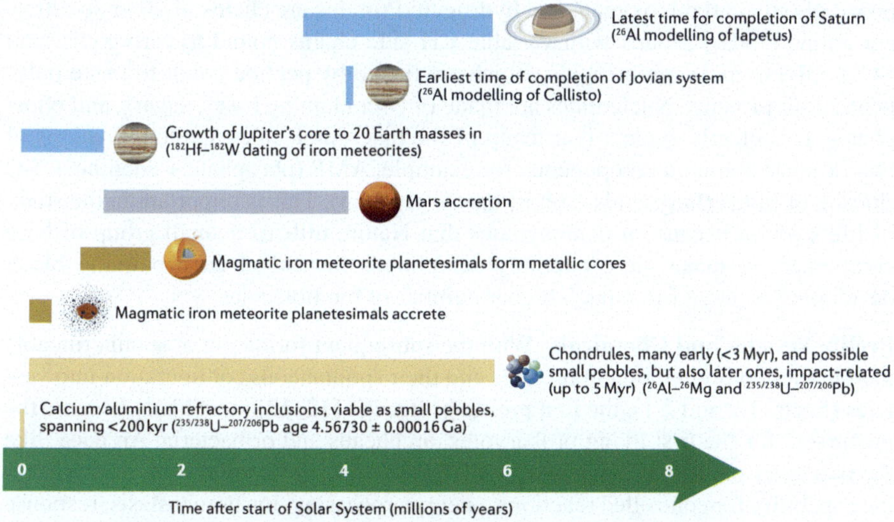

Fig. 3.2 Chronology of the earliest Solar System, as determined from various isotopic systems: The oldest Solar System solids are the Ca–Al refractory inclusions (CAIs) forming <200 Kyr after the start of the Solar System. These were followed shortly after by formation of chondrules that persisted for several million years. The timing of planetesimal accretion (lasting ~300–500 Kyr) and differentiation (lasting ~800 Kyr to 1.5 Myr) is inferred from the age of iron meteorites. Mars's accretion occurred early, probably within the first 4 Myr after the Solar System formed. Jupiter's accretion occurred even faster in the presence of the gas-rich nebula, with its core growing to 20 Earth masses in the first million years. Therefore, accretion of planetary objects, from dust to gas giants, was completed within less than 10 Myr. In contrast to these early events, Earth's accretion was completed much later around 70–120 Myr after the formation of CAIs. (Cartoon with text (Fig. 2) and its legend resourced with permission from copyright owner of Halliday and Canup [9] (review))

ionizing radiation from space by formation of the ozone layer, and consistent energy from the sun, lightning, and volcanic activity.

One report proposes that some organic molecules that were utilized for the onset of lifeforms were made during the Hadean period before 3.8 Bya while the Earth was still being bombarded with some cosmic debris [11]. The theory is based on evidence that silica molecules like quartz sand (SiO_2) can induce formation of mineral self-organizing structures (MISOS) that could function as catalysts for prebiotic chemical reactions and form simple and complex carbon molecules on the MISOS crystal platforms. These reactions to form "silica/carbonate biomorphs" and "silica/metal oxyhydroxides" could have occurred in a nascent, silica-rich, high pH (pH units 10–11) ocean.

Chemical Groups of Life Broadly, the molecules of life today are considered as four complex bioorganic, groups: lipids (fatty acids), carbohydrates (sugars), proteins (amino acids), and nucleotides (e.g., RNA, DNA, ATP). Lipids are largely H-C bonds with some ^{-}OH groups (glycerol has 3 ^{-}OH groups) and phosphates. Carbohydrates are short, or long chained, branched, or unbranched sugars composed of only carbon, oxygen, and hydrogen. Proteins are chains of 20 or so different amine ($^{-}NH_2$) groups with variable CH side chains bound to carboxylic acid (-CO=OH) to make amino acids, which are linked by peptide bonds to make polypeptides or proteins. Nucleotides are made of N-containing bases, sugars, and phosphates. It is clearly evident that many of the macromolecules are combinations of two or more common components, for example, AMP (phosphate + adenine + 5-C ribose), or lipids (fatty acids + phosphate + glycerol). This is important in the study of life's origin because it demonstrates that Nature utilizes a small group of base compounds to make more complex compounds instead of each building block developing de novo for a much higher number of biomolecules.

Prelife Vesicles and Chemicals With the subsequent formation of semipermeable membranes composed of amphiphiles and their spontaneous configuration into vesicles (Sects. 1.2 and 2.1), the first prelife protocells [12, 13] most likely became the precursors for the first living prokaryotes, archaeans and or bacteria. Archaea, like the two other proposed domains of life (Bacteria and Eukaryota) (Fig. 3.3) possess the capability for controlled reactions, energy processing, ion homeostasis, response and adaption to a changing environment, repair/replacement, replication, and very importantly, the ability to transfer genetic instructions to progeny cells, that is, heredity [15]. More recent work in taxonomy using phylogenomic analysis proposes that there are only two Domains of life, Archaea and Bacteria (Sect. 3.3) [16, 17].

The groundbreaking experiments by Miller [18, 19], carried out in 1952, tested for a chemical origin of life by simulating conditions thought to exist in the

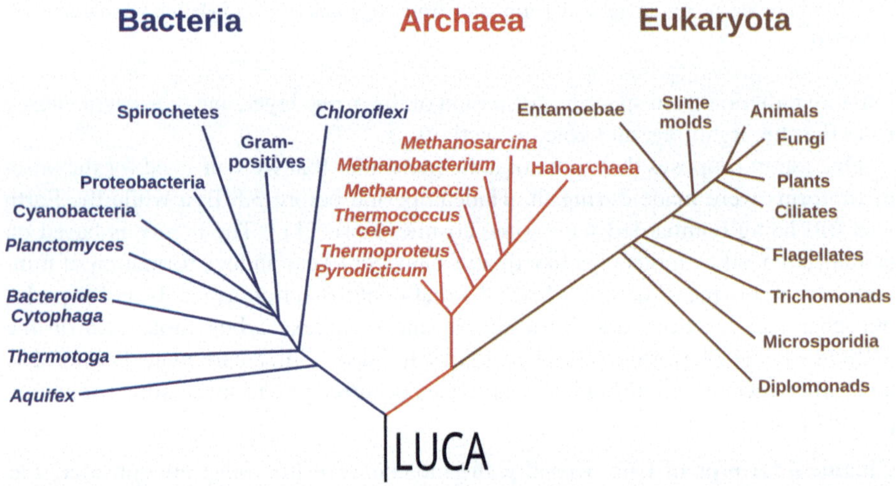

Fig. 3.3 The three domains of life: Archaea as a third domain is based in part on differences in ribosomal structure in Archaea vs. Bacteria and Eukaryota proposed by Woese and Fox [14]. (Drawing (https://creativecommons.org/licenses/by-sa/4.0/) available from Wikipedia: https://en.wikipedia.org/wiki/Three-domain_system)

atmosphere of the early prebiotic Earth. The simple experiments mixed H_2O, CH_4, NH_3, and H_2 in a heated flask with sparks from an electric arc. At first five amino acids were found. Eventually, products were shown to include more natural amino acids. Later studies [20] found that the nucleotide base adenine could be made from hydrogen cyanide (HCN) and NH_3 in a water solution. Many amino acids also were found to be formed from HCN and NH_3 under these conditions [21]. Other experiments showed that RNA and DNA nucleobases could be formed through simulated prebiotic chemistry with an atmosphere rich in CH_4 and NH_3 but devoid of O_2 [22]. This may be important because the higher electronegativity of O over N would have allowed O to form covalent bonds with C, N, and P (Sect. 1.3), which would impede formation of prebiotic amino acids and nucleosides.

Utilizing the select and essential elements, H, C, O, N, P, the first simple molecules and compounds (Fig. 3.4), were synthesized into the essential building blocks of sugars, amino acids, fatty acids, and nucleotides [23]. Condensation and polymerization reactions strung them together into chains to make polysaccharide structures, protein structures and enzymes, lipid membranes, and DNA/RNA strands [13, 24], as shown in Fig. 3.5, for storing and releasing chemical information from one cell to another. The reaction schema of Fig. 3.6 depicts a putative reaction pathway of prebiotic polymerization in which a volcanic gas (carbonyl sulfide, COS) exothermically reacts with an α-amino acid to form the α-amino acid thiocarbamate, a molecule that might be capable of mediating the chemical evolution and polymerization of both peptides and nucleosides [23].

Early Chemical Pathways Plausible, prebiotic key chemical pathways that are based on nascent coexisting biochemical subsystems are summarized in Fig. 3.7 [13]. This simple diagram proposes that life's origins were preceded by self-organizing reaction networks of different biochemical subsystems in a metabolism first approach [25]. It is envisioned that molecules of the different, original

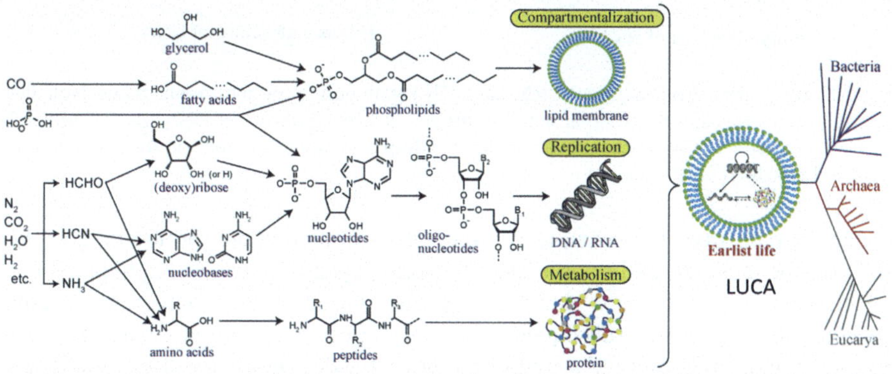

Fig. 3.4 Molecular groups: Structures, components, and abiotic synthetic pathways of biomacromolecules operating three fundamental functions of life (replication, compartmentalization, and metabolism). (Schema (Fig. 2) resourced with license (https://creativecommons.org/licenses/by-nc-nd/4.0/) from Kitadi and Maruyama [23] (review))

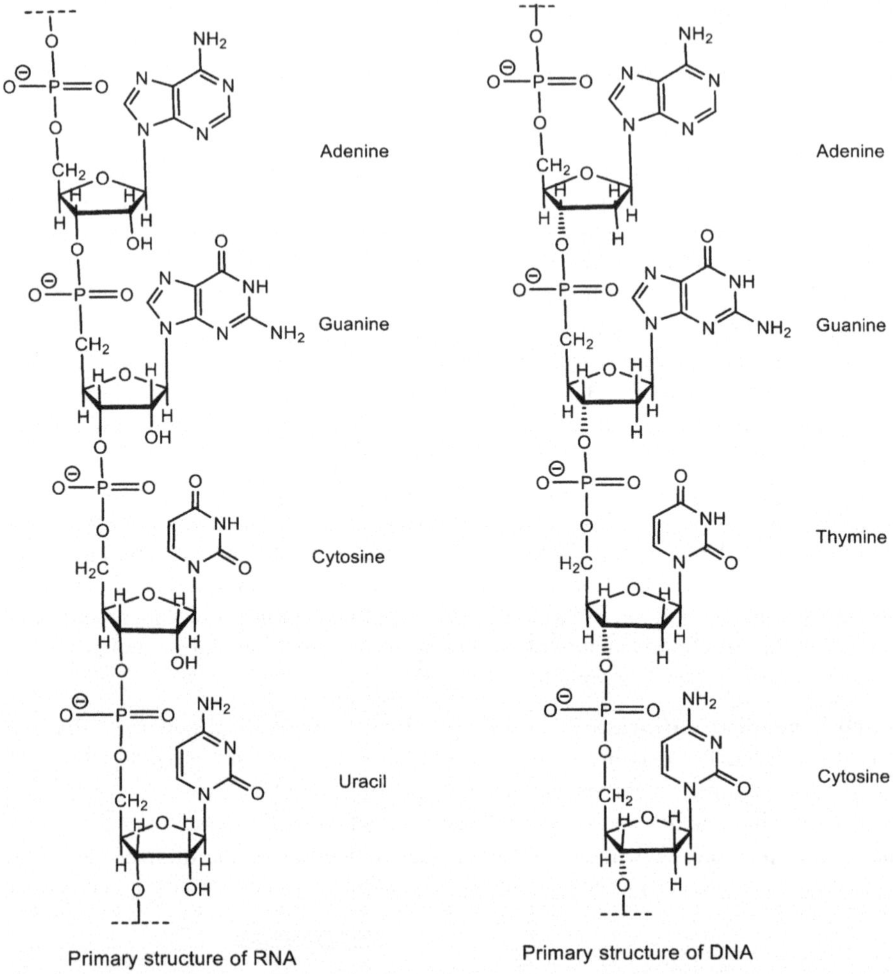

Fig. 3.5 Representative structures of RNA and DNA with purine and pyrimidine bases: Note that ribose on RNA has an OH group at the 2′ position and a deoxyribose on DNA has a H at the 2′ position. (Structures (Fig. 4) resourced with license (https://creativecommons.org/licenses/by/4.0/) from Kolodiazhnyl [24] (review))

metabolic pathways, for example, those involved in membrane formation, glycolysis, gluconeogenesis, and protein and nucleotide synthesis, were coordivnated in a nonenzymatic, "bottom-ups" building approach based on the interactions of converging sub-systems that subsequently became increasingly complex. This approach predicts a multitude of similar and different molecules into networks of metabolic pathways viewed as working together as an ensemble [25], which possibly laid the foundation for polymerization of like molecules.

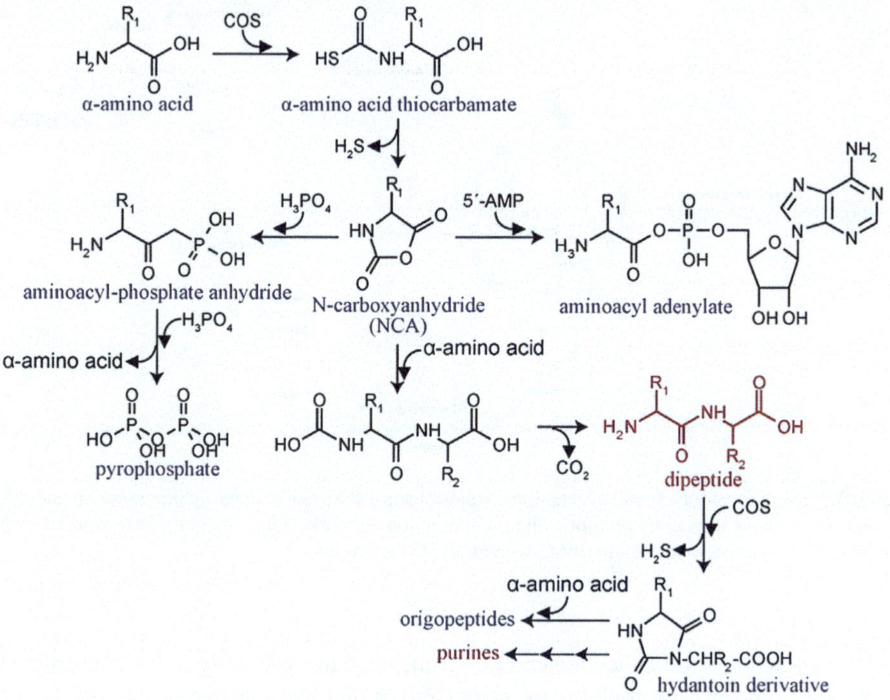

Fig. 3.6 Prebiotic polymerization: A simple volcanic gas, carbonyl sulfide (O=C=S) is formed from the condensation of CO and H_2S. COS can readily react (exothermically) with the amino group of an α-AA to give the α-AA thiocarbamate, a precursor molecule proposed to mediate the chemical evolution and polymerization of both peptides and nucleotides. The COS-mediated condensation is substantially enhanced in the presence of metal cations, e.g., Pb^{2+}, and Fe^{2+}, and oxidizing agents. These connections between peptides and nucleic acids via COS suggest that the linkage between replication of protein enzyme and DNA/RNA is a consequence of their co-evolution from the very dawn of lifeforms. (Schema (Fig. 5) and its modified legend resourced with license (https://creativecommons.org/licenses/by-nc-nd/4.0/) from Kitadi and Maruyama [23] (review))

Acetate, the Carbon Builder The first, single C hydrocarbon, methane (CH_4), and 1-C formaldehyde (CH_2O), are found throughout the Universe (Sects. 1.1 and 3.1). H_2 and CO_2 are also ubiquitous. The synthesis of larger C-coupled compounds was required for prebiotic sources of energy and molecular structure. The 2-C molecule acetate ($C_2H_3O_2$, or CH_3COO^-) is important as a substrate for building more complex biochemical structures. An acetyl group consists of a carbonyl group (C=O) bound to a methyl group (CH_3) as in (C=O)-CH_3. But before the last universal common ancestor (LUCA) existed as the template for all future lifeforms, how was 2-C acetic acid and similar higher carbon molecules made, like citric acid (also known as 2-hydroxypropane-1,2,3 tricarboxylic acid) that has 6-Cs ($C_6H_8O_7$), also written as $HOC(CO_2H)(CH_2CO_2H)_2$)?

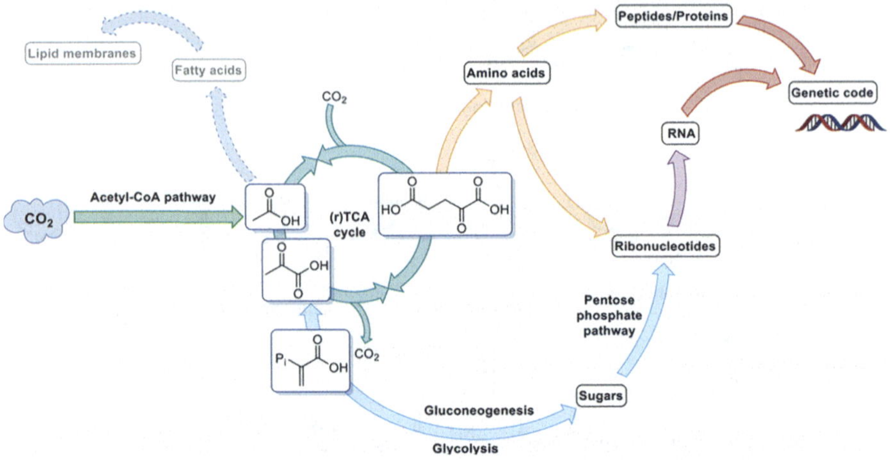

Fig. 3.7 First prebiotic chemical reactions: Relationships between coexisting biochemical subsystems in a "metabolism first" prebiotic chemical reaction network. (Diagram (Fig. 8) resourced with permission from copyright owner of Gozen et al. [13] (review))

It is now known that acetate can be synthesized inorganically in the presence of a single, valence-neutral alloy, awaruite (Ni_3Fe) that likely arrived on Earth during the late Heavy Bombardment 4 Bya [26]. In this reaction, H_2 and CO_2 can form acetate as well as pyruvate ($C_3H_4O_3$; or CH_2COCOO^-), under the mild conditions of a hydrothermal vent (dark, mildly alkaline, hydrothermal conditions (100 °C, 24 bar, 16 h incubation) [27]. The Ni_3Fe-catalyzed pathway is the only CO_2 fixing pathway that operates via carbon monoxide (CO) as an intermediate step [28]; this pathway generates carboxyl groups from carbonyl, rather than by reducing carboxyl groups, which is the key to its exergonic nature. Other pathways expend energy to reduce carboxyl groups [29].

Acetate synthesis is an example of a naturally occurring exergonic reaction catalyzed by transition metal minerals in which CO_2 is reduced by H_2 by chemisorption [27]. This reaction supports the principle of forward (autotrophic) evolution as an early prebiotic event 4 Bya in the absence of light and O_2, and long before photosynthetic bacteria transformed energy from sunlight. This means that the first cells could have lived off H_2 and CO_2 to form organic molecules in the geochemical setting of hydrothermal vents [30]. Some anaerobic (fermenting) bacteria today are acetogenic (can make acetate) and methanogenic (can make CH_4) enzymatically by conversion of fatty acids (FA) in β-oxidation or from CO_2 in the presence of H_2 [27]. Autotrophic chemical evolution does not exclude the probability of side-by-side heterotrophic evolution discussed earlier in this section.

Carbon building, as in the reverse tricarboxylic acid (TCA) cycle, starting from acetate or pyruvate (2 or 3-C to 6-C) (Sect. 5.1), may have been the main source of

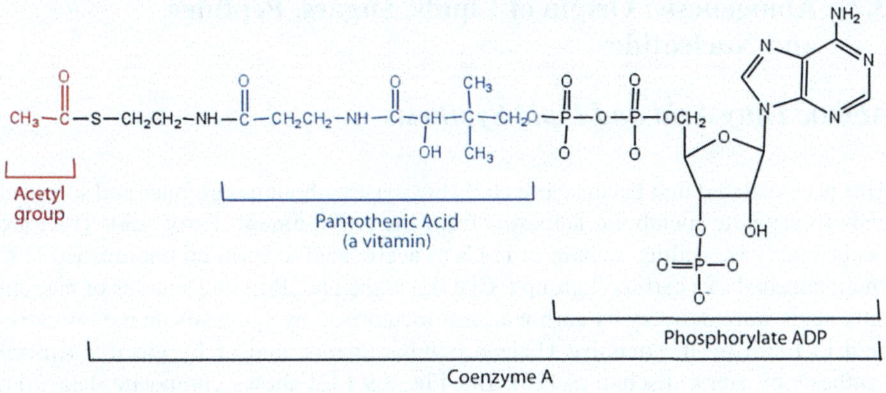

Fig. 3.8 Acetyl-CoA. (Structure resourced with license (https://creativecommons.org/licenses/by-nc-sa/4.0) from Sarah Knapp, Biology Dictionary (2020). Available from https://biologydictionary.net/acetyl-coa/)

carbon skeletons for biosynthesis. It is plausible that subsequent carbon compounds did not each start from CO_2, as discussed above, but first from 2-C acetate or 3-C pyruvate. Thus, biochemical pathways may have evolved from CO_2 outward via a process of "patterned pathway evolution", in which naturally occurring anabolic chemical reactions preceded reactions that later came to be catalyzed by enzymes rather than by metal catalysts [27]. However, as reviewed in Sect. 1.4, 2.2, 2.3, and 6.1, many metals remain as the essential catalytic center of many enzymes.

Coenzyme A The next evolutionary step in carbon building utilized an organic catalyst, coenzyme A (Fig. 1.16), that is highly involved in multiple chemical reactions of all life forms today. Its major function is to deliver a 2-C acetyl group to the TCA cycle to be oxidized for energy production (Sects. 3.3 and 5.1). In the TCA cycle today CoA-SH also attaches to α-ketoglutarate (αKG) to make NADH and succinyl-CoA, which then releases the CoA-SH to form GTP and succinate. Acetyl-CoA (Fig. 3.8) is a linkage of ADP with phosphor-pantothenic acid (a derivative of vitamin B_5) and with a cysteamine (β-mercaptoethylamine) tail where the acetyl group, attaches to, and detaches from, the sulfur atom of cysteamine.

Acetyl-CoA has been called the primary, intermediate building block of life [31]. There is strong chemical evidence that the acetyl-CoA pathway is older than the genes that encode the enzymes that catalyze acetyl-CoA today [27]. As an essential molecule, co-enzyme A participates in many biochemical reactions, in addition to the TCA cycle, that regulate protein, carbohydrate, and lipid metabolism.

3.2 Abiogenesis: Origin of Lipids, Sugars, Peptides, and Nucleotides

Abiotic Fatty Acid and Lipid Synthesis

This is considered first because life could not exist without membranes and spheroid cells to separate metabolic activities from the environment. Fatty acids (FA) are made simply by adding a chain of H-Cs to acetic acid to form an unbranched H-C chain attached to a carboxyl group ($^{-}$COOH) at the end. Possible sources of nascent fatty acids were delivery by carbonaceous meteorites, by synthesis on metals delivered to Earth by the massive Hadean bombardments, and or by electrochemical synthesis by spark discharges [23]. But Fig. 3.9 [32] shows composite data from multiple studies where FAs of at least 8-C can be synthetically formed. These results led to a hypothesis that FA synthesis, occurring on Fe-rich surfaces derived from the rocky bombardments in contact with an impact-generated reducing atmosphere, could have contributed 1000 times more total mass of FAs than subsequent delivery by either carbonaceous meteorites or electrochemical synthesis [32].

Membrane lipids of eukaryotes today are predominantly phospholipid molecules, the most common of which is phosphatidylcholine (H-C + phosphate + glycerol + choline) as in Fig. 3.10b [13]. The phosphate is bound to glycerol and

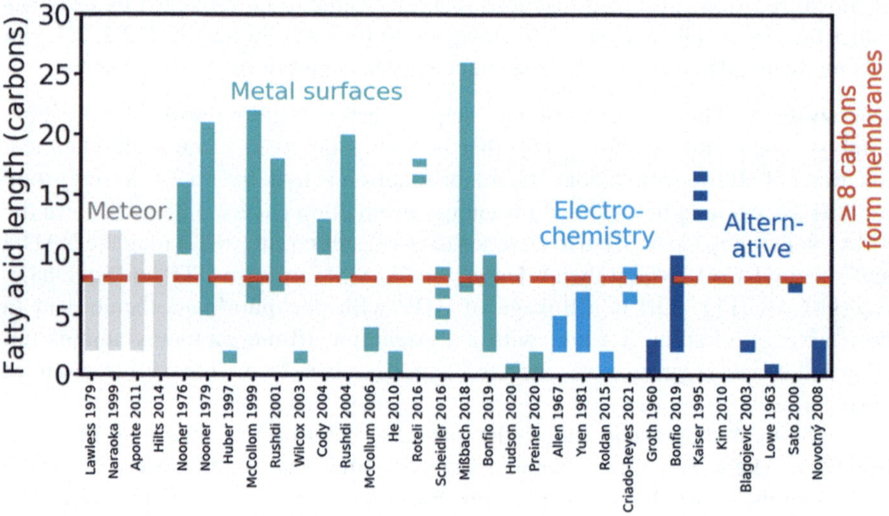

Fig. 3.9 Development of fatty acids: Length of fatty acids delivered by meteorites vs. length produced experimentally by abiotic synthesis, including metal catalysis and electrochemistry. Fatty acids with at least eight carbons (dashed red line) can assemble into membranes. The fatty acids produced are saturated and unbranched. Note that detection of a fatty acid with a certain length may not have been attempted during every experiment. Refer to article for references underlying experiments and explorations. (Plot (Fig. 5) resourced with license (https://creativecommons.org/licenses/by/4.0/) from Cohen et al. [32] (review))

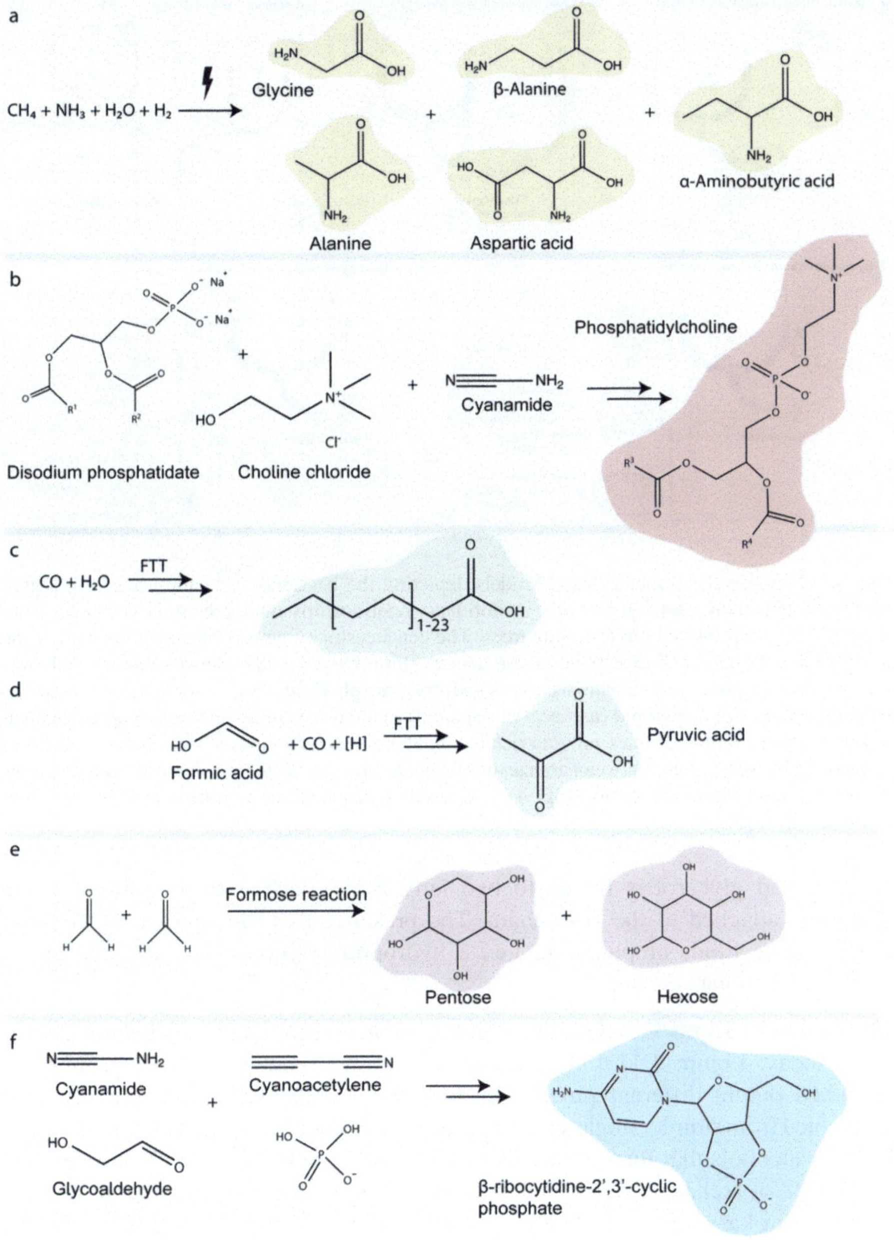

Fig. 3.10 Prebiotically plausible key chemical reactions: (**a**) Miller-Urey synthesis of amino acids. (**b**) Phospholipid synthesis. (**c, d**) Fatty acid synthesis. (**e**) Formose reaction and synthesis of ribose sugars. (**f**) Ribonucleotide synthesis. FTT = Fischer-Tropsch-type synthesis reaction (C-H chains from H_2 and CO). (Structures and schema (Fig. 3) resourced with permission from copyright owner of Gozen et al. [13] (review))

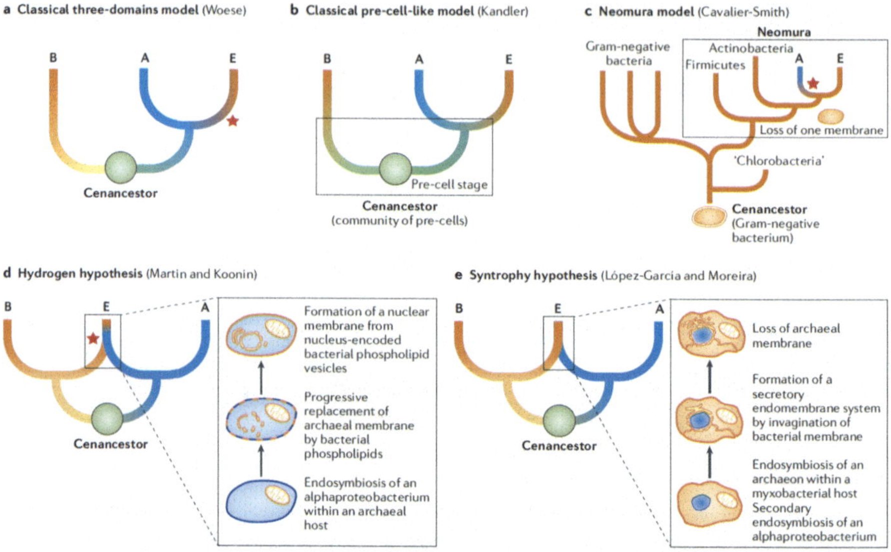

Fig. 3.11 Phospholipid membranes: Models depicting the bacterial-like nature of phospholipid membranes in eukaryotes: Views of the evolutionary relationships among the three domains of life are depicted as simplified phylogenetic trees. The cenancestor (green) is placed at the root of the trees. Orange branches correspond to organisms with bacterial-like phospholipids, and blue branches correspond to organisms with archaeal-like phospholipids. Red stars indicate transitions from one type of phospholipid (archaeal or bacterial) to the other. Insets in the hydrogen and syntrophy (symbiosis) hypotheses provide details about lipid evolution after the chimeric origin of eukaryotes by a symbiosis between archaeans and bacteria. See article for references to the various phospholipid models. (Cartoons (Fig. 4) resourced with permission from copyright owner of Lombard et al. [33])

choline, and glycerol is bound to two fatty acids, each with a carboxyl group (-COOH) attached to the H-C chain. The presence of O in lipids ($^{-}$COOH and PO_4^{3-}) make lipids amphiphilic, i.e., a hydrophobic (non-polar) HC end and a hydrophilic (polar) O end.

Formation of Lipid Vesicles The first membranes were likely much simpler than those today. Figure 3.11 [33] summarizes several models for how membranes branched out in different phyla. The first cell membranes were probably self-assembled from simple, single-chain amphiphilic molecules, such as monocarboxylic acids or alcohols that formed monolayered micelles [33]. The way that the plasma membrane of archaeans developed is likely distinctly different than that of modern bacteria (prokaryotes) and eukaryotes as envisioned in Fig. 3.12 [34]. It remains unclear if LUCA had both archaea and bacterial type membranes or had evolved to the bacterial type membrane [17]. Development of vesicles exposed hydrophilic groups to extend to both the exterior and interior of their bilayer boundary, that is, the nascent bilayer became capable of encapsulating a volume of solution that allowed nascent molecules to become trapped, but protected, within the membranes.

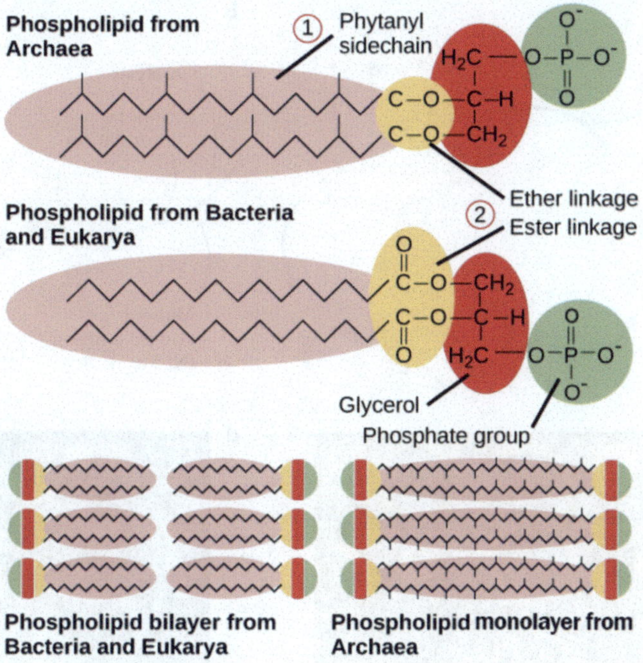

Fig. 3.12 Evolution of lipids: Comparison of lipid structure between Archaea vs. Bacteria and Eukaryota: Major differences in Archaea are ether bonds between phospholipid and glycerol (L-isomer), lipid side chains, and a monolayer membrane. In contrast, Bacteria and Eukaryota have ester bonds between phospholipid and glycerol (D-isomer), no lipid side chains, and a two-layer (lipid bilayer) membrane. (Cartoons resourced from OpenStax CNX, Structure of Prokaryotes. http://cnx.org/contents/9e7c7540-5794-4c31-917d-fce7e50ea6dd@11)

Depending on bilayer permeability and molecular size, the membranes facilitated the creation of gradients of individual molecules and ions. Moreover, vesicles in the prebiotic period could likely grow and divide spontaneously. From an origin-of-life perspective, these are interesting self-assembly properties (Fig. 3.13) [33] that made vesicles of amphiphilic compounds most important for isolating and concentrating ions and molecules within a self-contained milieu and for facilitating polymerization. The importance of lipids in forming membrane ion gradients is introduced in Sect. 2.1. Some common ion concentration gradients between extracellular fluid and cells, and between cells and their mitochondria are given in Sect. 1.2.

Prebiotic and Early Membranes of Lifeforms The first prebiotic membranes were likely assembled spontaneously from simple amphiphilic molecules [35]. The simplest membrane is a fatty acid (FA) with 10 or more hydrocarbons in a chain attached to a carboxylic acid head group; this allows them to form into monolayers on the surface of a liquid and to form vesicles in a solution composed of bimolecular membranes. The self-assembly of membranes is affected by concentrations of mon-

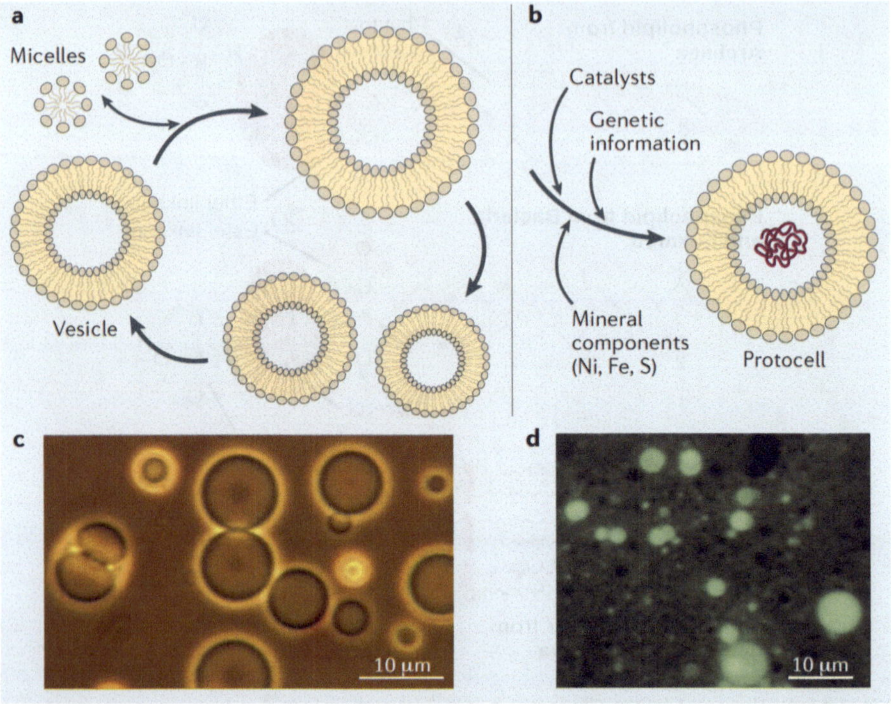

Fig. 3.13 Self-assembled vesicles: Formation, growth, and division of vesicles (**a**) and formation of protocells (**b**). Micrographs show self-assembled vesicles formed from amphiphilic C3–C11 carboxylic acids and polycyclic hydrocarbon derivatives found in the Murchison meteorite (**c**) and those formed from decanoic acid (**d**). The vesicles formed from decanoic acid show short, fluorescently labelled DNA fragments incorporated during wet–dry–wet cycles. (Micrographs (**c**, **d**) courtesy of D. Deamer (University of California, Santa Cruz, USA). Cartoons (Fig. 1a, b) resourced with permission from copyright owner of Lombard et al. [33])

ovalent NaCl and KCl inside and outside the membrane because osmosis of H_2O can shrink or expand membrane vesicles. Divalent cations (mostly Ca^{2+} and Mg^{2+}) (Sect. 2.3) can modify the membrane because of binding with carboxyl and phosphate groups [36]. The fluidity of these membranes between 0 and 100 °C depends on the presence of unsaturated C=C bonds, that is, not saturated with H^+ bonds, and C-C chain branching.

Making Polymers on Land and Sea A key question is whether amphiphilic compounds would form more favorably in surface hydrothermal fields than next to deep ocean, hydrothermal vents [37]. Before large land masses formed due to plate tectonics, cellular life may have originated in water pools around volcanic islands where H_2O vapor and steam came down as freshwater. It is proposed that the undulating interfaces of minerals and water, minerals and atmosphere, and water and atmosphere, resulted in cycles of hydration and dehydration [37]. In this way,

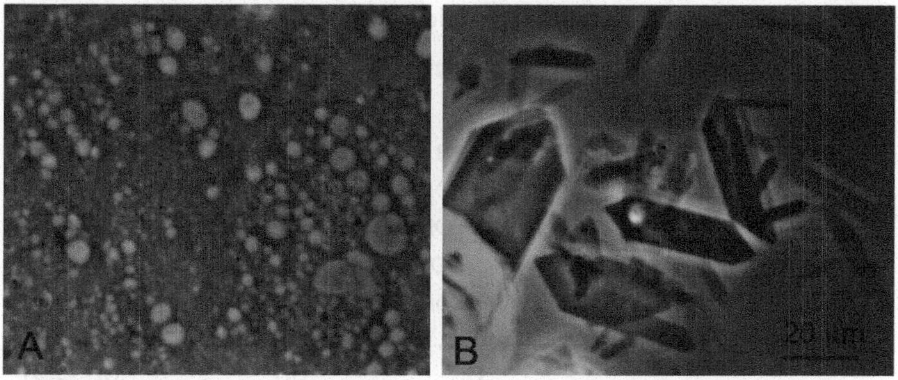

Fig. 3.14 Effect of seawater on self-assembly of lipid bilayer membranes: (**a**) Vesicles composed of decanoic acid readily form in dilute ionic solutions at pH 7.2; (**b**) In seawater at the same pH, crystals of a calcium-magnesium decanoate form instead of vesicles, thereby preventing the self-assembly of bilayer membranes. Bar shows 20 μm in both images. (Images (Fig. 3) with modified legend resourced with license (https://creativecommons.org/licenses/BY-NC/4.0/) from Deamer [36] (review))

organic solutes became concentrated by condensation on mineral surfaces that acted as catalytic co-factors for organic synthesis and polymerization of molecules.

In addition to not having wet-dry cycles that promote polymerization, a disadvantage for the premise of life forming near ocean vents is the composition of seawater. Seawater has much higher concentrations of Mg^{2+} and Ca^{2+} than freshwater. Figure 3.14 [36] shows that decanoic acid readily forms self-assembled vesicles in dilute seawater (akin to freshwater), whereas in seawater decanoic acid forms Mg^{2+}-Ca^{2+} crystals rather than vesicles. Since there is evidence supporting both hydrothermal vents and surface hydrothermal fields, this suggests that both environments had a time and condition-dependent role in life's beginning.

The nonpolar interior of a bilayer membrane prevents ions and polar molecules from passing through, that is, it forms a permeability barrier, which creates charge (voltage) and chemical gradients that are essential as energy sources for excitability and selective transport of ions and polar compounds. The processes of photosynthesis and oxidative phosphorylation (OxPhos) would be impossible without encasing chlorophyll and *e-* transport systems into vesicles with the capability of providing proton transfer and ATP synthesis.

The incorporation of phosphate anion and sterol linkages within amphiphilic membranes today allow phospholipids to assemble into bilayers at μM concentrations rather than at the mM concentrations of single chain lipids, for example, FAs required to form membranes in the absence of phosphates and sterols [36]. This implies that membrane evolution with these linkages was an evolutionary advantage over single chain lipids in forming membranes.

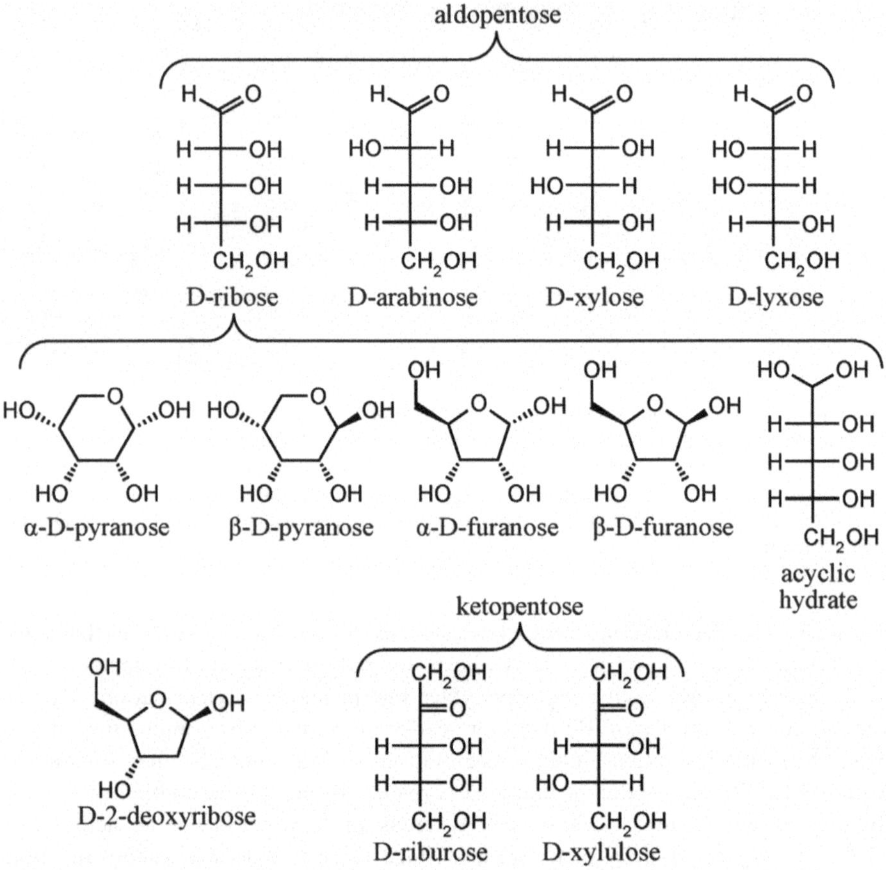

Fig. 3.15 Synthesis of the sugar D-ribose: There are some others pentoses with similar structure. (Structures and schema (Fig. 6) and its legend resourced with license (https://creativecommons.org/licenses/by-nc-nd/4.0/) from Kitadi and Maruyama [23] (review))

Abiotic Sugar Synthesis

Sugars and carbohydrates made photosynthetically today by algae and plants are food sources of chemical energy. But one sugar, ribose (Fig. 3.15) [23], is also a major component linking molecules together, particularly adenine nucleotides and RNA. Monosaccharides sugars, e.g., glyceraldehyde ($C_3H_6O_3$), ribose ($C_5H_{10}O_5$), and glucose ($C_6H_{12}O_6$), and polysaccharides ($C_x(H_2O)_y$), can be synthesized under prebiotic conditions using C based molecules containing O, such as 1-C formaldehyde (CH_2O), with two molecules forming 2-C glycolaldehyde ($HO{=}CH_2$-CHO); this is called the formose reaction. These 2-C molecules are found in interstellar matter. Once formed and placed in aqueous solutions, 5-C and 6-C sugars become arranged in rings as in Fig. 3.16h) [38].

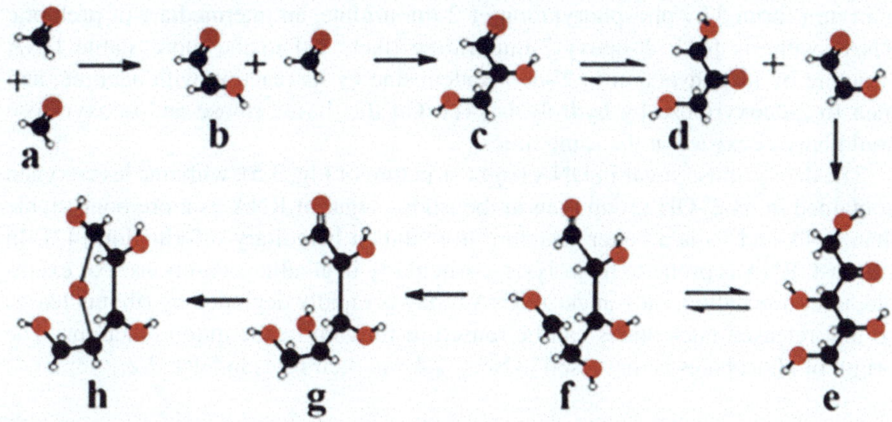

Fig. 3.16 D-ribose synthesis in the formose reaction: Two formaldehyde molecules (**a**) condense to glycolaldehyde (**b**). The subsequent aldol condensation forms glyceraldehyde (**c**), which undergoes a ketose-aldose isomerization (**d**). Glyceraldehyde reacts with glycolaldehyde to produce pentulose (**e**). Selective isomerization takes place between pentulose (**e**) and D-ribose (**f**), favoring the nucleophilic addition reaction of D-ribose (**g**) and ring formation to-D-ribose (**h**). Atoms C = black; O = red; H = clear. (Structures (Fig. 1) and its legend resourced with license (https://creativecommons.org/licenses/by/4.0/) from Banfalvi [38])

Simple sugars like ribose and glucose are joined together to make disaccharides and polysaccharides by dehydration synthesis; again, this is a process in which H_2O is removed (condensation) to join the sugar molecules together in a glycosidic linkage [38]. These links can occur spontaneously in the absence of enzymes and so may represent early, prebiotic, molecular self-replicating systems. Such replicating systems likely operated not only for ribose sugars, but also for the synthesis of polypeptides, fatty acids, phospholipids, and ribonucleotides by polymerization as depicted schematically in Fig. 3.10 [13, 39].

Importance of Ribose The high yield of ribose synthesis at high temperature and its protection from decomposition could have driven the abiotic polymerization of the ribose-phosphate "backbone" of RNA as well as the coupling of nucleobases into the backbone of RNA as shown in the left structure of Fig. 3.5. Also, ribose may have been evolutionarily selected as the sugar for the backbone of RNA because of its much higher permeability coefficient (up to 660 fold) than other sugars through vesicles made of phospholipids, for example, myristoleate, an FA ester, or single chain fatty acids [40].

RNA can function both as a template and as a catalytic "enzyme" (ribozyme, Sect. 3.3). RNA is proposed to have served as the first genetic template because to make DNA today from conversion to deoxyribose (H at position 2′) from ribose (OH at position 2′) requires a Fe-containing enzyme (ribonucleotide reductase), which, of course, had not yet been synthesized. However, it was found that 2′-deoxy-2-thiouridine, 2′-deoxyadenosine, and 2′-deoxyribose, can be formed under prebiotic conditions by photoreduction of 2,2′-anhydro-2-thiouridine, which is in turn formed by phosphorylation of 2-thiouridine, an intermediate of prebiotic RNA synthesis [41]. 2′-deoxy-2-thiouridine likely led to the more stable DNA structure by its conversion to 2′-deoxyadenosine by its reaction with adenine, and then to 2-deoxyribose by hydrolysis [41]. On this basis, ribose and deoxyribose could have coexisted at the same time.

The deoxyribose sugar in DNA (right structure of Fig. 3.5), with one less oxygen contained in its 2′-OH group than in the ribose sugar of RNA, is more bond stable than RNA and so is a better structure to maintain hereditary information [42]. In contrast, RNA is prone to hydrolysis, particularly in alkaline conditions. For example, after translation, messenger mRNA today is rapidly degraded by ribonucleases so that released nucleotides can be reused in forward transcription reactions. The origin of phosphates is discussed in Sect. 1.4 and their roles in Sect. 2.2.

Abiotic Amino Acid and Protein Synthesis

The distinct properties of the only 20+ or so natural amino acids are based on their unique "R" groups generally designated as nonpolar (hydrophobic, aliphatic or aromatic), polar (hydrophilic), acidic, and basic. Their unique characteristics, selected for incorporation in nascent lifeforms, set up the stage for diversity in structure and function, for example, ribosomes, metabolic enzymes, and ion channels, that are based on the unique sequence and length of the linked amino acids, as well as the distinct secondary and tertiary structures formed via folding (conformation). Up to 22 of the natural amino acids have been made synthetically in the laboratory [43], for example, in Fig. 3.10a [13]. and more than 500 different amino acids have been found in meteorites after hitting Earth [44, 45]. Most of these amino acids are L- (utilized today) rather than D- enantiomers as in sugars. Aromatic amino acids may have formed on Earth through a set of reactions that attaches substituents to an aromatic ring (Friedel–Crafts reactions) when catalyzed by an Fe-rich saponite clay, as demonstrated in the Mid-Atlantic oceanic ridge [46].

Hydrothermal Chemical Synthesis The demonstration of fluid-rock interactions in the oceanic lithosphere to generate amino acids abiotically gives credibility to the hydrothermal theory for generation of some molecules leading to the origin of life. Such findings may shed light on ancient pre-enzymic metabolism and the functioning of the present-day deep biosphere. Clay-forming hydrothermal alteration of oceanic rocks [46] may play a fundamental role in the synthesis and stabilization of complex organic compounds, such as aromatic amino acids, with mechanisms that can also protect the clay-trapped amino acids from hydrolysis, as also proposed in carbonaceous meteorites [47].

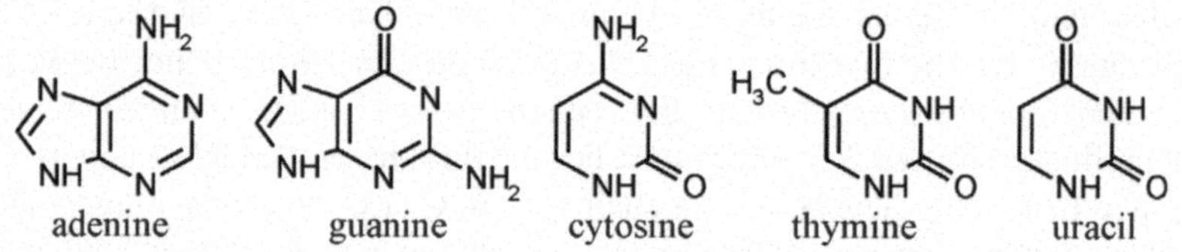

Fig. 3.17 Structures of the primary nucleobases found in DNA and RNA. (Structures (Fig. 8) and its legend resourced with license (https://creativecommons.org/licenses/by-nc-nd/4.0/) from Kitadi and Maruyama [23] (review))

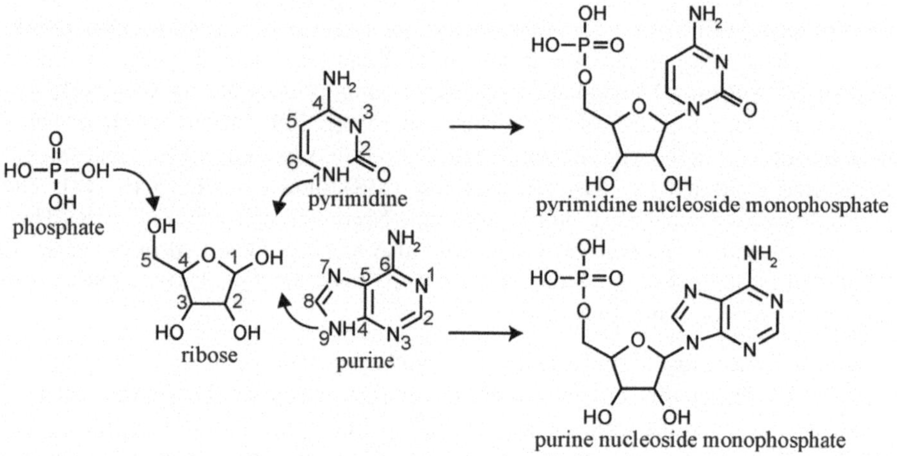

Fig. 3.18 Structure of nucleotides: Nucleotide components are ribose, nucleobase, and phosphate with their carbon atoms labeled. (Structures (Fig. 11) and the legend resourced with license (https://creativecommons.org/licenses/by-nc-nd/4.0/) from Kitadi and Maruyama [23] (review))

Abiotic Nucleoside Synthesis

In all lifeforms since LUCA, genetic information has been stored in the precise sequence of nitrogen bases in the sugar-phosphate skeletons of DNA, and in some instances, RNA. This information is translated into distinct amino acid sequences that form enzymes and protein structures. Nucleotide based molecules like nucleoside phosphates (e.g., AMP, GMP) and nucleoside coenzymes (e.g., NADH, $FADH_2$) contain only purine bases. Nucleic acids (e.g., RNA, DNA) (Fig. 3.5) comprise a D-ribose or 2-deoxy-D-ribose sugar that is linked by C-5′ to the phosphate unit, and by C-1′ to a nitrogen of one of six heterocycles, of which three are pyrimidines: cytosine (C), thymine (T), and uracil (U), and three are purines: adenine (A), guanine (G), as displayed in Fig. 3.17 [23], and hypoxanthine (H) (not shown).

The nucleotides are linked into chains by phosphodiester bonds between one phosphate group attached to the 5′ carbon of the sugar on one nucleotide with the free hydroxyl on the 3′ carbon of the next nucleotide as shown in Fig. 3.18 [23]. In

DNA double helix chains, the allowed pairings are A-T and C-G, and in RNA, they are A-U and C-G. The two strands of DNA are linked by two H-bonds between the bases A and T and by three H-bonds between the bases G and C. U and H are deamination products of C and A, respectively. For the third base of an RNA codon, U and C are interchangeable. The H base can pair with A, C or U, and so is mutagenic [48]. But what were the sources of these few bases, why were they selected, and how did they link up on the prebiotic Earth?

Origin of Nucleotide Bases A and G each have two N-containing rings with different functional groups; they may have arisen spontaneously from five interacting HCN molecules with metal catalysts and an energy source. However, these reactions to form purines are very endothermic and so are not likely to occur naturally [49]. Alternatively, purines, which have been found on meteorites, may have been formed in the cosmos by irradiation of ices containing H_2O, CH_3OH, CO_2, NH_3, and CH_4, and carried to Earth on meteorites and comets [49]. About 96 nucleobases have been found in the Murchison meteor that fell in Australia in 1969; an example is guanine identified by mass spectroscopy (MS) shown in Fig. 3.19 [47]. The pyrimidine bases today, C, T, and U, are single N-containing rings (Fig. 3.18). They too can be synthesized but since they were also recently identified in the cosmos, this could be their original source [47]. All five terrestrial nucleobases have now been found in meteorites [47].

Prebiotic Synthesis of Nucleotides The synthesis of purine based molecules into AMP and GMP, before enzymes were developed, was also likely carried out by condensation reactions [50]. Drying of the three components: purine, ribose, and phosphate, together on a surface with no H_2O, but in the presence of a metal catalyst like Ni, could have allowed the components to bind together with release of H_2O [51]. Figure 3.20 summarizes the binding possibilities of phosphates with either of the two sugars and one of the five bases. The phosphodiester bonds, as in AMP, RNA and DNA are between the phosphate groups and hydroxyl groups of two sugar groups as shown again in Fig. 3.5. These ester bonds form the backbone of nucleic acids and are crucial for stabilizing the structures of RNA and DNA. The discovery of phosphates [52] in the oceans of Saturn's moon Enceladus, and of HCN, CH_2O, C_2H_2, CH_3OH, and C_3H_6 in its volcanic plumes are portrayed in Fig. 3.21 [53]; these compounds could potentially support biologically mediated redox metabolism or polymerize to form nucleic and amino acid precursors. These findings furnish support for the possible existence of prebiotic or extraterrestrial life in our solar system.

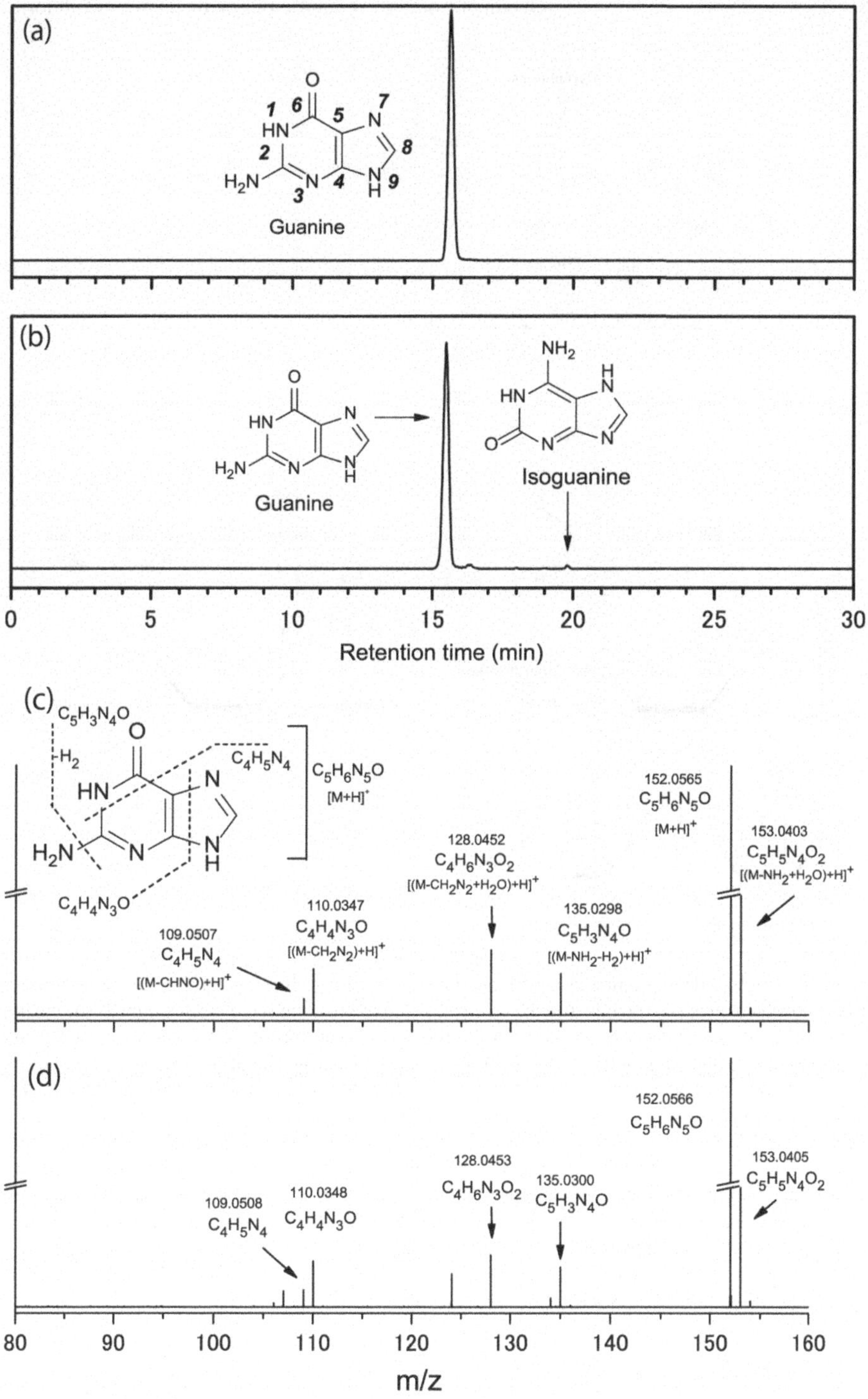

Fig. 3.19 Early nucleobases: Many nucleobases (96) have been identified in the Murchison (carbonaceous chondrite class) meteorite that fell to Earth in 1969: (**a**) Mass chromatograms (Fig. 1) detecting guanine ($C_5H_5N_5O$) and other purines. (MS plots and structures resourced with license (https://creativecommons.org/licenses/by/4.0/) from Oba et al. [47])

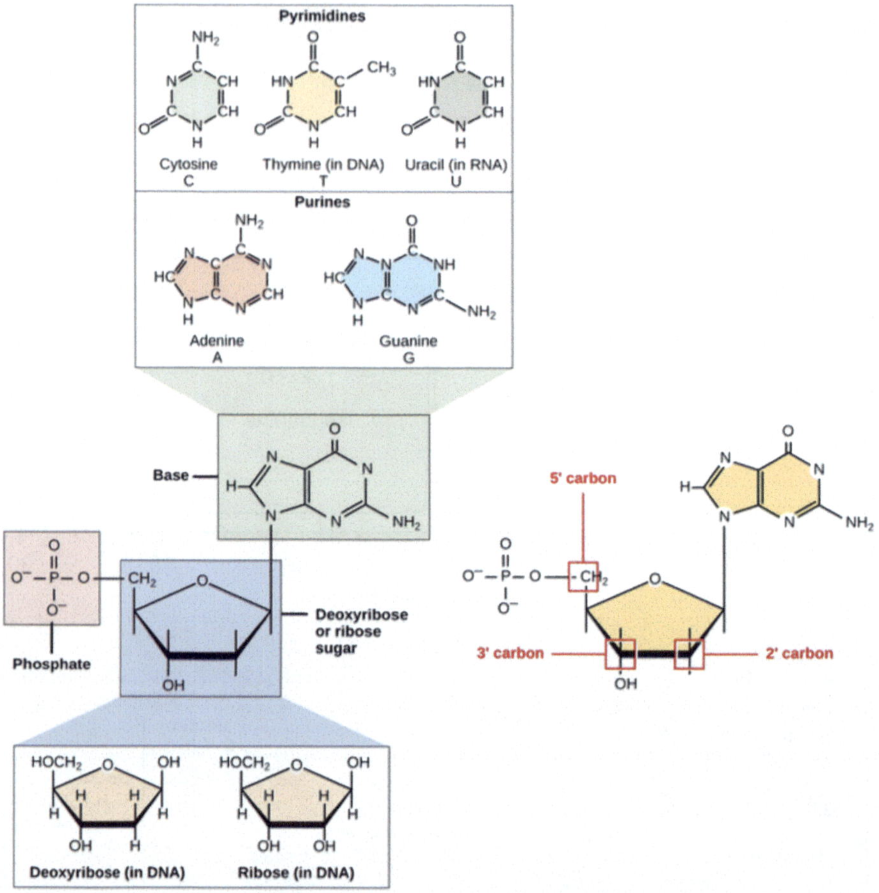

Fig. 3.20 Nucleotide Structure: A nucleotide has a nitrogenous base, a pentose sugar, and one or more phosphate groups. Carbon residues in the pentose are numbered 1′ through 5′ (the prime distinguishes these residues from those in the base, which are numbered without using a prime notation). The base is attached to the 1′ position of the ribose, and the phosphate is attached to the 5′ position. When a polynucleotide is formed, the 5′ phosphate of the incoming nucleotide attaches to the 3′ hydroxyl group at the end of the growing chain. Two types of pentose are found in nucleotides, deoxyribose (found in DNA) and ribose (found in RNA). Deoxyribose is similar in structure to ribose, but it has an H instead of an OH at the 2′ position. Bases can be divided into two categories: purines and pyrimidines. Purines have a double ring structure, and pyrimidines have a single ring. (Grouped structures (Fig. 3.14.1) and text resourced with license (https://creativecommons.org/licenses/by-nc-sa/4.0) from LibreTexts at: https://bio.libretexts.org/Bookshelves/Introductory_and_General_Biology/Book%3A_General_Biology_%28Boundless%29/03%3A_Biological_Macromolecules/3.14%3A_Nucleic_Acids_-__Types_of_RNA)

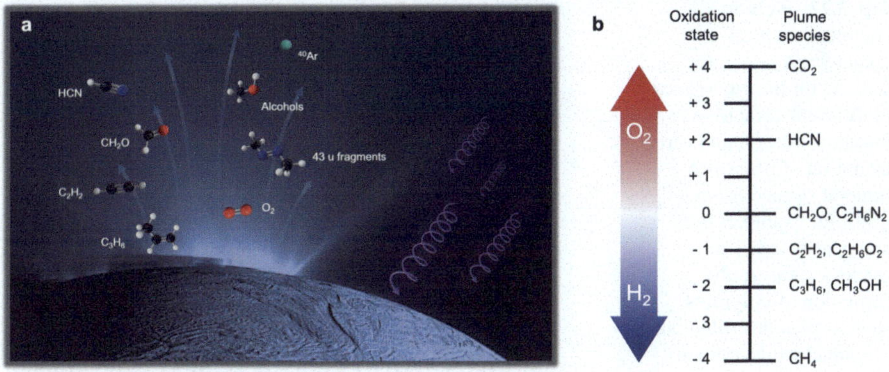

Fig. 3.21 Ejecta from Enceladus: New compounds identified in the Enceladus plume indicate a potentially habitable environment: (**a**) Jets emanating from ice fissures in Enceladus' South Polar Terrain feed a plume of ejected material containing organic molecules with varying oxidation states. Electron bombardment of the surface might help facilitate the production of oxidants and prebiotic feedstock molecules observed in the plume. These compounds could potentially support biologically mediated redox metabolisms or polymerize to form nucleic and amino acid precursors leading to the origin of life. (**b**) The average oxidation state of carbon for organic compounds confirmed or suspected in the plume. Plume-derived H_2 and O_2 could act as strong reducing and oxidizing agents, respectively, and may be responsible for the diverse redox chemistry seen at Enceladus. (Cartoon (Fig. 1) and its legend resourced with permission from copyright owner of Peter et al. [53])

3.3 Abiogenesis: Evolution of Complex Biochemical Compounds and Structures

Metals as Catalysts Before Enzymes

Before there were ribozymes, nascent riboproteins, and subsequently, diverse cellular enzymes to facilitate reactions, the early catalysts for organic reactions were metals, for example, Mg, Mn, Fe, S, Zn, Ni, and Cu, or metal combinations such as $(Fe, Ni)_3P$ and Fe_3Ni (Sect. 3.1), and FeS (Sect. 2.3 and Sects. 5.1, 5.2, and 5.3). This points out the essential roles of metals in prebiotic times, as well as their continued importance as cofactors in reactions in all lifeforms (Sects. 2.3 and 2.4). These early nonenzymic reactions may have required a harsh environment suitable for the metal-catalyzed reactions to take place, especially for endothermic reactions that require more energy for the activation than the energy produced. The capability of metals as catalysts is due principally to their various valance states that would allow *e-* transfer or sharing from one molecule to another (change of valence), accompanied by addition or subtraction of atoms from the substrate yielding a different molecule as the product.

Figure 3.22 [54] depicts a general scheme proposing the emergence of different cofactors during the origin and early evolution of life based on the pre and post

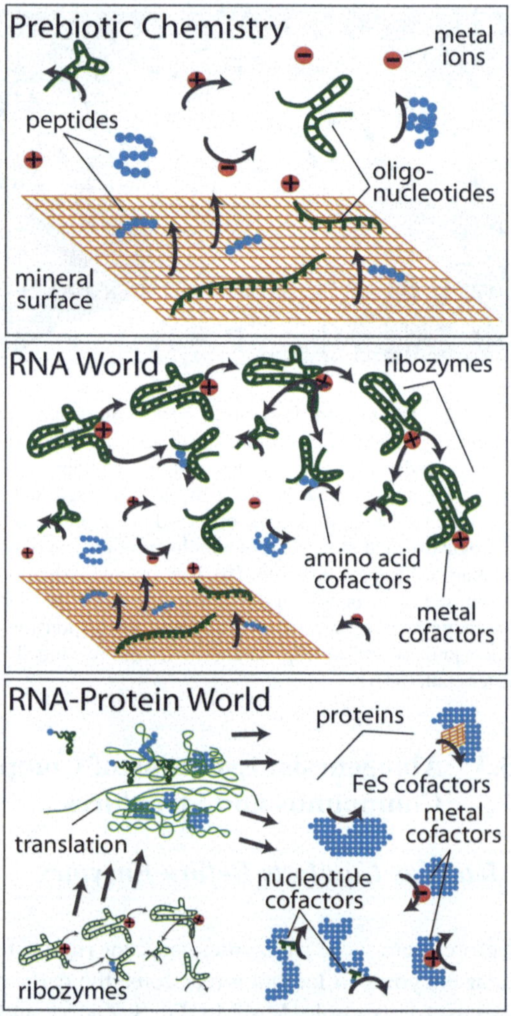

Fig. 3.22 Prebiotic chemical reaction cofactors: A general scheme for the emergence of different cofactors during the origin and early evolution of life: (**a**) A mineral surface (orange grid), for example, FeS, and nearby ions (red circles), catalyze the formation of polymers such as peptides (blue) and oligonucleotides (green), which themselves may have catalytic properties. (**b**) The "RNA world" scenario emerges from within this geochemical context. (**c**) Protein synthesis by translation evolved within the RNA-based genetic system and the resulting proteins retained cofactors that reflect the preceding RNA-based metabolism as well as the geochemical setting in which it first originated and evolved. (Cartoons (Fig. 2) and its legend resourced with license (https://creativecommons.org/licenses/by/4.0/) from Goldman and Kacar [54] (commentary))

"RNA world" concept. Evidence supports that peptides and oligonucleotides first condensed and polymerized on metal ion-rich surfaces in dry-wet cycles [54]. Corresponding oligonucleotide bases became H-bonded to each other to form hairpin loops, and metal cofactors permitted RNA bases to steer interaction with certain amino acids to form the ribozyme, which copies sequences of peptides. Following the evolution of protein synthesis by translation, the surrounding structural scaffold of these ancient ribozymes was supported by ribosomal protein but leaving behind the active site RNA as a coenzyme to link amino acids together [54].

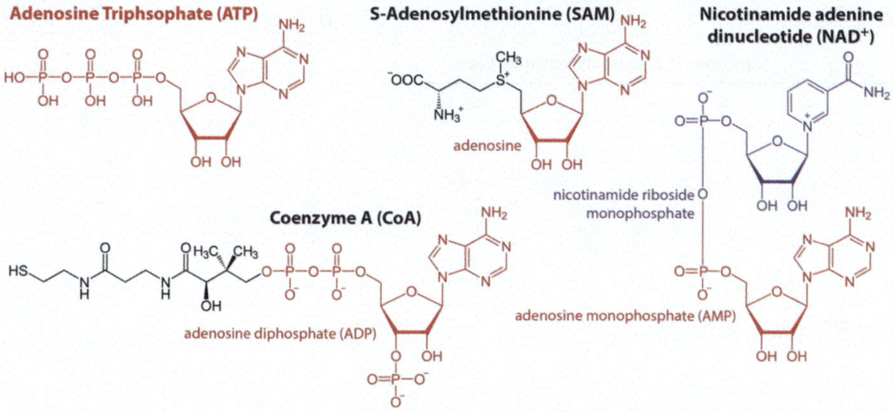

Fig. 3.23 Several prominent group transfer cofactors composed of or derived from nucleotides: ATP is both a group transfer cofactor and the adenosine monomer added to RNA. NAD^+ is a dinucleotide composed of adenosine monophosphate and the non-nucleic acid nucleotide, nicotinamide riboside monophosphate. CoA is composed of an adenosine diphosphate nucleotide attached to a cysteamine group (derived from cysteine amino acid) with pantothenate in between. SAM is composed of an adenosine monophosphate attached to a methionine amino acid. (Chemical structures (Fig. 1) and legend resourced with license (https://creativecommons.org/licenses/by/4.0/) from Goldman and Kacar [54] (commentary))

Organic Cofactors as Catalysts

More than 40 years ago it was proposed that certain newly synthesized organic compounds that were composed of smaller molecules, could then act together as organic cofactors to catalyze reactions [55]. Today these are called coenzymes, and in addition to ribosomal (r) RNA and transfer (t) RNA, most are composed of nucleobases or nucleoside moieties like coenzyme A (CoA) (Figs. 1.16 and 3.8) (Sects. 1.4, 3.1, 3.2, 5.1 and 6.3) and S-adenosyl methionine (SAM) (Fig. 3.23) [54]; alternatively, they are actual nucleotides like ADP (Fig. 1.17), ATP (Figs. 1.14 and 1.17), NADH (Fig. 3.24), and $FADH_2$ (see Fig. 5.5) [54]. Many of these coenzymes pick up one functional group as part of one reaction and release the same functional group in a subsequent, separate reaction; thus, they are also known as "group transfer cofactors" [54]. A display of common group transfer cofactors in Fig. 3.23 [54] shows the utility of molecules derived from adenine nucleotides to function as coenzymes in many reactions. The essential role of NADH and $FADH_2$ in transfer of energy from foodstuffs to ATP is addressed comprehensively in Sects. 3.2, 3.3, 5.1, 5.2 and 5.3.

Why ATP? The quintessential group transfer cofactor is ATP. But why was ATP naturally selected as the universal currency of energy? Or more precisely, why adenine? All nucleoside triphosphates are high in energy. Today purine synthesis requires 6 phosphorylation steps so why was the purine base guanine not selected over adenine, or why not use a pyrimidine? The establishment of ATP as nearly the

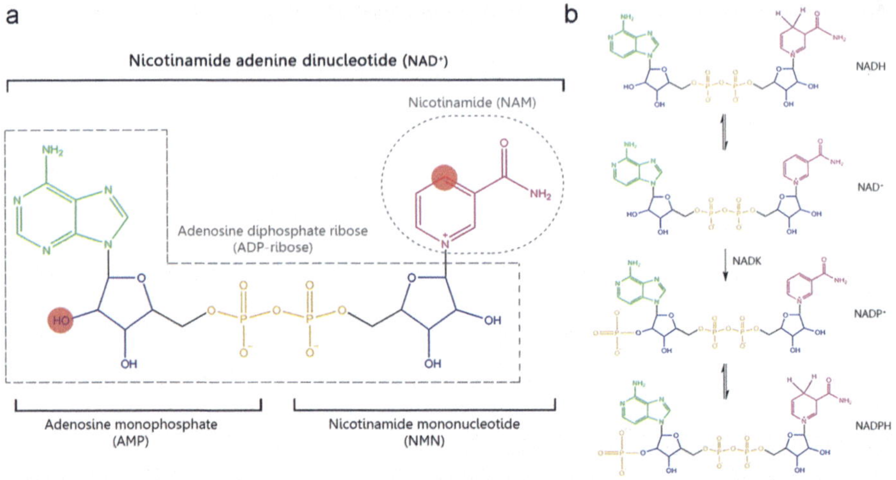

Fig. 3.24 Structure of nicotinamide adenine dinucleotide (NAD^+): NAD^+ is built from AMP linked to nicotinamide mononucleotide (NMN): (**a**) AMP is formed by adenosine (in green), ribose ring (in blue) whose hydroxyl (in red) can be phosphorylated resulting in $NADP^+$ and a phosphate group (in orange). NMN is formed by a ribose ring, phosphate group, and nicotinamide (NAM) (in pink) that is responsible for redox NAD^+ functions. Thus, the atom carbon (marked in red) accepts a hydride anion (H^+, 2*e*-) leading to the reduced form NADH. In addition to the classical redox functions, NAD^+ acts as a substrate of multiples enzymes donating the group adenosine diphosphate ribose (ADP-ribose) and releasing NAM as the catabolic reaction product. (**b**) Structure of the different states of NAD^+. NAD^+ can be phosphorylated to $NADP^+$ by the NADK enzyme. Thus, NAD^+ and $NADP^+$ can be reduced to NADH or NADPH, respectively. *NADH* nicotinamide adenine dinucleotide, *NAD^+* nicotinamide adenine dinucleotide, *$NADP^+$* nicotinamide adenine dinucleotide phosphate, *NADPH* reduced nicotinamide adenine dinucleotide phosphate. (Structures (Fig. 1) and its abridged legend resourced with license (https://creativecommons.org/licenses/by/4.0/ from Navas and Carnero [56] (review))

sole cellular energy source indicates its conservation in all life forms since LUCA is a result of a prebiotic equivalent of ATP [57, 58]. Evidence is provided that one route to ATP was acetyl phosphate (AcP) [58].

AcP is a bridge between high energy thioester and phosphate metabolism (Sect. 1.4) [57]; AcP can be synthesized in water at ambient condition from thioacetate ($CH_3(C{=}O)SH$). It links acetyl-CoA to the substrate-level phosphorylation of ADP in glycolysis and the TCA cycle. In the presence of Fe^{3+} ions in water at pH 5.5–6, 30 °C, AcP can phosphorylate ADP to ATP with an energy of hydrolysis greater than that of ATP. No other nucleoside diphosphate, e.g. guanine (GDP), cytosine (CDP), or uracil (UDP), can be phosphorylated via AcP [58]. From this it is proposed that ATP is universally conserved across lifeforms as the currency of energy because its formation is chemically favored in aqueous conditions. The chemical structure of adenine appears to be important [59]. Another possibility is that adenine, which is found in space and in meteorites alongside many other nitrogenous bases [47, 49], is more resistant to ionizing radiation than other nucleotide bases

[60]. Recall that ATP (and other nucleotides), once synthesized, is unstable because of the third phosphate group. In seeking out a reactant, ATP releases energy which can drive forward energetically unfavorable reactions and mediate multiple metabolic processes.

Cofactors Required for Chemiosmosis Equally ancient and ubiquitous are the dinucleotide containing cofactors in the electron transport system (ETS), that is, NADH, $FADH_2$, and flavin mononucleotide (FMN), as discussed in Sects. 5.1, 5.2 and 5.3. These two-ring, adenine-containing (Fig. 3.17) cofactors are essential for *e*- transfer along a series of molecules by redox reactions that culminates in production of ATP to drive other metabolic reactions. The two other group transfer cofactors, CoA and SAM (Fig. 3.23), transfer the 2-C acyl (acetate) and 3-C methyl groups, respectively. Each of these also has a nucleoside group (adenosine + ribose) and other side groups.

Other cofactors are histidine, an amino acid with its imidazole side chain that mimics a purine base, and thiamine, which has a pyrimidine (Fig. 3.17) moiety. Histidine has an important role in the acid–base chemistry of many enzymes and it is the most common amino acid found in the active site of enzymes [61]. Also the class of porphyrins (Fig. 2.5), which includes hemes (Sect. 2.2) can be synthesized abiotically; these are also cofactors in large part by virtue of the Fe at their center. Also ZnS promotes prebiotic chemistry by facilitating light-driven CO_2 reduction to formate ($HC{=}OO^-$) [62].

Abiotic Polypeptide Synthesis

Before ribozymes, ribosomes, and protein-based enzymes, polypeptides may have arisen on warm, dry surfaces by dehydration so that the ^-COOH group of one amino acid linked up with the $^-NH_2$ group of another amino acid to form a peptide bond that connects amino acids together as depicted in Fig. 3.25 [63]. One such surface could be clay, which contains elemental metal atoms such as Cu, Fe, S, Ni, and Zn, that could function as catalysts in this dehydration process of self-replication. This suggests that both dry land and water were necessary for forming and replicating the chemical structures that preceded proteins and enzymes in early lifeforms. Hydrothermal fields around volcanoes and tidal basins may have been a prime location for these reactions to occur. Because these peptides were random and non-coded, they may have had no useful function. Once ribozymes and protein synthesizing enzymes were developed, specific amino acid sequences were capable of being translated by ribosomes. Ribosomes take specific instructions from messenger RNA contained in the sequence of the three base codons carried by messenger RNA after transcription from DNA.

The TIM (triose-phosphate isomerase) barrel, with its α/β barrel protein architecture is proposed as an early bridge between the RNA world concept and protein mediated metabolism [54, 64]. During the transition from RNA as an enzyme to

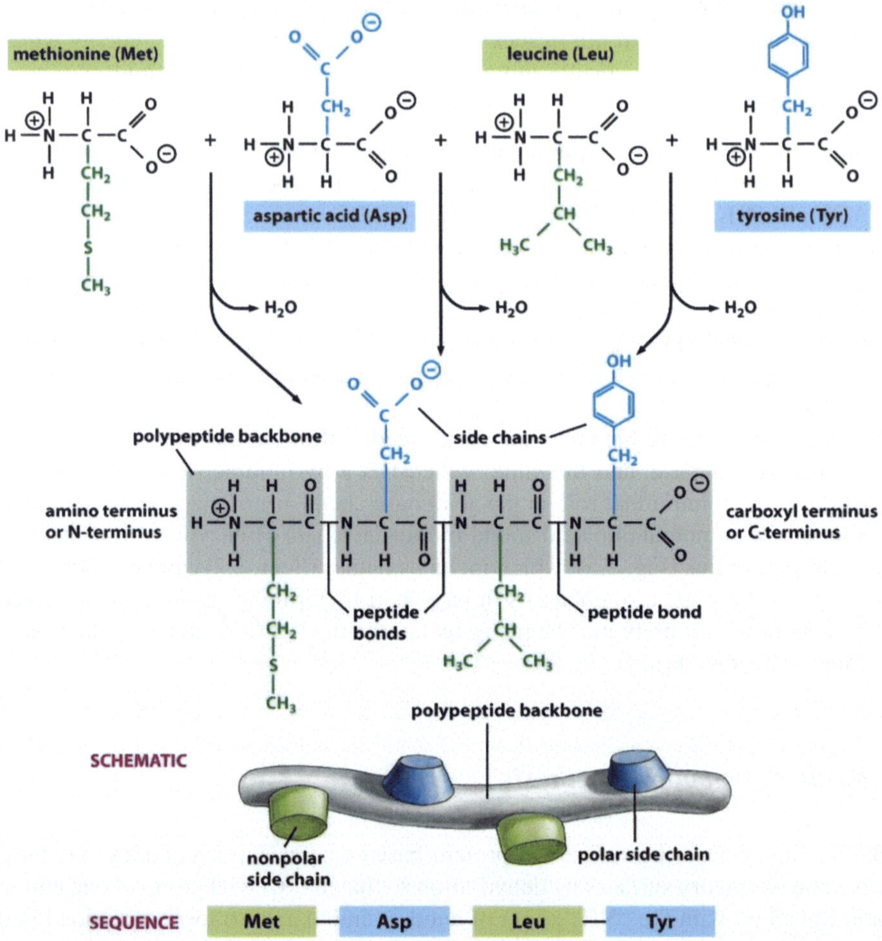

Fig. 3.25 Structural components of a protein: A protein consists of a polypeptide backbone with attached amino acid side chains. Each protein is distinct due to its unique sequence and number of amino acids. The end carrying the free amino acid group (NH_3^+) is the N-terminus, and the end carrying the free carbonyl group (COO^-) is the C-terminus. Amino acid sequences are always read in the N-to-C direction. (Structures and scheme (figure 3.2) and abridged and modified legend resourced with permission from copyright owner of The Shape and Structure of Proteins. Alberts et al. [63] (book chapter))

protein catalyst-mediated metabolism, TIM barrel proteins may have provided a universal active site into which different ancient cofactors could be placed through evolutionary processes, thus transferring their catalytic properties from ribozymes (or earlier, minerals) into protein enzymes [54, 64]. After the discovery of ribosomes, it was postulated that co-enzymes may have been the first catalytic sites for ancient ribozymes [54]. Their importance as co-enzymes throughout evolution in each domain of life attests to their essential roles not only in the origin life but also in its continuation.

Abiotic RNA and DNA Synthesis

During the long prebiotic period, many diverse polymeric molecules must have been randomly derived and assembled from organic precursors that were available on the early Earth. A few must have been capable of storing information and reproducing themselves as discussed above. The most efficient of these replicators subsequently must have evolved into RNA and DNA, the primary carriers of hereditary information in all lifeforms [65]. Over billions of years, DNA with its four different subunits (nucleotide bases), encoded in the linear sequence of these nucleosides, became the carrier of genetic information. Eventually, the interactions between two strands of DNA with complementary sequences (double helix) satisfied the process of stable DNA replication within and across all species. Much has been learned in the past 70 years since DNA and its function were discovered only in 1869 and 1953, respectively [66].

Over time, DNA replication, RNA transcription, and protein translation, were organized and catalyzed by enzymes, for example, DNA polymerase, RNA polymerases for messenger (m), transfer (t), and ribosomal (r) RNA, and peptidyl transferase in ribosomes. With genetic templates, and enzymes to catalyze the reactions, organisms could replicate into like organisms. But errors in replication (genetic mutations) and transcription or translation (phenotype mutations) alter the phenotype of an organism. If a mutation was advantageous for survival, it could be adapted in the progeny of the organism so that over approximately 3 Byr, a great diversity of lifeforms has evolved.

Different Roles for RNA and DNA The processes of replication, transcription, and translation are largely well understood today, but much remains to be known about how these systems were first generated and how they have evolved. It is theorized that RNA was synthesized first [38, 42]. But RNA and DNA could have coevolved. RNA bases can bind with other bases on the same molecule because of its flexible structure (Fig. 3.26) [42]. RNA has vital functions both in the transcription of DNA (information storage) and in the translation of nucleotide sequences into structural proteins and enzymes, whereas DNA contains the vital hereditary information of an organism to develop, survive and reproduce. Similar to RNA, with its bases coupled onto a strand of ribose-phosphate linkages, experiments indicate that DNA, with its bases coupled onto a strand of deoxyribose-phosphate linkages, could have been synthesized from organic compounds present in a prebiotic environment [67].

One key to understanding the storage of genetic information is how the coupling of deoxyribose, the backbone sugar of DNA, occurred. Figure 3.27 [67] presents experimentally verified, selective pathways to all deoxy-sugar nucleosides formed by what is called the "nucleobase-activated pathway". It may be that deoxyribose was not linked to the base in a single step. Instead, deoxyribose may have been built up on a preformed base by a short sequence of reaction steps involving simple organic molecules such as acetaldehyde and glyceraldehyde [67]. It is also possible that primitive DNA originally had a different sugar than deoxyribose as the

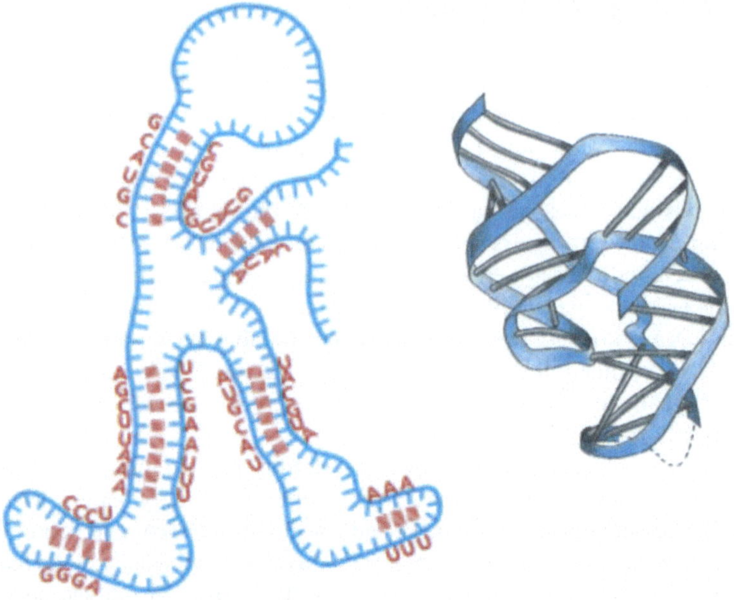

(A) (B)

Fig. 3.26 Conformation of an RNA molecule: (**a**) Nucleotide sequence pairing between different regions of the same RNA polymer chain causes the molecule to adopt a distinctive shape. (**b**) 3D structure of an RNA molecule from hepatitis delta virus that catalyzes RNA strand cleavage. (Structures (Fig. 1.6) resourced with permission from copyright owner of The Universal Features of Cells on Earth. Alberts et al. [15] (book chapter))

sugar-phosphate structural backbone [67]. From these selective pathways to only deoxy-sugar nucleosides formed by the nucleobase activated pathways [67], it is suggested that the earliest DNA molecules could have appeared in parallel with RNA 4 Bya, or shortly thereafter, when there were at least some pools of surface H_2O.

Prebiotic Condensation and Polymerization

It is intriguing to understand how chains of amino acids and nucleotides were formed into peptides and nucleic acids in the absence of ribozymes and polymerases in prebiotic conditions. Polymerization is endothermic and so requires an energy source. In the absence of H_2O, anhydrous condensation can form ester and peptide bonds at a rate constant higher than the hydrolysis that occurs in H_2O [36]. It is proposed that geothermal energy (heat) from hydrothermal fields in the presence of metal catalysts in dry clay provided the conditions for dehydration and concentration of monomers to condense into chains [68–71].

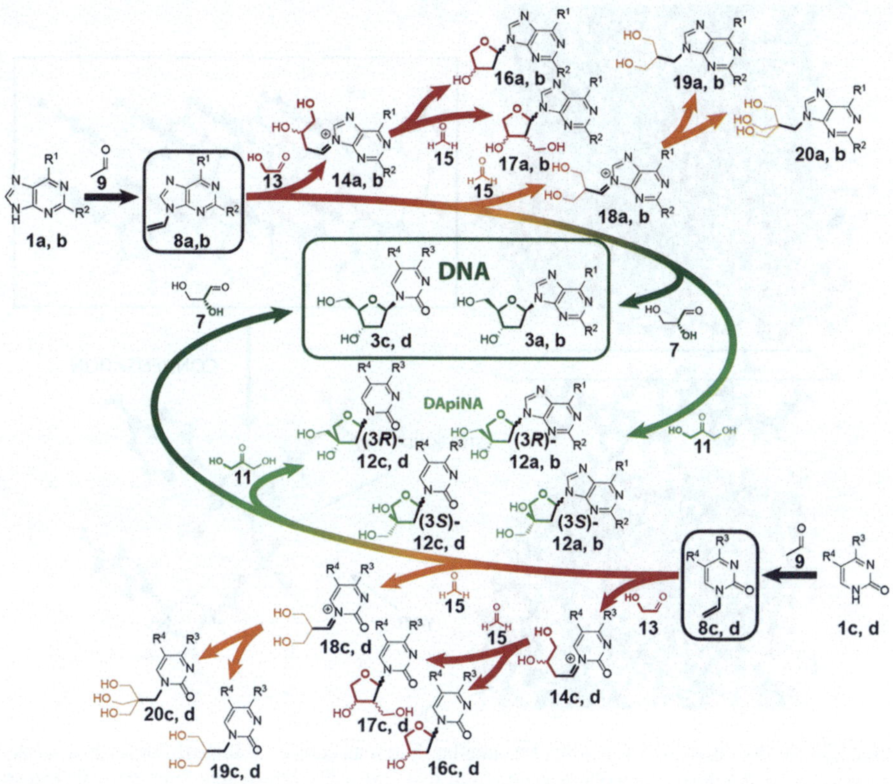

Fig. 3.27 Deoxy-sugar nucleoside formation: Experimentally verified selective pathways to all deoxy-sugar nucleosides formed by the nucleobase activated pathway. The initial step is formation of 8 from the 1 and 9 (black pathway). Reaction with 13 and 15 leads to16 (dTNA) and 17 (HMENA), respectively, which are unable to polymerize into a defined structure (red pathway). In a side-reaction under Cannizzaro conditions with 15, two deoxy-nucleosides are formed,19 and 20 (orange pathway). Reaction of 8 with 11 results in the DNA progenitor 12 (DApiNA nucleosides, light-green pathway), and with 7, 3 is obtained (dark-green pathway). (a) $R^1 = NH_2$, $R^2 = H$; (b) $R^1 = OH$, $R^2 = NH_2$, (c) $R^3 = NH_2$, $R^4 = H$; (d) $R^3 = OH$, $R^4 = CH_3$. (Reaction structures and schemes (Fig. 3) and its legend resourced with permission from copyright owner of Teichert et al. [67])

Early Energy Bonds Synthesis of the covalent phosphodiester bond (Figs. 1.16 and 1.17) can occur experimentally in cycles of heating, drying, and rehydration on clay but only a few disorganized dimers and trimers are formed in this way. However, if mononucleotides are present in a solution of amphiphiles during dehydration, the solutes become concentrated and organized within a two-dimensional matrix where polymerization can occur [72]. Figure 3.28 [36] shows how a lamellar lipid matrix can arrange monomers to form chains. Further work demonstrates that RNA synthesis can be initiated in a simulated environment of a hydrothermal field, with 5′-AMP and 5′-UMP as monomers in a solution of lysophosphatidyl-choline during rehydration [69]. Figure 3.29 [36] gives a proposed mechanism for phosphoester bond for-

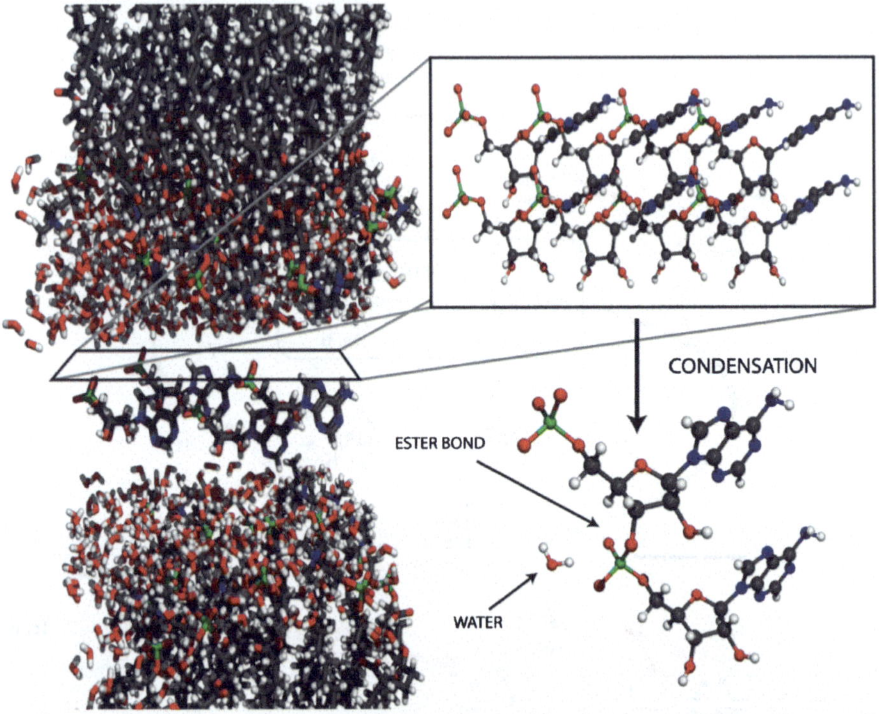

Fig. 3.28 X-ray diffraction structure of a lamellar lipid matrix that has imposed order on a solute, in this case 5′-adenosine monophosphate (from text Ref. [73]). (Structure (Fig. 6) with legend resourced with license (https://creativecommons.org/licenses/BY-NC/4.0/) from Deamer [36] (review))

mation [75]. This proposal suggests that the RNA world view was not a singularity, but likely codeveloped with membrane vesicles as platforms, not only for RNA chain synthesis, but also for the synthesis of repeat bonding of other molecules into chains.

These discoveries support the energy requiring prebiotic synthesis of nucleic acids and peptides. However, the reverse processes of hydrolysis of ester and peptide bonds, including separation of a nucleotide base from its sugar, also occur spontaneously. So it is possible that under prebiotic conditions, the synthesis of polymers was faster than their degradation by hydrolysis [36]. Today, degradation is mostly executed by dehydrogenases.

Evolution of Ribozymes, Ribosomes, and Enzymes

On completion of the abiotic synthesis of a multitude of different amino acids and nucleic acids, of which only a few were selected and utilized for LUCA, these amino and nucleic acids were likely polymerized by condensation to polypeptides

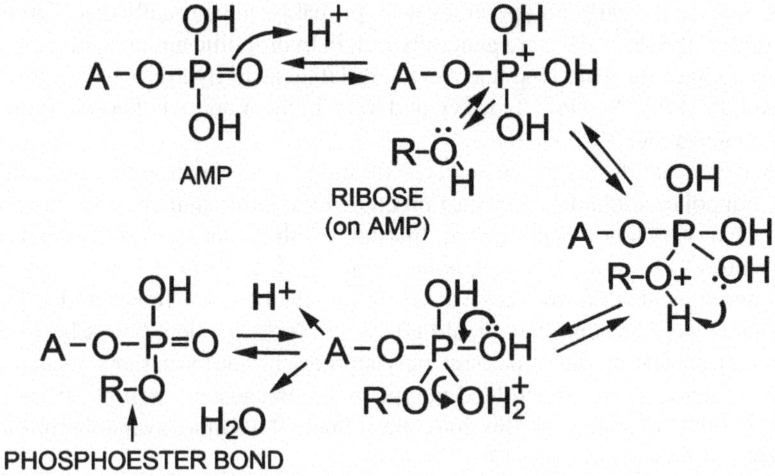

Fig. 3.29 Proposed mechanism for phosphoester bond formation in a condensation reaction driven by simulated hydrothermal conditions (text Ref. [74]). (Reaction scheme (Fig. 7) with legend resourced with license (https://creativecommons.org/licenses/BY-NC/4.0/) from Deamer [36] (review))

and nucleotides in wet-dry cycles by their interaction with the minerals in clay [76–79], as presented above. Thiodepsipeptide, a molecule that contains both a peptide and a high energy thioester bond (ΔG^{o} −7.7 kcal/mol) is proposed as a plausible route for proto-peptide formation [73]. The simultaneous polymerization of peptides and RNA [80] may have allowed the conditions for heralding the dawn of the simplest early lifeforms [80, 81].

Multifunctional RNA? The "RNA world" concept revolves around the catalytic properties of RNA-based enzymes, i.e., ribozymes, before the advent of proteins as enzymes [74]. The RNA world concept postulates that the most prominent molecular constituent for the development of life was RNA, or something chemically like RNA [55, 82], that could: (1) self-replicate; (2) catalyze reactions (in place of or in addition to metal catalysts) as in pre-ribosomal rRNA; and (3) chemically transfer coding information from nucleotides to peptides [83]. The first part was arguably self-replication of RNA; isolation of an RNA enzyme with RNA replicase activity supports this theory [84]. Next, strands of spontaneously formed polypeptides may have interacted with strands of RNA nucleotides to form proto-ribosomes (ribozymes). Indeed, nucleotide bases and amino acid residues readily interact, as in RNA binding proteins (RBPs), mostly by H-binding and van der Waals interactions [85, 86].

Early Disordered Peptides During the evolution of the first proteins in abiotic conditions, the first functional peptides were likely short and linear [79]. They could have formed secondary sheet structures in H_2O due to H-bonding, but they were thought incapable of folding into complex tertiary structures (conformations) [87].

This is because the early polypeptides were probably "intrinsically disordered peptides (IDPs)" [88, 89]. IDPs are generally rich in hydrophilic amino acids (e.g., Arg, Asp, Cys, Gln, Glu, His, Lys, and Ser) and deficient in hydrophobic amino acids (e.g., Ala, Ile, Leu, Phe, Pro, Trp, Tyr, and Val). In the modern eukaryote proteome, there remains about 25–30% IDPs.

IDP means that the sequence and type of amino acids (hydrophilic polar, hydrophobic nonpolar, aliphatic, aromatic) precluded the individual amino acid residues (R side chains) in a polypeptide from linking together, that is, as in charged amino acid residues to form ionic bonds, polar amino acids to form H-bonds, and hydrophobic amino acids to form weak van der Waals interactions. Importantly, if cysteine residues were present, they would interact to form covalent bonds between the sulfurs to fold and stabilize nascent polypeptides in conformations so that RNA could not interact with these folded polypeptides. Because of their plasticity in not forming strict conformations, this could have made IDPs more available for binding with RNA as proto-ribosomes [79].

RNA binding could have catalyzed the selection of short, disordered polypeptides with patterned sequencing. But if the peptide did not result in a functional use, it would not be selected. The selection of an IDP with very few, new sequence changes could convert an unstructured IDP into a more stable helix–hairpin–helix motif capable of bonding suitable amino acid residues [79] to form an ordered amino acid arrangement. Thus, the role of IDPs in the origin of selective folded domains is key to the evolution of specific, stable protein conformations that allow an enzyme to perform a specific task. RNA that was able to associate with small stabilizing peptides, and then fold them, likely preceded the ribosome. The transition from a flexible peptide to a stable protein structure with a specific function must have been key to origination of the first proto-ribosomes [90].

Non-Coding RNA Non-canonical RNA base pairing is different from canonical base pairing as found in the Watson–Crick double helix DNA with adenine (A) forming a pair with uracil (U) using two H-bonds (weaker), and guanine (G) forming a pair with cytosine (C) using three H-bonds (stronger). Noncanonical RNA bases are held together by multiple hydrogen bonds that are mostly planar and stable. Noncanonical RNA does not encode proteins but plays an essential role in shaping the structure and function of different functional RNA molecules, like those found today in transfer (t) and ribosomal (r) RNAs [74, 91]. These base pairing are considered to be ancient in the RNA world concept as they are able to establish peptide synthesis directly on RNA [91]. The decorating of complex peptide-sequences onto rRNA to make new chimeric molecules supports the early existence of an RNA-peptide world [74].

Ribozymes to Ribosomes The proposed evolution of ribonucleoprotein (RNP) from a pure ribozyme (rRNA only) is gleaned (a) from evidence of comparative analysis of non-coding RNAs that perform the same function in phylogenetically disparate organisms, and (b) from biochemical and structural analysis of the separate contributions of the RNAs and proteins in various RNP enzymes [92]. There are big gaps in understanding the evolution of the original RNP. But it is widely agreed

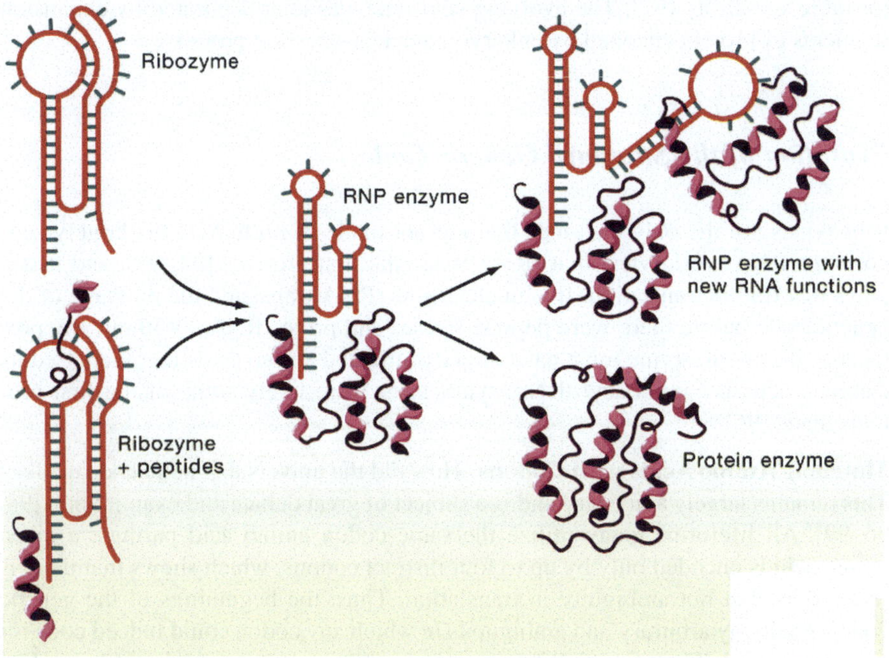

Fig. 3.30 Model for the evolution of biological catalysis: (Upper left) Primordial ribozymes were capable of purely RNA catalysis, but (lower left) given random peptides in their environment, selection for fitness may have occurred at the level of the RNP. (Middle) A more advanced stage in which systems reproducibly synthesized peptides and polypeptides that enhanced RNP stability and function, ultimately including translation of mRNAs on ribosomes. (Upper right) RNP enzymes have continued to evolve by adding (and subtracting) RNA elements that either contribute directly at the RNA level or bind to additional protein subunits. (Lower right) In most cases, catalysis has been taken over by protein enzymes. (Cartoon (Fig. 1) and modified legend resourced with permission from copyright owner of Cech [92] (review))

that the first peptides were translated at the peptidyl transferase center (PTC) site of the ribozyme (RNA world concept). Before LUCA the ribosome was essentially a PTC ribozyme supported by proteins made at the ribozyme that (a) furnished functional stabilization of the ribosome, (b) contributed to the correct folding of the RNA structure, and (c) improved the efficiency and accuracy of translation [93].

The First RNA Made Peptides Figure 3.30 [92] depicts how phylogenetically conserved rRNA might have first made peptides on rRNA that furnished stability and function to the nascent RNP and eventually translated mRNA. A key development was reproducible synthesis of useful proteins. As observed in different present-day phyla, the rRNA continued to evolve by adding and subtracting RNA elements to give it new function. Eventually, the protein synthesized at the PTC site was released and catalysis was taken over by protein enzymes, likely because protein catalysis had better functionality such as hydrophobic and hydrophilic side groups for more versatility [92]. The evolving ribosome was later decorated by expansion segments of protein encoded by eukaryote nuclear-encoded proteins.

Evolution of the Standard Genetic Code

Ribozymes are the only enzymes that are not made of proteins. This largely supports the RNA world view that rRNA made the first proteins [94, 95], and it still plays that role in translation. It is unclear how rRNA translated the message of the genetic code before there were peptide ligases and proto-tRNAs. Without any proteins at all, the ribozyme must have directly attracted amino acids that increased its catalytic activity; aminoacyl-tRNA synthetases were likely some of the first proteins made [96].

Matching Amino Acids with Codons How did the universal genetic code evolve? This remains largely a mystery and is a subject of great debate and examination [94, 96–99]. All lifeforms today utilize the same codon amino acid parings; a given amino acid is encoded only by up to four distinct codons, which shows that there is redundancy but not ambiguity in translation. Thus, the beginnings of the genetic code were likely arbitrary and ambiguous in which any codon could indeed code for any amino acid. The mix of different polypeptides that were produced are called statistical proteins. The evolving genetic code from codon sequence to amino acid sequence had two problems to fix, code ambiguity and translation errors [100].

An earlier stereochemical theory posited that the standard genetic code (SGC) was determined by the stereochemical affinities between codons and amino acids. But experiments concluded that any codon can be associated with any amino acid. It is now widely accepted that the SGC is the result of natural selection in which mutations and mistranslations of the nascent code eventually led to a minimization of errors in phenotypes and stabilization of the phenotype [100].

The top theories today are based on a process called "ambiguity reduction" by which evolution took place with a mechanism that systematically reduced the emerging code's ambiguity and eventually removed it such that one or more codons became capable of coding only one amino acid [100]. Once the genetic code had achieved universality by the time of LCUA it must have been self-optimized to minimize errors in transcription and translation.

Equilibrium fitness would be attained when more sets of phenotypes had a genetic code that always made the same protein [101]. Higher fitness means a long-term natural selection between a population of protein-coding genes that becomes more and more reproducible while more and more of its codons matched up to the same amino acid product [101]. It is proposed that for this to occur there must be horizontal gene transfer (HGT) (Sect. 4.2) at the level of an individual in a single generation [101, 102].

One mathematical model [102] takes the 64 codons ambiguously (randomly) encoding the 20 amino acids to test for universality, meaning that over time, what conditions would result in a given codon only coding for one amino acid? The researchers used the mechanism of HGT to demonstrate that convergence of the genetic code to a near universal solution using dynamic simulations is directly linked to asexual (horizontal) mutation rates. HGT is evidenced by endosymbiosis in prokaryotes (Sect. 4.2) but it also had an important role eukaryotic genome evolution [103] where it could have accelerated delineation of the eventual final SGC.

Chemical Templates for the Evolution of Life From the above background, it is proposed that a functioning ribosome must have developed long before the appearance of the last universal common ancestor (LUCA) (Figs. 2.1, 3.3, 3.10, and 3.22) [104]. The conservation of rRNA is so prominent that it has allowed establishment of phylogenetic trees to divide lifeforms into the two or three proposed life domains, Archaea, Bacteria, and Eukaryota [14, 83, 105]. Despite many advances in ribosome research, it remains quite controversial how the ribosomal subunits, composed mostly of rRNA and a few polypeptides, transfer tRNA, and the enzymes associated with breaking and connecting strands, were collectively put to use [79, 106–110]. Moreover, advances in molecular genetics raise new questions about the origins of the three domains and the participants in endosymbiosis [111–113].

Current evidence suggests that some form of cellular organization arose well before the time of the last universal common ancestor (LUCA). Standard phylogenetic analyses have shown that several protein families associated with membrane translocation, membrane transport, and membrane bioenergetics were very likely present in the proteome of LUCA. Despite these cellular systems emerging prior to the LUCA, extant archaea, bacteria, and eukaryotes have significant differences in cellular infrastructure and the molecular functions that support it, leading some researchers to argue that true cellularity did not evolve until after LUCA [17, 111, 112, 114].

For example, comparisons of reconstructed partial proteomes of LUCA, the last archaeal common ancestor (LACA), and the last bacterial common ancestor (LBCA) demonstrate that a broad set of functions associated with cellular organization and function were already present by the time of LUCA. However, the proteomes of LACA and LBCA related to cellular organization continued to change after the divergence of LUCA into the proposed Domains of life [114].

One model for the chemical basis of life is diagramed in Fig. 3.31 [80]. This model proposes first the abiotic formation of polypeptides by polymerization with cofactors, with the later development of ribosome components to carryout biosynthesis [80]. The model also posits that peptides and nucleic acids coevolved together through collaboration at successive stages of sophistication. A recent update considers some of the controversies over how early ribosomes developed [79]. Recent attention has been given to the evolution and elongation of ribosomes, maturation of the RNA rich peptidyl transferase center (PTC), where the synthesis of proteins takes place, and the formation and extension of the ribosomal tunnel where translation and modifications occur before the ribosomal units separate from the just formed polypeptide sequence [79].

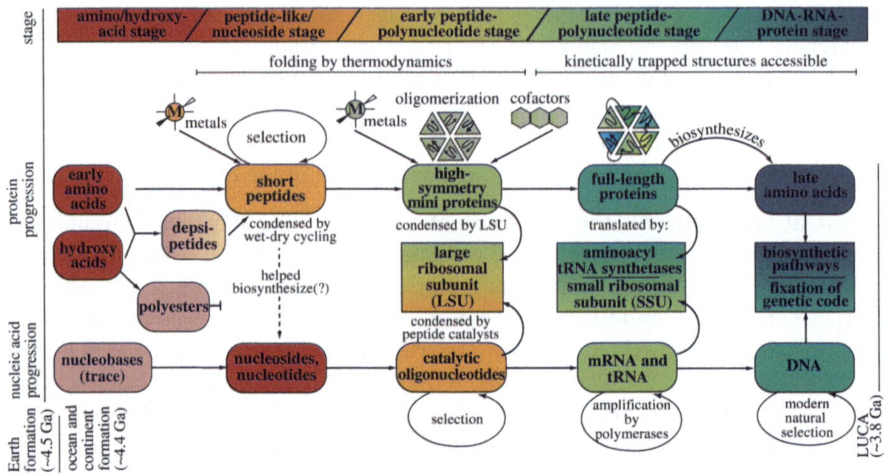

Fig. 3.31 Proposed origin of life model: A model informed by the early accessibility of various monomeric organic molecules leading to the formation of short peptide-like molecules and emphasizing the various ways in which peptides and nucleic acids co-evolved through collaborating at successive stages of sophistication. (Cartoons and scheme (Fig. 1) and its legend resourced with license (https://creativecommons.org/licenses/by/4.0/ from Fried et al. [80] (review))

LUCA The hypothetical ancestor to all lifeforms today is based on a hypothetical cell line that diverged between 3.5 and 3.9 Bya with common characteristics shared by each of the three proposed domains, Archaea, Bacteria, and Eukarya [14, 83, 115–117]. Shared by all three domains are dsDNA and m, r, and tRNAs, with DNA and RNA polymerases, ribosomes, and the nucleobases and amino acids utilized today. A complete characterization of LUCA remains much debated. LUCA is proposed to have had genes and a genetic code that could be replicated, transcribed, and translated into peptides and proteins with catalytic activity. This would make it a genote in which a fixed genotype propagates the phenotype.

A controversy over a membrane makeover after LUCA would make it a progenote, where substantial changes could yet occur [17]. LUCA likely had a lipid bilayer membrane that tended to exclude Na^+ and concentrate K^+ in its watery cytoplasm by ion ATP-ase driven pumps and ion exchangers. Energy was likely converted to ATP by the chemiosmosis process. Cells likely divided by fission and there was genetic recombination. LUCA is preceded by a first universal common ancestor (FUCA), a proposed first lifeform that did not have all these characteristics and existed millions of years before LUCA.

Figure 3.32 depicts what the ancestral LUCA protokaryote might have consisted of [117]. Because all life forms on Earth today use the same genetic heredity code and protein translation code, it is assumed that DNA was transferred on cell division and ribosomes were present to replicate proteins essential for energy transduction and repair [116, 117]. The investigators examined phylogenetic trees for 6.1 million genes from sequenced prokaryotic genomes to reconstruct the microbial ecology of LUCA. They identified 355 protein families that trace for LUCA by phylogenomic

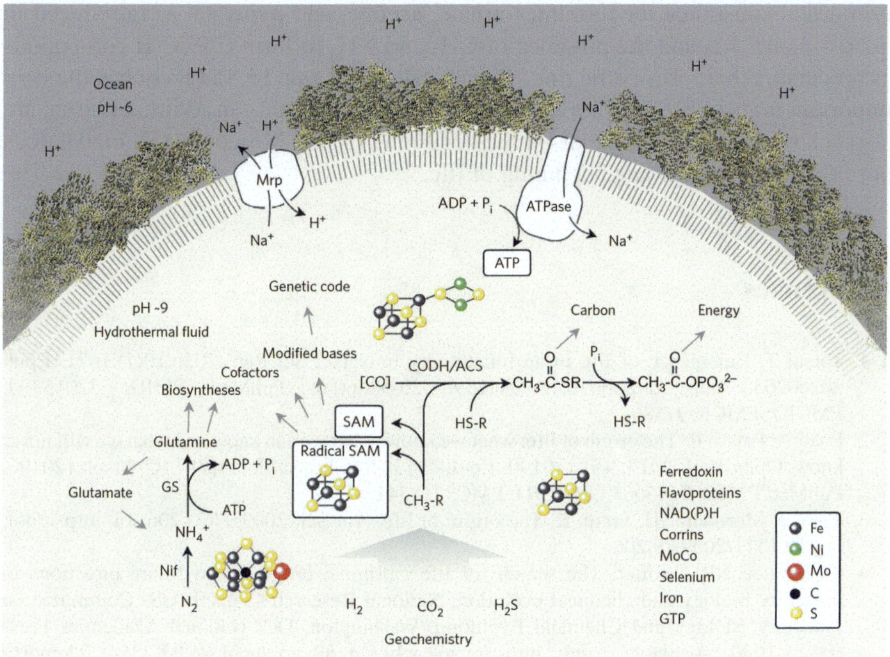

Fig. 3.32 LUCA reconstructed from genome data: Summary of the main interactions of LUCA with its environment, a vent-like geochemical setting (text Refs. [12, 13, 18, 20]), as inferred from genome data (Supplementary Table 2 in text). Note the proposed geochemical precursors: N_2, H_2, CO_2, H_2S. Abbreviations: *CODH/ACS* carbon monoxide dehydrogenase/acetyl CoA-synthase, *Nif* nitrogenase, *GS* glutamine synthetase, *Mrp* MrP type Na^+/H^+ antiporter, *CH3-R* methyl groups, *HS-R* organic thiols. The components listed on the lower right are present in LUCA, in addition to the cofactors listed in Supplementary Table 1 in text. In modern CODH/ACS complexes, CO is generated from CO_2 and reduced ferredoxin (text Ref. [22]). The figure does not make a statement regarding the source of CO in primordial metabolism (uncatalyzed via the gas water shift reaction or catalyzed via transition metals), symbolized by [CO]. A Na^+/H^+ antiporter could transduce a geochemical pH gradient (indicated on the left) inherent in alkaline hydrothermal vents (text Refs. [13, 18]) into a more stable Na^+ gradient to feed a primordial Na-dependent ATP synthase (text Ref. [20]). LUCA undisputably possessed genes, because it had a genetic code; the question of which genes it possessed has hitherto been more difficult to address. The transition metal catalysts at the nitrogenase active site and the CODH/ACS active site as well as a 4Fe–4S cluster as in ferredoxin are indicated. (Cartoon (Fig. 3) and modified legend resourced with permission from copyright owner of Weiss et al. [117])

criteria. Their conclusions depicted LUCA as anaerobic, CO_2-fixing, N_2-fixing, thermophilic H_2-dependent, with LUCA utilizing the Wood-Ljungdahl pathway, which fixes carbon by the reductive acetyl-CoA pathway (Sects. 3.1 and 5.1). There was evidence from sequence homology for proteins that harbor FeS clusters and free radical reactions, flavins, SAM, CoA, and ferredoxin, among others. These investigators suggest that LUCA originated in and inhabited a geothermal active environment rich in H_2, CO_2, and Fe; their data support the theory of an autotrophic origin of life [117]. This theory is based on the early presence of H_2 and CO_2 as primordial substrates for forming formate, acetate, and pyruvate, as introduced in Sect. 1.3 and 3.1, and the presence of CH_4 and NH_3 to form CN^-/C_2H compounds as precursors that form cyclic rings (Sects. 1.4, 3.1, 5.1 and 5.3). Of course, the very important roles of P, S, and single and multi-valanced metals, in addition to iron, are also acknowledged as essential for the beginning of life. Moreover, this model does not rule out a heterotropic beginning of life.

References

1. Totani T. Emergence of life in an inflationary universe. Sci Rep. 2020;10(1):1671. Epub 20200203. https://doi.org/10.1038/s41598-020-58060-0. PubMed PMID: 32015390; PMCID: PMC6997386.
2. Pross A, Pascal R. The origin of life: what we know, what we can know and what we will never know. Open Biol. 2013;3(3):120190. Epub 20130306. https://doi.org/10.1098/rsob.120190. PubMed PMID: 23466673; PMCID: PMC3718341.
3. Trefil J, Morowitz HJ, Smith E. The origin of life. Am Sci. 2009;97(3):206–14. https://doi.org/10.1511/2009.78.206.
4. Committee NRC, editor. The search for life's origins: progress and future directions in planetary biology and chemical evolution, National Research Council (US) Committee on Planetary Biology and Chemical Evolution. Washington, DC: National Academies Press (US); 1990. Available from: https://www.ncbi.nlm.nih.gov/books/NBK235415/?report=printable.
5. Rasmussen S, Chen L, Deamer D, Krakauer DC, Packard NH, Stadler PF, Bedau MA. Evolution: transitions from nonliving to living matter. Science. 2004;303(5660):963–5. https://doi.org/10.1126/science.1093669. PubMed PMID: 14963315.
6. Haldane JBS. The origin of life. 1929. Available from: https://en.wikipedia.org/wiki/Primordial_soup#cite_note-:0-2.
7. Oparin AI. The origin of life. 1924. Available from: https://en.wikipedia.org/wiki/Primordial_soup#cite_note-1.
8. Mitton S. A short history of panspermia from antiquity through the mid-1970s. Astrobiology. 2022;22(12) https://doi.org/10.1089/ast.2022.0032.
9. Halliday A, Canup RM. The accretion of planet Earth. Nat Rev Earth Environ. 2023;4:19–36. https://doi.org/10.1038/s43017-022-00370-0.
10. Mojzsis SJ, Harrison TM, Pidgeon RT. Oxygen-isotope evidence from ancient zircons for liquid water at the Earth's surface 4,300 Myr ago. Nature. 2001;409(6817):178–81. https://doi.org/10.1038/35051557. PubMed PMID: 11196638.
11. Garcia-Ruiz JM, van Zuilen MA, Bach W. Mineral self-organization on a lifeless planet. Phys Life Rev. 2020;34–35:62–82. Epub 20200113. https://doi.org/10.1016/j.plrev.2020.01.001. PubMed PMID: 32303465.
12. Henderson JCI. Prebiotic chemistry, the primordial replicatorm and modern protocells. In: Rasmussen S, Bedau MA, Chen L, Deamer D, Krakauer DC, Packard NH, Stadler PF, editor. Protocells: bridging nonliving and living matter. The MIT Press; 2022. 9780262282093. https://doi.org/10.7551/mitpress/9780262182683.001.0001.
13. Gozen I, Koksal ES, Poldsalu I, Xue L, Spustova K, Pedrueza-Villalmanzo E, Ryskulov R, Meng F, Jesorka A. Protocells: milestones and recent advances. Small. 2022;18(18):e2106624. Epub 20220323. https://doi.org/10.1002/smll.202106624. PubMed PMID: 35322554.
14. Woese CR, Fox GE. Phylogenetic structure of the prokaryotic domain: the primary kingdoms. Proc Natl Acad Sci USA. 1977;74(11):5088–90. https://doi.org/10.1073/pnas.74.11.5088. PubMed PMID: 270744; PMCID: PMC432104.

15. Alberts B, Johnson A, Lewis J, Raff M, Roberts K, Walter P. The universal features of cells on earth. In: Molecular biology of the cell. 4th ed. New York: Garland Science; 2002. Available from: https://www.ncbi.nlm.nih.gov/books/NBK26871/.
16. Williams TA, Cox CJ, Foster PG, Szollosi GJ, Embley TM. Phylogenomics provides robust support for a two-domains tree of life. Nat Ecol Evol. 2020;4(1):138–47. Epub 20191209. https://doi.org/10.1038/s41559-019-1040-x. PubMed PMID: 31819234; PMCID: PMC6942926.
17. Di Giulio M. The origins of the cell membrane, the progenote, and the universal ancestor (LUCA). Biosystems. 2022;222:104799. Epub 20221027. https://doi.org/10.1016/j.biosystems.2022.104799. PubMed PMID: 36330883.
18. Miller SL, Urey HC. Organic compound synthesis on the primitive earth. Science. 1959;130(3370):245–51. https://doi.org/10.1126/science.130.3370.245. PubMed PMID: 13668555.
19. Miller SL. A production of amino acids under possible primitive earth conditions. Science. 1953;117(3046):528–9. https://doi.org/10.1126/science.117.3046.528. PubMed PMID: 13056598.
20. Oro J, Kimball AP. Synthesis of purines under possible primitive earth conditions. I. Adenine from hydrogen cyanide. Arch Biochem Biophys. 1961;94:217–27. https://doi.org/10.1016/0003-9861(61)90033-9. PubMed PMID: 13731263.
21. Oró J, Kamat SS. Amino-acid synthesis from hydrogen cyanide under possible primitive earth conditions. Nature. 1961;190:442–3. https://doi.org/10.1038/190442a0. PubMed PMID: 13731262.
22. Oró J. In: Fox S, editor. Origins of prebiological systems and of their molecular matrices. Wakulla Springs: New York Academic Press; 1967. p. 137. ISBN 978-1-4832-2861-7. https://doi.org/10.1016/C2013-0-12108-3.
23. Kitadi N, Maruyama S. Origins of building blocks of life: a review. Geosci Front. 2017;9(4):1117–53. https://doi.org/10.1016/j.gsf.2017.07.007.
24. Kolodiazhnyi OI. Phosphorus compounds of natural origin: prebiotic, stereochemistry, application. Symmetry. 2021;13:889. https://doi.org/10.3390/sym13050889.
25. Muchowska K, Varma SJ, Moran J. Nonenzymatic metabolic reactions and life's origin. Chem Rev. 2020;120(15):7708–44. https://doi.org/10.1021/acs.chemrev.0c00191.
26. Martin WF. Hydrogen, metals, bifurcating electrons, and proton gradients: the early evolution of biological energy conservation. FEBS Lett. 2012;586(5):485–93. Epub 20111001. https://doi.org/10.1016/j.febslet.2011.09.031. PubMed PMID: 21978488.
27. Martin WF. Older than genes: the acetyl CoA pathway and origins. Front Microbiol. 2020;11:817. Epub 20200604. https://doi.org/10.3389/fmicb.2020.00817. PubMed PMID: 32655499; PMCID: PMC7325901.
28. Ragsdale SW. Life with carbon monoxide. Crit Rev Biochem Mol Biol. 2004;39(3):165–95. https://doi.org/10.1080/10409230490496577. PubMed PMID: 15596550.
29. Xavier JC, Preiner M, Martin WF. Something special about CO-dependent CO_2 fixation. FEBS J. 2018;285(22):4181–95. Epub 20181008. https://doi.org/10.1111/febs.14664. PubMed PMID: 30240136; PMCID: PMC6282760.
30. Lane N, Allen JF, Martin W. How did LUCA make a living? Chemiosmosis in the origin of life. BioEssays. 2010;32(4):271–80. https://doi.org/10.1002/bies.200900131. PubMed PMID: 20108228.
31. Ljungdahl LG. A life with acetogens, thermophiles, and cellulolytic anaerobes. Ann Rev Microbiol. 2009;63:1–25. https://doi.org/10.1146/annurev.micro.091208.073617. PubMed PMID: 19575555.
32. Cohen ZR, Todd ZR, Wogan N, Black RA, Keller SL, Catling DC. Plausible sources of membrane-forming fatty acids on the early Earth: a review of the literature and an estimation of amounts. ACS Earth Space Chem. 2023;7(1):11–27. Epub 20221222. https://doi.org/10.1021/acsearthspacechem.2c00168. PubMed PMID: 36704178; PMCID: PMC9869395.

33. Lombard J, Lopez-Garcia P, Moreira D. The early evolution of lipid membranes and the three domains of life. Nat Rev Microbiol. 2012;10(7):507–15. Epub 20120611. https://doi.org/10.1038/nrmicro2815. PubMed PMID: 22683881.
34. van Wolferen M, Pulschen AA, Baum B, Gribaldo S, Albers SV. The cell biology of archaea. Nat Microbiol. 2022;7(11):1744–55. Epub 20221017. https://doi.org/10.1038/s41564-022-01215-8. PubMed PMID: 36253512; PMCID: PMC7613921.
35. Deamer D, Dworkin JP, Sandford SA, Bernstein MP, Allamandola LJ. The first cell membranes. Astrobiology. 2002;2(4):371–81. https://doi.org/10.1089/153110702762470482. PubMed PMID: 12593777.
36. Deamer D. The role of lipid membranes in life's origin. Life (Basel). 2017;7(1). Epub 20170117. https://doi.org/10.3390/life7010005. PubMed PMID: 28106741; PMCID: PMC5370405.
37. Deamer DW, Georgiou CD. Hydrothermal conditions and the origin of cellular life. Astrobiology. 2015;15(12):1091–5. https://doi.org/10.1089/ast.2015.1338. PubMed PMID: 26684507.
38. Banfalvi G. Prebiotic pathway from ribose to RNA formation. Int J Mol Sci. 2021;22(8). Epub 20210408. https://doi.org/10.3390/ijms22083857. PubMed PMID: 33917807; PMCID: PMC8068141.
39. Liu Y, Sumpter DJT. Mathematical modeling reveals spontaneous emergence of self-replication in chemical reaction systems. J Biol Chem. 2018;293(49):18854–63. Epub 20181003. https://doi.org/10.1074/jbc.RA118.003795. PubMed PMID: 30282809; PMCID: PMC6295724.
40. Sacerdote MG, Szostak JW. Semipermeable lipid bilayers exhibit diastereoselectivity favoring ribose. Proc Natl Acad Sci U S A. 2005;102(17):6004–8. Epub 20050414. https://doi.org/10.1073/pnas.0408440102. PubMed PMID: 15831588; PMCID: PMC556129.
41. Xu J, Green NJ, Gibard C, Krishnamurthy R, Sutherland JD. Prebiotic phosphorylation of 2-thiouridine provides either nucleotides or DNA building blocks via photoreduction. Nat Chem. 2019;11(5):457–62. Epub 20190401. https://doi.org/10.1038/s41557-019-0225-x. PubMed PMID: 30936523; PMCID: PMC6597365.
42. Alberts B, Johnson A, Lewis J, Raff M, Roberts K, Walter P. The RNA world and the origins of life. In: Molecular biology of the cell. 4th ed. New York: Garland Science; 2002. Available from: https://www.ncbi.nlm.nih.gov/books/NBK26904/.
43. Ambrogelly A, Palioura S, Soll D. Natural expansion of the genetic code. Nat Chem Biol. 2007;3(1):29–35. https://doi.org/10.1038/nchembio847. PubMed PMID: 17173027.
44. Cronin JR, Pizzarello S. Amino acids in meteorites. Adv Space Res. 1983;3(9):5–18. https://doi.org/10.1016/0273-1177(83)90036-4. PubMed PMID: 11542462.
45. Glaven D, McLain HL, Dworkin JP, Parker ET, Elsila JE, Aponte JC, Simkus DN, PozaryckI CI, Graham HV, Nittler LR, Alexander CM. Abundant extraterrestrial amino acids in the primitive CM carbonaceous chondrite Asuka 12236. Meteorit Planet Sci. 2020;55(9):1976–2006. https://doi.org/10.1111/maps.13560.
46. Ménez B, Pisapia C, Andreani M, Jamme F, Vanbellingen QP, Brunelle A, Richard L, Dumas P, Réfrégiers M. Abiotic synthesis of amino acids in the recesses of the oceanic lithosphere. Nature. 2018;564(7734):59–63. https://doi.org/10.1038/s41586-018-0684-z.
47. Oba Y, Takano Y, Furukawa Y, Koga T, Glavin DP, Dworkin JP, Naraoka H. Identifying the wide diversity of extraterrestrial purine and pyrimidine nucleobases in carbonaceous meteorites. Nat Commun. 2022;13(1):2008. Epub 20220426. https://doi.org/10.1038/s41467-022-29612-x. PubMed PMID: 35473908; PMCID: PMC9042847.
48. Lin T, Zhang L, Wu M, Jiang D, Li Z, Yang Z. Repair of hypoxanthine in DNA revealed by DNA glycosylases and endonucleases from hyperthermophilic archaea. Front Microbiol. 2021;12:736915. Epub 20210831. https://doi.org/10.3389/fmicb.2021.736915. PubMed PMID: 34531846; PMCID: PMC8438529.
49. Bera PP, Stein T, Head-Gordon M, Lee TJ. Mechanisms of the formation of adenine, guanine, and their analogues in UV-irradiated mixed NH_3:H_2O molecular ices containing purine.

Astrobiology. 2017;17(8):771–85. Epub 20170714. https://doi.org/10.1089/ast.2016.1614. PubMed PMID: 28708419; PMCID: PMC5734622.

50. Yadav M, Kumar R, Krishnamurthy R. Chemistry of abiotic nucleotide synthesis. Chem Rev. 2020;120(11):4766–805. Epub 20200109. https://doi.org/10.1021/acs.chemrev.9b00546. PubMed PMID: 31916751.
51. Fernandez-Garcia C, Coggins AJ, Powner MW. A chemist's perspective on the role of phosphorus at the origins of life. Life (Basel). 2017;7(3). Epub 20170713. https://doi.org/10.3390/life7030031. PubMed PMID: 28703763; PMCID: PMC5617956.
52. Postberg F, Sekine Y, Klenner F, Glein CR, Zou Z, Abel B, Furuya K, Hillier JK, Khawaja N, Kempf S, Noelle L, Saito T, Schmidt J, Shibuya T, Srama R, Tan S. Detection of phosphates originating from Enceladus's ocean. Nature. 2023;618(7965):489–93. Epub 20230614. https://doi.org/10.1038/s41586-023-05987-9. PubMed PMID: 37316718; PMCID: PMC10266972.
53. Peter J, Nordheim TA, Hand KP. Detection of HCN and diverse redox chemistry in the plume of Enceladus. Nat Astron. 2023; https://doi.org/10.1038/s41550-023-02160-0.
54. Goldman AD, Kacar B. Cofactors are remnants of life's origin and early evolution. J Mol Evol. 2021;89(3):127–33. Epub 20210206. https://doi.org/10.1007/s00239-020-09988-4. PubMed PMID: 33547911; PMCID: PMC7982383.
55. White HB 3rd. Coenzymes as fossils of an earlier metabolic state. J Mol Evol. 1976;7(2):101–4. https://doi.org/10.1007/BF01732468. PubMed PMID: 1263263.
56. Navas LE, Carnero A. NAD^+ metabolism, stemness, the immune response, and cancer. Signal Transduct Target Ther. 2021;6(1):2. Epub 20210101. https://doi.org/10.1038/s41392-020-00354-w. PubMed PMID: 33384409; PMCID: PMC7775471.
57. Whicher A, Camprubi E, Pinna S, Herschy B, Lane N. Acetyl phosphate as a primordial energy currency at the origin of life. Orig Life Evol Biosph. 2018;48(2):159–79. Epub 20180303. https://doi.org/10.1007/s11084-018-9555-8. PubMed PMID: 29502283; PMCID: PMC6061221.
58. Pinna S, Kunz C, Halpern A, Harrison SA, Jordan SF, Ward J, Werner F, Lane N. A prebiotic basis for ATP as the universal energy currency. PLoS Biol. 2022;20(10):e3001437. Epub 20221004. https://doi.org/10.1371/journal.pbio.3001437. PubMed PMID: 36194581; PMCID: PMC9531788.
59. Shapiro R. The prebiotic role of adenine: a critical analysis. Orig Life Evol Biosph. 1995;25(1–3):83–98. https://doi.org/10.1007/BF01581575. PubMed PMID: 11536683.
60. Barbatti M, Ullrich S. Ionization potentials of adenine along the internal conversion pathways. Phys Chem Chem Phys. 2011;13(34):15492–500. Epub 20110729. https://doi.org/10.1039/c1cp21350d. PubMed PMID: 21804965.
61. Ribeiro AJM, Tyzack JD, Borkakoti N, Holliday GL, Thornton JM. A global analysis of function and conservation of catalytic residues in enzymes. J Biol Chem. 2020;295(2):314–24. Epub 20191203. https://doi.org/10.1074/jbc.REV119.006289. PubMed PMID: 31796628; PMCID: PMC6956550.
62. Mulkidjanian AY. On the origin of life in the zinc world: 1. Photosynthesizing, porous edifices built of hydrothermally precipitated zinc sulfide as cradles of life on Earth. Biol Direct. 2009;4:26. Epub 20090824. https://doi.org/10.1186/1745-6150-4-26. PubMed PMID: 19703272; PMCID: PMC3152778.
63. Alberts B, Johnson A, Lewis J, Raff M, Roberts K, Walter P. The shape and structure of proteins. In: Molecular biology of the cell. 4th ed. New York: Garland Science; 2002. Available from: https://www.ncbi.nlm.nih.gov/books/NBK26830/.
64. Goldman AD, Beatty JT, Landweber LF. The TIM barrel architecture facilitated the early evolution of protein-mediated metabolism. J Mol Evol. 2016;82(1):17–26. Epub 20160105. https://doi.org/10.1007/s00239-015-9722-8. PubMed PMID: 26733481; PMCID: PMC4709378.
65. Forterre P, Filee J, Myllykallio H. Origin and evolution of DNA and DNA replication machineries. In: The genetic code and the origin of life. Austin: NCBI Bookshelf. A service of the National Library of Medicine, National Institutes of Health. Madame Curie Bioscience

Database [Internet]. Landes Bioscience; 2007. p. 145–68. https://doi.org/10.1007/0-387-26887-1_10. Available from: https://www.researchgate.net/publication/226781413.
66. Pray L. Discovery of DNA structure and function: Watson and Crick. Nat Educ. 2008;1:100.
67. Teichert JS, Kruse FM, Trapp O. Direct prebiotic pathway to DNA nucleosides. Angew Chem Int Ed Engl. 2019;58(29):9944–7. Epub 20190624. https://doi.org/10.1002/anie.201903400. PubMed PMID: 31131499.
68. Da Silva L, Maurel MC, Deamer D. Salt-promoted synthesis of RNA-like molecules in simulated hydrothermal conditions. J Mol Evol. 2015;80(2):86–97. Epub 20141209. https://doi.org/10.1007/s00239-014-9661-9. PubMed PMID: 25487518.
69. DeGuzman V, Vercoutere W, Shenasa H, Deamer D. Generation of oligonucleotides under hydrothermal conditions by non-enzymatic polymerization. J Mol Evol. 2014;78(5):251–62. Epub 20140513. https://doi.org/10.1007/s00239-014-9623-2. PubMed PMID: 24821106.
70. Deamer D. Liquid crystalline nanostructures: organizing matrices for non-enzymatic nucleic acid polymerization. Chem Soc Rev. 2012;41(16):5375–9. Epub 20120523. https://doi.org/10.1039/c2cs35042d. PubMed PMID: 22623014.
71. Powner MW, Sutherland JD. Prebiotic chemistry: a new modus operandi. Philos Trans R Soc Lond Ser B Biol Sci. 2011;366(1580):2870–7. https://doi.org/10.1098/rstb.2011.0134. PubMed PMID: 21930577; PMCID: PMC3158916.
72. Toppozini L, Dies H, Deamer DW, Rheinstadter MC. Adenosine monophosphate forms ordered arrays in multilamellar lipid matrices: insights into assembly of nucleic acid for primitive life. PLoS One. 2013;8(5):e62810. Epub 20130507. https://doi.org/10.1371/journal.pone.0062810. PubMed PMID: 23667523; PMCID: PMC3646914.
73. Frenkel-Pinter M, Bouza M, Fernandez FM, Leman LJ, Williams LD, Hud NV, Guzman-Martinez A. Thioesters provide a plausible prebiotic path to proto-peptides. Nat Commun. 2022;13(1):2569. Epub 20220511. https://doi.org/10.1038/s41467-022-30191-0. PubMed PMID: 35562173; PMCID: PMC9095695.
74. Muller F, Escobar L, Xu F, Wegrzyn E, Nainyte M, Amatov T, Chan CY, Pichler A, Carell T. A prebiotically plausible scenario of an RNA-peptide world. Nature. 2022;605(7909):279–84. Epub 20220511. https://doi.org/10.1038/s41586-022-04676-3. PubMed PMID: 35546190; PMCID: PMC9095488.
75. Tessera M. Research program for a search of the origin of Darwinian evolution: research program for a vesicle-based model of the origin of Darwinian evolution on prebiotic early Earth. Orig Life Evol Biosph. 2017;47(1):57–68. Epub 20160311. https://doi.org/10.1007/s11084-016-9482-5. PubMed PMID: 26968859.
76. Forsythe JG, Yu SS, Mamajanov I, Grover MA, Krishnamurthy R, Fernandez FM, Hud NV. Ester-mediated amide bond formation driven by wet-dry cycles: a possible path to polypeptides on the prebiotic Earth. Angew Chem Int Ed Engl. 2015;54(34):9871–5. Epub 20150715. https://doi.org/10.1002/anie.201503792. PubMed PMID: 26201989; PMCID: PMC4678426.
77. Paecht-Horowitz M. Clays as possible catalysts for peptide formation in the prebiotic era. Orig Life. 1976;7(4):369–81. https://doi.org/10.1007/BF00927932. PubMed PMID: 1023136.
78. Campbell TD, Febrian R, McCarthy JT, Kleinschmidt HE, Forsythe JG, Bracher PJ. Prebiotic condensation through wet-dry cycling regulated by deliquescence. Nat Commun. 2019;10(1):4508. Epub 20191004. https://doi.org/10.1038/s41467-019-11834-1. PubMed PMID: 31586058; PMCID: PMC6778215.
79. Farias-Rico JA, Mourra-Diaz CM. A short tale of the origin of poteins and ribosome evolution. Microorganisms. 2022;10(11). Epub 20221026. https://doi.org/10.3390/microorganisms10112115. PubMed PMID: 36363706; PMCID: PMC9694802.
80. Fried SD, Fujishima K, Makarov M, Cherepashuk I, Hlouchova K. Peptides before and during the nucleotide world: an origins story emphasizing cooperation between proteins and nucleic acids. J R Soc Interface. 2022;19(187):20210641. Epub 20220209. https://doi.org/10.1098/rsif.2021.0641. PubMed PMID: 35135297; PMCID: PMC8833103.

81. Sankaran N. The RNA world at thirty: a look back with its author. J Mol Evol. 2016;83(5–6):169–75. Epub 20161119. https://doi.org/10.1007/s00239-016-9767-3. PubMed PMID: 27866234.
82. Higgs PG, Lehman N. The RNA World: molecular cooperation at the origins of life. Nat Rev Genet. 2015;16(1):7–17. Epub 20141111. https://doi.org/10.1038/nrg3841. PubMed PMID: 25385129.
83. Woese CR, Kandler O, Wheelis ML. Towards a natural system of organisms: proposal for the domains Archaea, Bacteria, and Eucarya. Proc Natl Acad Sci USA. 1990;87(12):4576–9. https://doi.org/10.1073/pnas.87.12.4576. PubMed PMID: 2112744; PMCID: PMC54159.
84. Joyce GF. Building the RNA world. Ribozymes Curr Biol. 1996;6(8):965–7. https://doi.org/10.1016/s0960-9822(02)00640-1. PubMed PMID: 8805318.
85. Kagra D, Jangra R, Sharma P. Exploring the nature of hydrogen bonding between RNA and proteins: a comprehensive analysis of RNA protein complexes. Chemphyschem. 2022;23(2):e202100731. Epub 20211117. https://doi.org/10.1002/cphc.202100731. PubMed PMID: 34747094.
86. Corley M, Burns MC, Yeo GW. How RNA-binding proteins interact with RNA: molecules and mechanisms. Mol Cell. 2020;78(1):9–29. https://doi.org/10.1016/j.molcel.2020.03.011. PubMed PMID: 32243832; PMCID: PMC7202378.
87. Greenwald J, Riek R. On the possible amyloid origin of protein folds. J Mol Biol. 2012;421(4–5):417–26. Epub 20120424. https://doi.org/10.1016/j.jmb.2012.04.015. PubMed PMID: 22542525.
88. Yruela I, Neira JL. Intrinsically disordered proteins in biology: one for all, all for one. Arch Biochem Biophys. 2020;684:108328. Epub 20200304. https://doi.org/10.1016/j.abb.2020.108328. PubMed PMID: 32145248.
89. Uversky VN. Protein intrinsic disorder and structure-function continuum. Prog Mol Biol Transl Sci. 2019;166:1–17. Epub 20190608. https://doi.org/10.1016/bs.pmbts.2019.05.003. PubMed PMID: 31521229.
90. Seal M, Weil-Ktorza O, Despotovic D, Tawfik DS, Levy Y, Metanis N, Longo LM, Goldfarb D. Peptide-RNA coacervates as a cradle for the evolution of folded domains. J Am Chem Soc. 2022;144(31):14150–60. Epub 20220729. https://doi.org/10.1021/jacs.2c03819. PubMed PMID: 35904499; PMCID: PMC9376946.
91. Rios AC, Tor Y. On the origin of the canonical nucleobases: an assessment of selection pressures across chemical and early biological evolution. Isr J Chem. 2013;53(6–7):469–83. https://doi.org/10.1002/ijch.201300009. PubMed PMID: 25284884; PMCID: PMC4181368.
92. Cech TR. Crawling out of the RNA world. Cell. 2009;136(4):599–602. https://doi.org/10.1016/j.cell.2009.02.002. PubMed PMID: 19239881.
93. Noller HF. Evolution of protein synthesis from an RNA world. Cold Spring Harb Perspect Biol. 2012;4(4):a003681. Epub 20120401. https://doi.org/10.1101/cshperspect.a003681. PubMed PMID: 20610545; PMCID: PMC3312679.
94. Papastavrou N, Horning DP, Joyce GF. RNA-catalyzed evolution of catalytic RNA. Proc Natl Acad Sci USA. 2024;121(11):e2321592121. Epub 20240304. https://doi.org/10.1073/pnas.2321592121. PubMed PMID: 38437533; PMCID: PMC10945747.
95. Forterre P. The two ages of the RNA world, and the transition to the DNA world: a story of viruses and cells. Biochimie. 2005;87(9–10):793–803. Epub 20050412. https://doi.org/10.1016/j.biochi.2005.03.015. PubMed PMID: 16164990.
96. Wolf YI, Koonin EV. On the origin of the translation system and the genetic code in the RNA world by means of natural selection, exaptation, and subfunctionalization. Biol Direct. 2007;2:14. Epub 20070531. https://doi.org/10.1186/1745-6150-2-14. PubMed PMID: 17540026; PMCID: PMC1894784.
97. Woese CR, Olsen GJ, Ibba M, Soll D. Aminoacyl-tRNA synthetases, the genetic code, and the evolutionary process. Microbiol Mol Biol Rev. 2000;64(1):202–36. https://doi.org/10.1128/MMBR.64.1.202-236.2000. PubMed PMID: 10704480; PMCID: PMC98992.

98. Fontecilla-Camps JC. Reflections on the origin and early evolution of the genetic code. Chembiochem. 2023;24(10):e202300048. Epub 20230413. https://doi.org/10.1002/cbic.202300048. PubMed PMID: 37052530.
99. Henras AK, Plisson-Chastang C, O'Donohue MF, Chakraborty A, Gleizes PE. An overview of pre-ribosomal RNA processing in eukaryotes. Wiley Interdiscip Rev RNA. 2015;6(2):225–42. Epub 20141027. https://doi.org/10.1002/wrna.1269. PubMed PMID: 25346433; PMCID: PMC4361047.
100. Barbieri M. Evolution of the genetic code: the ambiguity-reduction theory. Biosystems. 2019;185:104024. Epub 20190906. https://doi.org/10.1016/j.biosystems.2019.104024. PubMed PMID: 31499094.
101. Sella G, Ardell DH. The impact of message mutation on the fitness of a genetic code. J Mol Evol. 2002;54(5):638–51. https://doi.org/10.1007/s00239-001-0060-7. PubMed PMID: 11965436.
102. Argyriadis JA, He YH, Jejjala V, Minic D. Dynamics of genetic code evolution: the emergence of universality. Phys Rev E. 2021;103(5–1):052409. https://doi.org/10.1103/PhysRevE.103.052409. PubMed PMID: 34134257.
103. Keeling PJ, Palmer JD. Horizontal gene transfer in eukaryotic evolution. Nat Rev Genet. 2008;9(8):605–18. https://doi.org/10.1038/nrg2386. PubMed PMID: 18591983.
104. Delaye L, Becerra A, Lazcano A. The last common ancestor: what's in a name? Orig Life Evol Biosph. 2005;35(6):537–54. https://doi.org/10.1007/s11084-005-5760-3. PubMed PMID: 16254691.
105. Davidovich C, Belousoff M, Wekselman I, Shapira T, Krupkin M, Zimmerman E, Bashan A, Yonath A. The proto-ribosome: an ancient nano-machine for peptide bond formation. Isr J Chem. 2010;50(1):29–35. https://doi.org/10.1002/ijch.201000012. PubMed PMID: 26207070; PMCID: PMC4508870.
106. Root-Bernstein M, Root-Bernstein R. The ribosome as a missing link in the evolution of life. J Theor Biol. 2015;367:130–58. Epub 20141210. https://doi.org/10.1016/j.jtbi.2014.11.025. PubMed PMID: 25500179.
107. Petrov AS, Gulen B, Norris AM, Kovacs NA, Bernier CR, Lanier KA, Fox GE, Harvey SC, Wartell RM, Hud NV, Williams LD. History of the ribosome and the origin of translation. Proc Natl Acad Sci USA. 2015;112(50):15396–401. Epub 20151130. https://doi.org/10.1073/pnas.1509761112. PubMed PMID: 26621738; PMCID: PMC4687566.
108. Fox GE. Origin and evolution of the ribosome. Cold Spring Harb Perspect Biol. 2010;2(9):a003483. Epub 20100609. https://doi.org/10.1101/cshperspect.a003483. PubMed PMID: 20534711; PMCID: PMC2926754.
109. Bowman JC, Petrov AS, Frenkel-Pinter M, Penev PI, Williams LD. Root of the tree: the significance, evolution, and origins of the ribosome. Chem Rev. 2020;120(11):4848–78. Epub 20200506. https://doi.org/10.1021/acs.chemrev.9b00742. PubMed PMID: 32374986.
110. Noller HF, Donohue JP, Gutell RR. The universally conserved nucleotides of the small subunit ribosomal RNAs. RNA. 2022;28(5):623–44. Epub 20220203. https://doi.org/10.1261/rna.079019.121. PubMed PMID: 35115361; PMCID: PMC9014874.
111. Raymann K, Brochier-Armanet C, Gribaldo S. The two-domain tree of life is linked to a new root for the Archaea. Proc Natl Acad Sci U S A. 2015;112(21):6670–5. Epub 20150511. https://doi.org/10.1073/pnas.1420858112. PubMed PMID: 25964353; PMCID: PMC4450401.
112. Hug LA, Baker BJ, Anantharaman K, Brown CT, Probst AJ, Castelle CJ, Butterfield CN, Hernsdorf AW, Amano Y, Ise K, Suzuki Y, Dudek N, Relman DA, Finstad KM, Amundson R, Thomas BC, Banfield JF. A new view of the tree of life. Nat Microbiol. 2016;1:16048. Epub 20160411. https://doi.org/10.1038/nmicrobiol.2016.48. PubMed PMID: 27572647.
113. Tsurumaki M, Saito M, Tomita M, Kanai A. Features of smaller ribosomes in candidate phyla radiation (CPR) bacteria revealed with a molecular evolutionary analysis. RNA. 2022;28(8):1041–57. Epub 20220610. https://doi.org/10.1261/rna.079103.122. PubMed PMID: 35688647; PMCID: PMC9297842.

114. Kailing F, Lieberman J, Wang J, Turner JL, Goldman AD. Evolution of cellular organization along the first branches of the tree of life. J Mol Evol. 2024. Epub 20240717. https://doi.org/10.1007/s00239-024-10188-7. PubMed PMID: 39020132.
115. Woese C. The universal ancestor. Proc Natl Acad Sci USA. 1998;95(12):6854–9. https://doi.org/10.1073/pnas.95.12.6854. PubMed PMID: 9618502; PMCID: PMC22660.
116. Weiss MC, Preiner M, Xavier JC, Zimorski V, Martin WF. The last universal common ancestor between ancient Earth chemistry and the onset of genetics. PLoS Genet. 2018;14(8):e1007518. Epub 20180816. https://doi.org/10.1371/journal.pgen.1007518. PubMed PMID: 30114187; PMCID: PMC6095482.
117. Weiss MC, Sousa FL, Mrnjavac N, Neukirchen S, Roettger M, Nelson-Sathi S, Martin WF. The physiology and habitat of the last universal common ancestor. Nat Microbiol. 2016;1(9):16116. Epub 20160725. https://doi.org/10.1038/nmicrobiol.2016.116. PubMed PMID: 27562259.

Chapter 4
Genesis of Mitochondria to Power Complex Lifeforms

Abstract The fermentation (glycolysis) of glucose with release of carbon dioxide generates only two net molecules of ATP, whereas complete oxidation of glucose to water generates 12-fold more molecules of ATP. In anaerobic conditions the final electron acceptor can be sulfur or sulfate, but with the advent of an O_2 gas-containing atmosphere on the Earth's surface and waters, life adapted to use the highly electronegative O_2 molecule as the final electron acceptor. This process relies on the chemiosmotic mechanism whereby electrons are transferred through reducing equivalents, for example, NADH, using mobile, lipid permeable quinones as intermediate charge carriers, while protons are pumped across an ion impermeable membrane using the energy of the Sun or substrates to establish a voltage (electro-)-chemical gradient. In turn, most ATP is generated when the protons reenter down this gradient. This transduction process was crucial for life to develop. The oxidation phosphorylation machinery existed in prokaryotes long before the tricarboxylic cycle and electron transport system became incorporated entirely from a prokaryote to another as an endosymbiont. The encapsulated, two-membrane endosymbiont is now called the mitochondrion, and its evolution is likely tied to encapsulation of a prokaryote's genome into another's, forming a defined nucleus within the first eukaryotes. Most of the endosymbiont's mitochondrial genome was transferred to the host nucleus or eliminated, while the host DNA now furnishes over 1000 nuclear-encoded proteins to assure efficient function and regulation of its mitochondria. Mitochondria exhibit great diversity within a tissue or organism (heterogeneity). They are very flexible in regulating energy output according to needs, for example, running or flying, and in adjusting to cell environmental changes, for example, ischemic stress. Modern-day cardiac myocyte mitochondria are exemplified here as high-energy mitochondria. Before the last common ancestor, the first proteins were made to facilitate the work of the ribozyme in precisely coding peptides with information stored in RNA after transcription from DNA. This required adaption of a final standard genetic code by which only four DNA nucleobases, arranged in a linear codon of only three nucleobases, encode only one amino acid out of the 20 amino acids utilized in lifeforms. The genetic code exhibits redundancy in the last base of the codon but coding is not ambiguous as a given codon only translates one distinct amino acid. Protein function is dependent not only on the linear sequence of amino acids but also on the

D. F. Stowe, *Evolution of Bioenergetics from Elements to Life*,
https://doi.org/10.1007/978-3-031-76668-8_4

folding of the protein by hydrogen bonding among the amino acids and water in the cytoplasm. Viruses likely coexisted with early prokaryote genomes.

Keywords Aminoacyl transfer RNA · Anaerobic bacteria · Archaean Eon · Arachaeon cells · Bioelectricity · Bioenergetics · Biogenesis · Biosynthesis · Chemiosmosis · Chloroplasts · Chomatophores · Coupled redox reaction · Dynamic kinetic stability · Electromagnetic current · Electrolytes · Electronegativity · Electron transport system · Encapsulation · Endosymbiont · Endosymbiotic hypothesis · Eukaryotes · Evolutionary bioenergetics · First eukaryote common ancestor · Fermentation · Heteroplasmy · Homeoplasmy · Horizontal gene transfer · Last eukaryote common ancestor · Last universal common ancestor · Matrilinear ancestry · Mitochondrial genome · Mitochondrial heterogeneity · Nucleobases · Nucleolus · Oxidative phosphorylation · Peptidyl transfer RNA · Peroxisomes · Phagocytosis · Photo-autotrophs · Photo-heterotrophs · Photosynthesis · Phylogenomic analyses · Polyanions · Prokaryotes · Protomitochondrion · Proton motive force · Quinones · Retrotranscribing viruses · Ribosomal RNA · Standard genetic code · Stereochemistry · Symbionts · Transfer RNA · Translocases

4.1 Origin, Biogenesis, and Heterogeneity of Mitochondria

Early bacteria could utilize fermentation to produce ATP, although inefficiently. In anaerobic bacteria where O_2 is scant or unavailable, as in deep-sea thermal vents, elemental sulfur (S^0) (or the polyanion SO_4^{2-}) was, and still is, utilized in chemiosmosis as the final *e*- acceptor with its reduction to H_2S. Much later ancient bacteria adapted to the increasing O_2 environment to adapt the respiratory chain proteins to utilize O_2 in chemiosmosis with ATP synthase to produce much more energy as ATP [1, 2]. The stronger electronegativity of O_2 made it an optimal *e*- acceptor over sulfur after O_2 became available.

Endosymbiosis Then, approximately 1.5 Bya, long after the appearance of LUCA [3, 4] (3.5–3.9 Bya), evidence was first presented to suggest that primitive versions of the protein respiratory complexes of purple non-sulfur bacteria were carried over to the cells that engulfed them to eventually become nascent mitochondria [2]. This process is called the "endosymbiotic hypothesis." It is possible that the symbionts were engulfed by food vacuoles of the hosts but that they escaped into the cytosol of the host, thus evading phagocytosis [5]. Apparent similarities between the archaeon bacteria that hosted the symbiont bacteria are circular DNA not enclosed by a nucleus, membrane bounded protoplasm, and reproduction by fission. Some details of these transitions are described below.

Purple non-sulfur bacteria are photosynthetic, two-membrane, facultative aerobic bacteria, and so are described as heterotrophic phototrophs, also known as photo-heterotrophs. That means they require carbon sources from foodstuffs rather

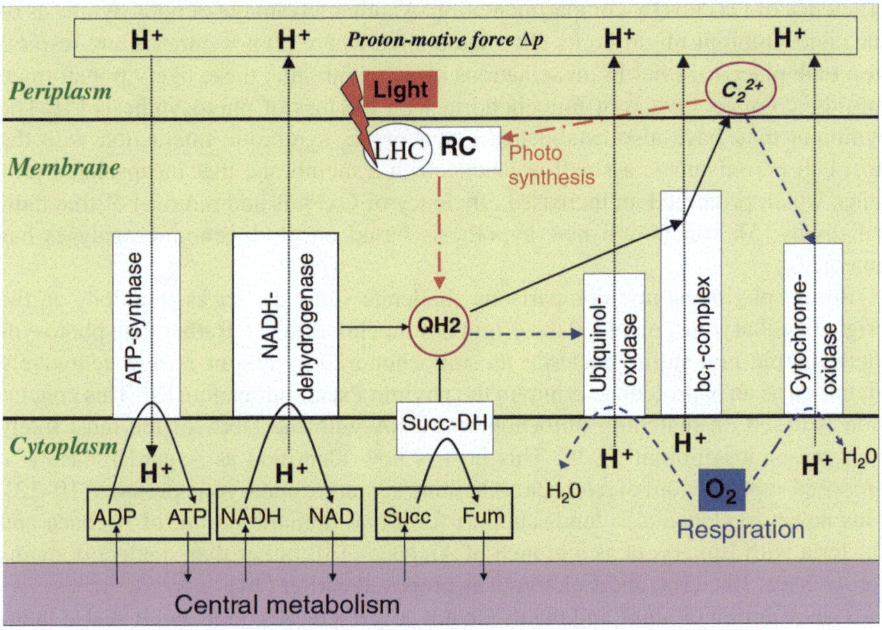

Fig. 4.1 Electron transport system in purple non-sulfur bacteria: Except for proton pumping (dotted lines), all arrows inside the membrane indicate *e-* transfers (flows). Purely photosynthetic and purely respiratory *e-* transfers are indicated by red dashed-dotted and blue dashed lines, respectively. Stoichiometries are not included. *LHC* light harvesting complexes, *RC* reaction centers, *Succ* succinate, *Fum* fumarate. (Diagram (Fig. 1) and its modified legend resourced with license (https://creativecommons.org/licenses/by/4.0/ from Klamt et al. [1])

than from CO_2 as plants (photoautotrophs) do. Figure 4.1 [1] depicts how photo-heterotrophs generate ATP using light in a bacteriochlorophyll or bacteriorhodopsin-based reaction center where the light excites molecules to cause transfer of *e-* through an ETS with protons pumped across their inner membrane and returned through the inner membrane (IM) to drive ATP synthesis [1]. In photo-heterotrophs, *e-* transfer occurs only in a cyclic pathway from the reaction center through the ETS, and then return to the reaction center (Fig. 4.1). In photo-heterotrophs, photosynthesis does not produce O_2, but they can reduce O_2 to H_2O in aerobic conditions.

It is well accepted that the engulfment and permanent entrapment of a primitive cell as symbionts in pro-eukaryote, archaeon host cells occurred as O_2 levels rose in Earth's atmosphere 1.5–2 Bya [1, 5, 6]. It was postulated that both the symbiont and host cells were facultative aerobic cells [5], with the symbiont likely a prokaryote, while the host could also have been a prokaryote, or perhaps a larger eukaryote. This means the symbiont as well as the host would have the capacity to utilize O_2 when available for OxPhos but could alternatively utilize photosynthesis to make ATP when deprived of O_2.

The ultimate selective advantage of the symbiont's respiratory components, genes, and its membranes into the host cytoplasm may have been a greater efficiency of OxPhos by compartmentation. Another advantage of photosynthesis of the endosymbiont might be its chromatophores where photosynthesis and respiration took place together in invaginations of its membrane; these likely transformed into the compact cristae of mitochondria with the loss of photosynthesis [5]. The symbiont may have also established a synergetic, symbiotic interaction with the host cell peroxisomes, a small organelle with a membrane that incapsulated proteins, which promoted an increased efficiency of OxPhos and removal of free radical toxins [5]. But now a new hypothesis based on phylogenomic analyses has emerged.

Recent phylogenomic comparisons challenge some earlier assumptions on the origin of eukaryotes, included the origin or mitochondria [7]. Rather than photosynthetic purple-non-sulfur bacteria, the mitochondrial symbiont is now tentatively identified as an α-proteobacterium in the phylum Pseudomonadota [8]. This conclusion is based on matching mitochondrial DNA with the DNA of the most likely bacterial endosymbiont [8, 9]. This host is now identified as Asgardarchaeota, a proposed superphylum of Archaea containing signature eukaryotic proteins [10–12]. This newer evidence also lends support for a two domain system of Archaea and Bacteria with Eukaryota as a branch of Archaea [13], rather than distinct domains for Archaea, Bacteria, and Eukaryota as proposed earlier [14].

The evolution of algae and plants are not covered in this book, but it is also interesting to note that a photosynthetic prokaryote could have been engulfed into a nascent algal or plant cell to compartmentalize the photosynthetic process within a new organelle, now called the chloroplast. This endosymbiotic event is theorized to have occurred later in cells that had lost photosynthesis [5].

Beginnings of Chemiosmosis Chemiosmotic coupling for generating chemical energy, usually to make ATP from substrates, is a universal facet of all lifeforms. Chemiosmosis requires a membrane passively impermeable to ions but containing proton pump(s), driven by light or by oxidizing reducible substrates. These pumps work as a battery to store energy (proton motive force, *pmf*) through a mechanism to utilize and transfer *e*- (current) that drive the proton pumps. This is coupled by transporting the energy stored in the battery by proton reuptake to make high energy ATP, which can then drive an enormous number of endergonic reactions throughout the cell. While this may appear as a complicated mechanism, it was the one apparently selected in evolution as the one most proficient with the macromolecules available.

It is moreover uncertain how chemiosmosis emerged at the dawn of life and its precise mechanism remains unclear. A mechanism for energy transduction from photons or oxidizable substrates to a stored chemical energy system had to be in place for complex metabolism in a cell to commence. Experimental evidence gives some clues [15]. A model liposome system was set up in which internal ferricyanide, $Fe^{3+}(CN)_6)$, served as the oxidant A and ascorbate, $C_6H_8O_6$, or dithionite, (DH, $S_2O_4^{-2}$), provided a source of electrons for transfer by quinones (Q) embedded in the liposomal bilayer membranes. The quinones link the donor ascorbate or dithionite *e*- across the membrane to the receptor ferricyanide [15]. This initiates a coupled redox reaction in which *e*- are transported across the membrane, while protons are released into the vesicle decreasing the interior pH. This produces a transmembrane chemical pH gradient, as well as a voltage gradient [15].

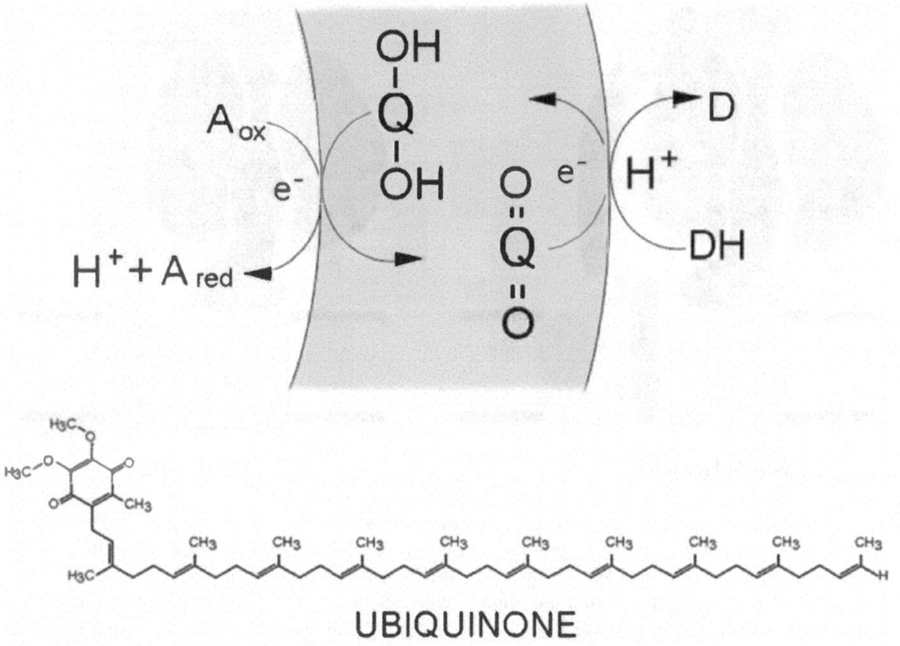

Fig. 4.2 Role of quinones in coupling electron transport to proton transport across bilayer membranes: An external donor source of reducing power (DH) transfers both electrons and protons to a quinone (Q) which diffuses to the other side of the membrane. The electrons reduce an oxidant (A_{ox}) such as an iron compound that does not accept protons in the reaction, and the protons are then released into the interior of the compartment. *Ox* oxidized, *Red* reduced. The structure of ubiquinone which functions in mitochondria as an electron transport cofactor is shown. (Cartoons (Fig. 1) and its modified legend resourced with license (https://creativecommons.org/licenses/BY-NC/4.0/) from Milshteyn et al. [15])

This chemiosmotic energy source in the form of an electrochemical proton gradient is managed by the quinones that acts as a lipid soluble H^+ shuttle from the external to the internal volume of the vesicle while the quinones shuttle (recycle) *e*- back and forth across the bilayer membrane. Note that in actual mitochondrial respiration the H^+ ions are pumped out via the IMS causing the matrix pH to increase. Figure 4.2 [15] is a cartoon to describe experimental evidence for chemiosmosis using ferricyanide as the oxidant (A) and dithionite as the donor (DH). These compounds are placed on either side of an artificial bilayer membrane containing quinones which diffuse from one side of the membrane to another. There is evidence that the quinones were present in carbonaceous meteorites [15], like the Murchison meteor, falling to Earth in the form of polycyclic aromatic hydrocarbon derivatives. Some of these Qs were tested in the experimental system [15].

Mitochondrial ATP The subunit composition of complex V (F_0F_1-ATP synthase) between cell-engulfed mitochondrial and true eubacteria (prokaryote with cell wall, often with pili and flagellum) has become more complex during evolution as dia-

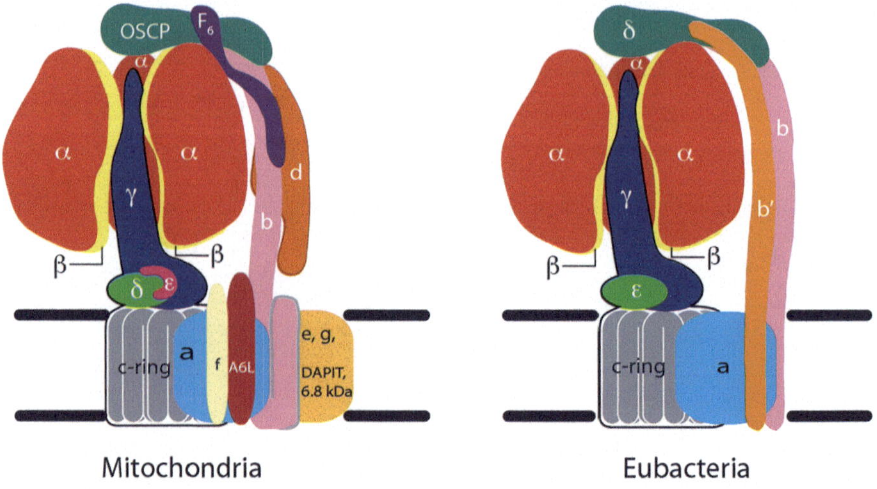

Fig. 4.3 Schematic subunit composition of modern mitochondrial vs. eubacterial ATP synthase: The bacterial and chloroplast ATP synthases have comparable subunit composition, with $\alpha_3\beta_3\gamma\delta\varepsilon$ constituting F_1 and F_0 comprising of ab_2c_{10-15}. Subunits γ and ε shape a primary stalk that non-covalently interacts with the membrane bound, hydrophobic c-ring. The proteolipidic oligomer c rotates in opposition to the peripheral stator that is made from ab_2 and δ. The stator restricts rotation of the $\alpha_3\beta_3$ hexamer, which provides six nucleotide binding sites (each located at $\alpha\beta$-subunit interface) for reversible ATP synthesis. Ions are translocated across channels at the a–c ring interface, generating a torque on the central stalk $\gamma\varepsilon$. The γ-subunit which forms α-helical coiled-coil structure is embedded in the F_1-hexamer. The torque of γ causes alternating conformational adjustments in the β-subunits, leading to differences in nucleotide binding affinities (open, tight, and free), which eventually bring about ATP synthesis or hydrolysis. Mitochondrial ATP synthase contains the bacterial core set of subunits but contains additional subunits in the F_0 domain (subunits DAPIT and 6.8 kDa proteolipid) and in the stator (subunits d and F6). OSCP (oligomycin sensitivity-conferring protein) is equivalent to the δ subunit of the bacterial enzyme. (Cartoon (Fig. 3) and its modified legend resourced with license (https://creativecommons.org/licenses/by/4.0/ from Neupane et al. [16] (review). Source: Mitochondrial Biology Unit, University of Cambridge (http://www.mrc-mbu.cam.ac.uk/projects/2679/subunit-composition))

gramed in Fig. 4.3 [16]. However, the initial selective advantage of the symbiont to remain in the host cell as a new organelle, the import of carriers (translocases) into the mitochondrial to transport substances via the outer mitochondrial membrane (OMM) and IMM (Fig. 4.4) [5, 17], the genetic pressure to eventually transfer most the symbionts' genes to the nascent cell nucleus, and the specific machinery of the symbiont that was transferred to the host cell remain debated [5, 18, 19]. It is possible that the adaption of the new organelle provided selective advantage for this new, larger, and compacted energy source in cells so that the mutual symbiotic relationship has continued throughout evolution. The more efficient transduction of foodstuffs into energy in the form of ATP with compartmentalization likely hastened the evolution of multicellular organisms into a wide variety of complex animals. Over time the most advantageous attributes of organelles were selected so that in eukaryotes today survival depends on their incorporated mitochondria.

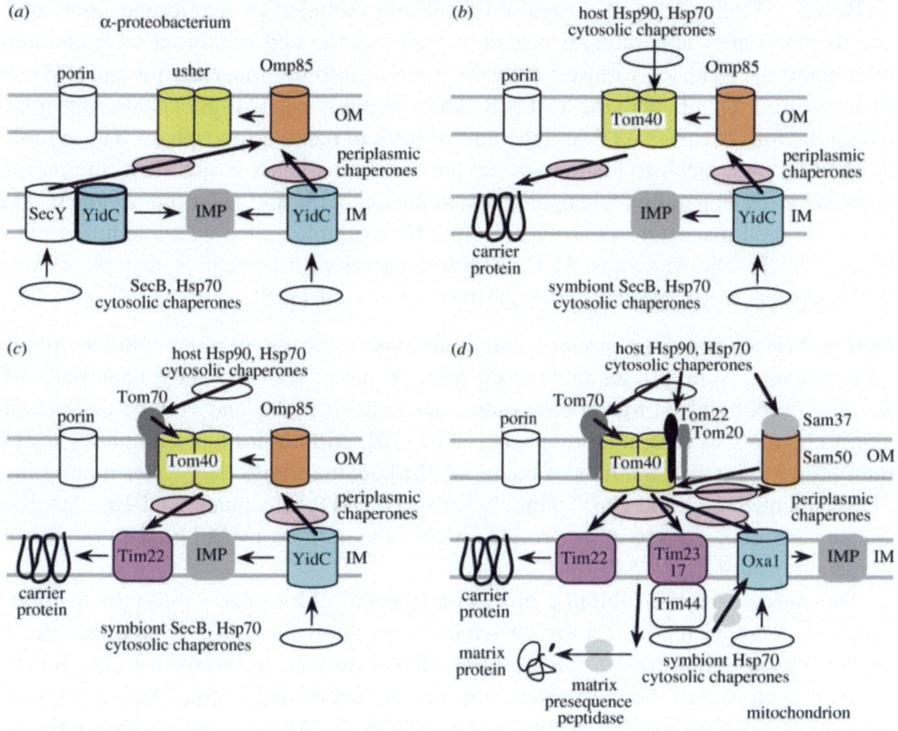

Fig. 4.4 Origin of mitochondrial protein-import systems (**d**) from the ancestral α-proteobacterial protein-export systems: (**a**) The outer mitochondrial membrane (OM) β-barrel proteins (Sam50, Tom40, porins) all evolved directly from symbiont OM proteins and Oxa1/2 evolved from the inner membrane (IM) protein YidC. (**b**) Enslavement was initiated by the insertion of novel carrier proteins able to enter through Tom40 with the help of pre-existing host and periplasmic chaperone proteins and insert into the IM. The first carrier, possibly a descendent of the peroxisomal ATP importer, then exported ATP or photosynthesate to the host and generated over 30 other carriers by gene duplication. Their import was made efficient by evolution of receptor Tom70, possibly of cyanobacterial origin by lateral gene transfer, inserting via its N-terminal membrane-spanning a-helix, and Tim22 entering via the Tom40 pore. (**c**) Following massive transfer of genes from symbiont to nucleus, the pre-sequence mechanism evolved by gene duplication of Tim22 to Tim23 and Tim17 to generate the inner membrane translocase and the addition of Tom20 and Tom22 (**d**) to recognize the hydrophobic and positively charged parts, respectively, of the pre-sequences of Tim44, to transfer the symbiont/matrix chaperone Hsp70 more efficiently onto the emerging pre-proteins and a matrix peptidase to remove their pre-sequences. Preexisting periplasmic chaperones diversified and adapted to prevent periplasmic aggregation and improve transfer between membrane–protein complexes, and various, less key proteins were added to each to increase stability and improve transfer rates and efficiency. SecB machinery was retained by the excavate jakobid protozoa (text in Gray et al. [8]) but was lost by other mitochondria. Acquisition of pre-sequences by approximately 1000 proteobacterial genes transferred to the nucleus; greater similarity of many of these to γ- not α-proteobacteria may be an artefact stemming from major divergence during mitochondriogenesis, which allowed loss of their symbiont versions and huge mitochondrial genome reduction. This raised efficiency by increasing space for matrix enzymes and sparing nutrients and energy previously wasted on multiple copies of their DNA. *IMP* inner membrane proteins other than carriers and translocons. (Cartoons and scheme (Fig. 2) and its modified legend resourced with permission from copyright owner of Cavalier-Smith [5])

Beside ATP synthase, the engulfed symbiont included its membranes, genome, and its respiratory apparatus. It was important that the endosymbiont's membranes were naturally selected to insert a carrier protein into the inner membrane (IM) to transport photosynthate to the host cell. An example is the ATP-phosphate carrier by which the host receives ATP, and the endosymbiont receives phosphate. The symbiont had its own DNA to encode its membranes so as DNA is inherited, the membrane heredity is parallel. Changes to the symbiont's IM and outer membrane (OM) likely allowed insertion of carrier proteins, for example,, adenine nucleotide translocase (ANT), also known as ACC (ADP/ATP carrier protein), to transport ATP out of the nascent mitochondria and ADP into mitochondria [5].

Cell to Mitochondrial Carriers Bacterial export carriers became cellular import carriers over several phases in which β-barrel proteins (porins) in the bacterial OM became mitochondrial translocase outer membrane (TOM) and voltage-dependent anion channels (VDAC), among others [19, 20]. Although more complex [5, 21] than presented here, channels of bacterial IM became translocase inner membrane (TIM) channels, K^+ and Ca^{2+} channels, and ADP/ATP, phosphate, and Na^+, H^+, K^+, Ca^{2+} exchangers, among other proteins within the modern IMM [19, 20], as shown in the cartoon of Fig. 4.5 [17].

The nascent endosymbionts, mitochondria and chloroplasts, brought in thousands of genes to the host, most of which were duplicates. It is argued that these similar genes could not be deleted until after they were transferred to the newly evolved nucleus and the pre-protein import machinery had evolved to import their proteins [5]. It is possible that the import of host TOM and TIM carriers into the endosymbiont was necessary to correct for the immense phenotypic damage that would result from translation of all the symbiont genes. During somatic cell

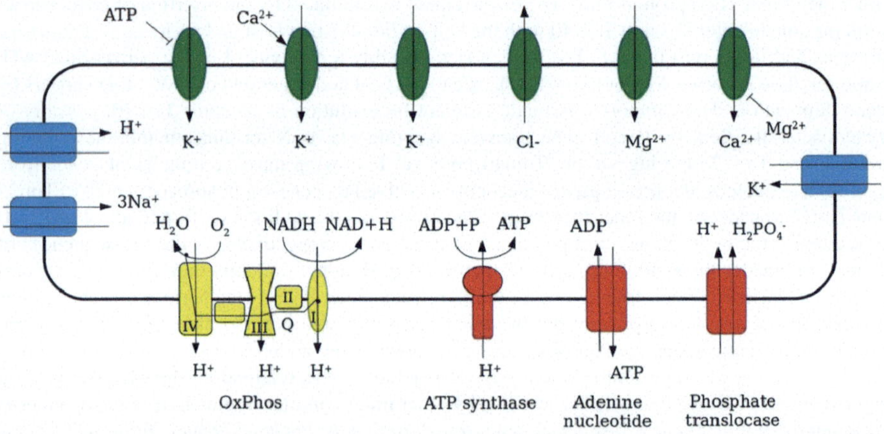

Fig. 4.5 Some enzymes of the inner mitochondrial membrane (IMM): Ion channels, exchangers, pumps, and oxidative phosphorylation chain; green = uniporters, blue = exchangers, yellow = respiratory complexes for electron transfer system, and red = ATP synthasome. There are several types of mitochondrial (m) mK_{Ca}, mK_v and mCl channels. (Cartoon (scheme 1) and its modified legend resourced with license (https://creativecommons.org/licenses/by/4.0/) from Jasielec et al. [17])

division horizontal gene transfer (HGT) via mitochondrial DNA can accelerate evolution across cells by bypassing the nuclear gene transfer mechanism. TIM IM translocons and OM pre-sequences are hypothesized to have evolved to retarget many proteins coded by the symbiont's gene back into the symbiont where, if not beneficial, they would be selected for deletion [5]. Gene loss from protomitochondria to the host nucleus was likely species variable, and as high as 99%. The reasons for these variable, one-way gene losses are unclear [5, 21, 22]. Because bacteria have no nucleus to contain genes, the eukaryote host cell membrane-bound nucleus may have been a more secure location for proto-mitochondrial genes [23].

New Platform for Chemiosmosis and Phosphorylation Encapsulation of the chemiosmosis mechanism of a cell into nascent mitochondria, but eventually using the nuclear genome to manage most all its genetic machinery, was proposed to greatly increase the efficacy to transduce energy in the host cell and with other specialized organelles allowed for the blossoming of an immense number of complex multicellular animals [24]. It is proposed [25] that an evolutionary increase in cellular complexity, as in endosymbiosis, does not necessarily demand an increase for more energy. Instead, the incorporation of new mitochondrial genes with weak and duplicitous functions likely imposes higher energetic costs to the host cell by overexpressing duplicate genes before their functions become evolutionarily refined [26]. In the long term, it is argued, the accumulation of new genes deviates resources away from growth and reproduction [25].

This opinion [25] is based in part on data showing how a cell's quantity of ATP increases as a function of a cell's volume, and how metabolic rate increases as a function of a cell's mass (e.g., a small bacterium and an larger eukaryotic cell) in diverse phyla in the absence or presence of mitochondria. It is posited that increasing cell complexity, like adding mitochondria as energy sources, deviates resources away from functions that provide for survival, growth, and reproduction [25]; this could occur, at least initially, until duplications in functions of the endosymbiont and host were eliminated. Thus, increased complexity of a cell may not be evolutionarily advantageous. It must be noted, however, that in complex, specialized cells, substrate requirements and energy output are much greater for the cell size. Metabolic function is compartmentalized in the cells and tissues of multicellular eukaryotes. Nerve and muscle cells have very large mitochondrial volumes and produce large amounts of energy. These cells provide for processing environmental information and for coordination and locomotion in highly metabolic animals such as mammals and birds. So eventually the enhanced diversity and complexity of some cells were eventually favored by natural selection.

Mitochondrial DNA The mitochondrial genome of humans now contains only 37 genes; 13 encode subunits of respiratory complexes I, III, IV, complex V, and the rest encode the 22 transfer tRNAs and 2 ribosomal rRNAs required for translation of the respiratory subunit mRNAs within the mitochondrial matrix as depicted in Fig. 4.6 [18]. Before the upcoming detailed discussion of energy transduction and metabolism in mitochondria and cells, note that glycolysis occurs only in the cytosol, whereas the tricarboxylic acid (TCA cycle) and the electron transfer system

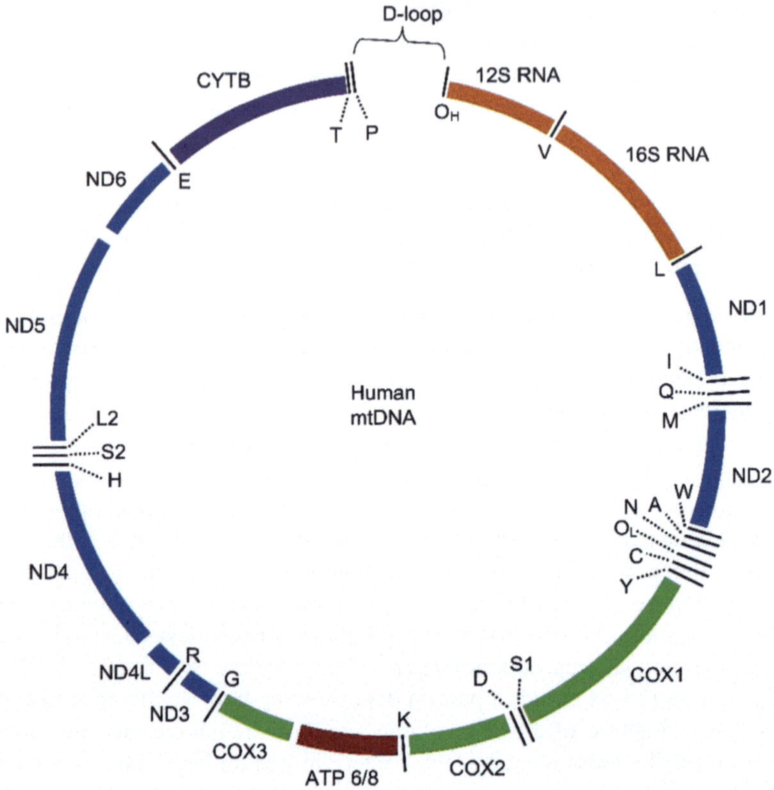

Fig. 4.6 Schematic representation of human mitochondrial DNA (mtDNA): The circular molecule (16,569 bp) contains genes coding for 7 subunits of complex I (ND1–ND6, NADH ubiquinone-oxidoreductase chain 1–6) (blue line), 1 subunit of complex III (CYTB, cytochrome B) (purple line), 3 subunits of complex IV (COX1–COX3, cyclooxygenase) (green line), and 2 subunits of complex V (ATP synthase 6 and 8) (red line). Moreover, mtDNA encodes 2 rRNAs (orange line) and 22 tRNAs for specific amino acids (marked with black lines with letter designation for specific amino acids) (based on Ref. [9]). (Cartoon (Fig. 1) and its legend resourced with license (https://creativecommons.org/licenses/by/4.0/) from Boguszewska et al. [18] (review))

(ETS) reactions occur only within the IMM and mitochondrial matrix. Figure 4.7 shows an electron micrograph of mammalian cardiac mitochondria with the large surface area of the IMM as observed by the densely folded cristae [27].

Mitochondrial DNA is unique in that it is transferred only from the ovum during formation of the zygote; That means that maternal mDNA is passed down to all offspring and represents matrilineal ancestry [28, 29]. Mitochondrial DNA testing in either males or females is used to trace the genealogy of maternal ancestors. This can be useful in tracing the heredity of eukaryote offsprings without the complication of haploid paternal DNA mixing with haploid maternal DNA as occurs during fertilization.

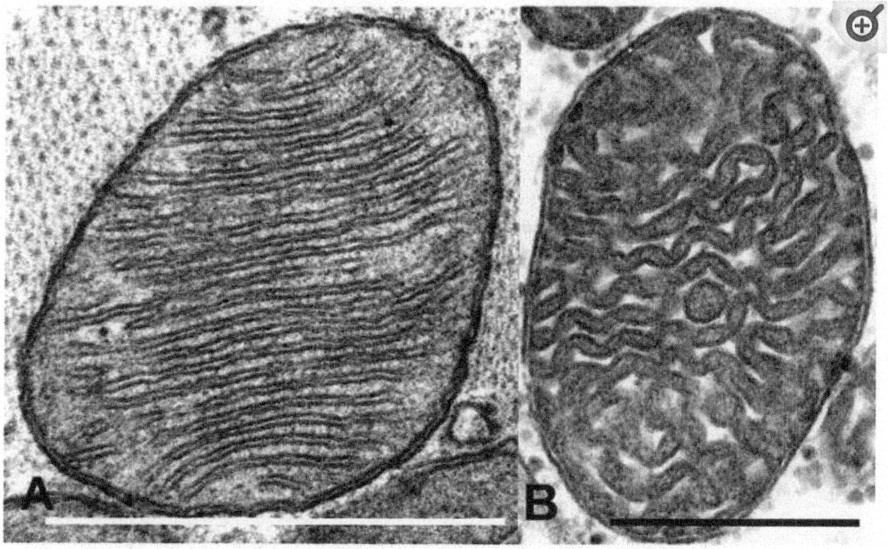

Fig. 4.7 Cardiac mitochondrial structure: (**a**) Rat; TEM. A typical cardiac mitochondrion with lamelliform cristae showing its unit membrane substructure to advantage. (**b**) Human; TEM. A cardiac mitochondrion with apparently tubular cristae. Specimen (archives of the Department of Pathology, Case Western Reserve University School of Medicine), was obtained by transjugular bioptome; the tissue sample was from the right side of the interventricular septum. (Transmission electron micrograph (TEM) (Fig. 1) and modified legend resourced with permission from copyright owner of Hoppel et al. [27])

Single deletions or point mutations in mDNA are known to cause deleterious phenotypes within skeletal and cardiac muscle and the nervous system, often manifested by cardiomyopathy, conduction defects, dystonia, seizures, encephalopathy, and visual and hearing impairment, among others [28, 30]. All of these conditions are though to arise from dysfunctional bioenergetics in the affected high energy cells. It is important to note that mutations in nuclear DNA encoding TCA cycle enzymes, respiratory subunit enzymes, and all others that participate in OxPhos, can also cause defects in bioenergetic function.

Normally, cells have mitochondria with multiple, identical copies of mDNA; this is called homoplasmy. Heteroplasmy is the presence of a mixture of mutated and wild-type mDNA within a cell. Unlike in the cell nucleus where mutations are inherited with more-or-less fixed frequences of approximately 75%, 50%, or 25%, depending on the parents' genotype, the degree of mDNA heteroplasmy can vary widely from 0% to 100%. Recent sequencing techniques show that the mutants generated may arise by clonal expansion of low level heteroplasmic variants as well as by somatic mutation [28, 30]. This means that once a threshold for phenotypic changes occurs, the phenotype may expand over time with aging causing increasing severity of disease. Figure 4.8 [28] shows the complex scenarios of an oocyte with mutated copies of mDNA that with fertilization and somatic cell division can lead

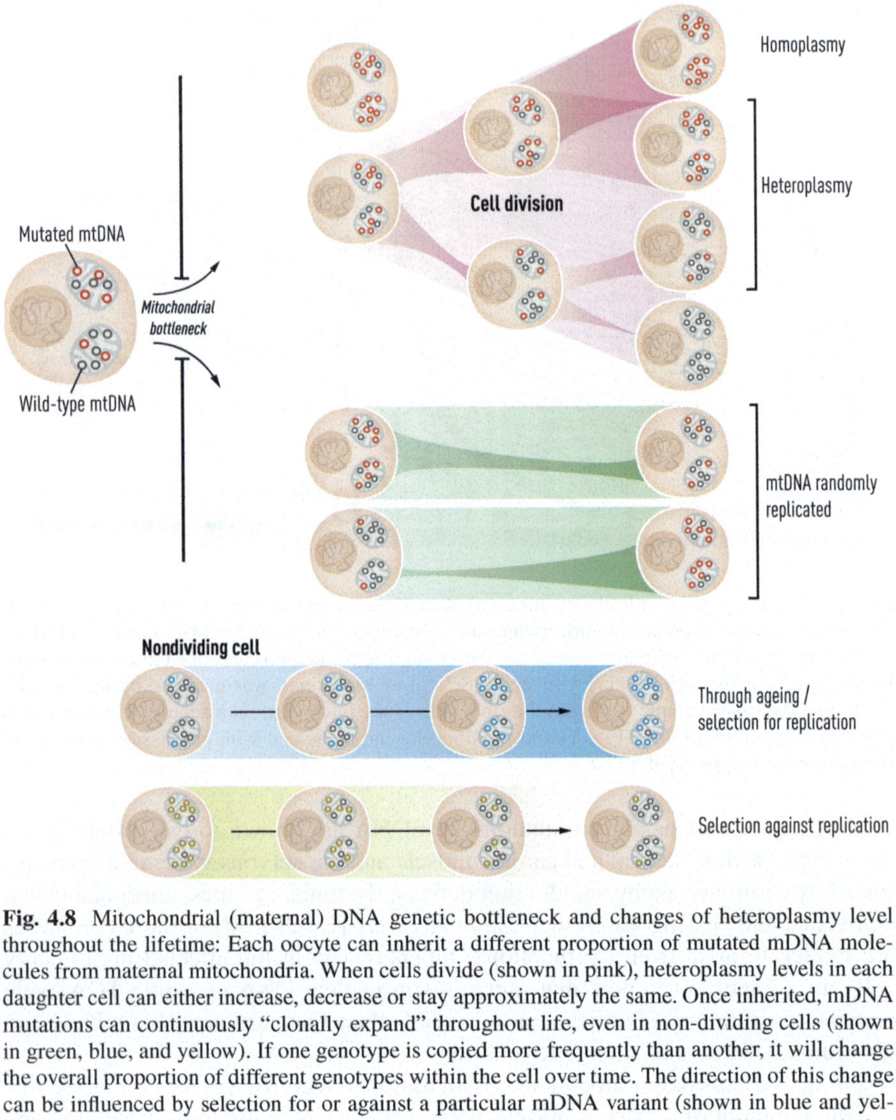

Fig. 4.8 Mitochondrial (maternal) DNA genetic bottleneck and changes of heteroplasmy level throughout the lifetime: Each oocyte can inherit a different proportion of mutated mDNA molecules from maternal mitochondria. When cells divide (shown in pink), heteroplasmy levels in each daughter cell can either increase, decrease or stay approximately the same. Once inherited, mDNA mutations can continuously "clonally expand" throughout life, even in non-dividing cells (shown in green, blue, and yellow). If one genotype is copied more frequently than another, it will change the overall proportion of different genotypes within the cell over time. The direction of this change can be influenced by selection for or against a particular mDNA variant (shown in blue and yellow). When a mutated mDNA molecule has a replicative advantage, the level will increase during life and possibly exceed the biochemical threshold, and thus contribute to the age-related pathologies or the ageing process (shown in the blue box). (Diagram (Fig. 1) and its modified legend resourced with license (https://creativecommons.org/licenses/BY-NC/4.0/) from Wei and Chinnary [28] (review))

to accumulation or loss of mDNA mutations. Even in non-dividing cells there can be selection for or against replication of mutations.

Mitochondrial Heterogeneity In addition to energy metabolism, modern mitochondria have large and variable roles to play in cellular function [31]. All mito-

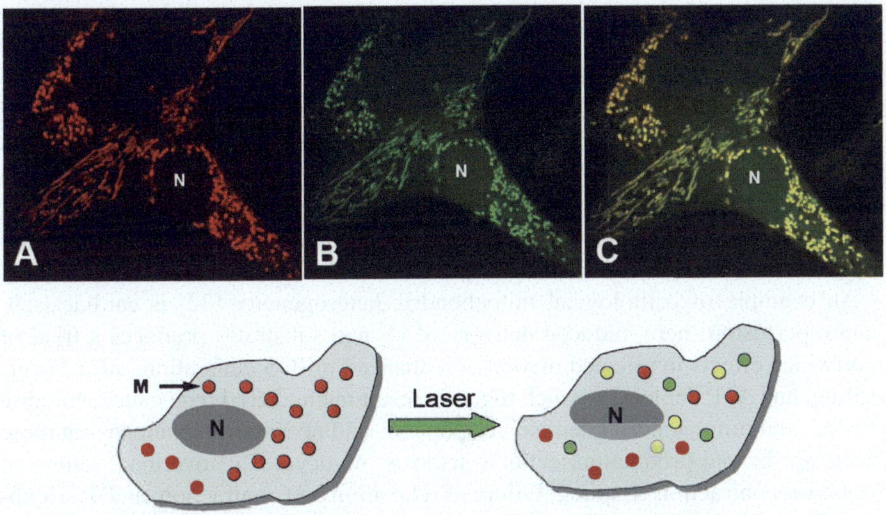

Fig. 4.9 Heterogeneity of mitochondrial membrane potential ($\Delta\Psi_m$) and ROS during photo-oxidative stress in HL-1 cells revealed by confocal fluorescent imaging: Upper panel: Simultaneous confocal imaging of red fluorescence of $\Delta\Psi_m$-sensitive probe TMRM, 0.1 μM (**a**), and green fluorescence of ROS-sensitive probe DCF, 20 μM (**b**). Fluorescence of TMRM and DCF is shown as a merge image (**c**). Some mitochondria with only green fluorescence indicate partially depolarized mitochondria with weaker TMRM signal and high levels of ROS production (ROS flashes). Lower panel: Red circles (M) show normally polarized $\Delta\Psi_m$ as monitored by TMRM fluorescence. Yellow circles show ROS-producing mitochondria (merged TMRM and DCF signal). Green circles are excessively ROS-producing and largely $\Delta\Psi_m$ depolarized mitochondria, which demonstrate high ROS (high DCF green signal) and, in parallel, the loss of $\Delta\Psi_m$ (low TMRM red signal). N shows nucleus. (Images (Fig. 3) and its modified legend resourced with license (https://creativecommons.org/licenses/BY-NC/4.0/) from Kuznetsov and Margreiter. [31] (review))

chondria are not alike in form or function. Within a given organ, or even within a single cell, mitochondria can have different morphologies and biochemical properties, and can interact differentially with other cellular structures as discussed in detail in Chap. 7. This includes potential differences in bioenergetics, reactive O_2 species (ROS) release, and mCa^{2+} handling. Figure 4.9 [31] shows a confocal fluorescent image as an example of heterogeneity of $\Delta\Psi_m$ and ROS during photo-oxidative stress in HL-1 cells (a modified, immortalized mouse cardiomyocyte cell line). This plasticity is called mitochondrial variation. Experimentation suggests that the heterogenicity of mitochondrial number, clustering, movement, form, and function, is dependent on the environment in which the mitochondria carryout their functions [32]. The heterogeneity of mitochondrial function is an example of an additional level of mitochondrial complexity within its cell that is interrelated with their interdependent, symbiotic functions to carry out metabolism, repair, and regeneration [31].

Unlike the mitochondrial genome with only 37 genes, the mitochondrial proteome contains at least 1100 proteins. Only 13 respiratory proteins are encoded by mDNA; the vast majority of proteins are imported into mitochondria after ribosomal translation in the cytosol. This reflects cell-mitochondria symbiosis over time. Mitochondria from the most diverse organisms contain approximately 75% of the same proteome imported from nuclear DNA [32], leaving possibly 25% of the proteome that is tailored for the function of heterogenetic mitochondria in different cell types. One such example mentioned in this book is the mitochondrial variants (SK3.1 and SK3.2) of the sBK_{Ca} channel identified in cardiac tissue [33, 34]. These are post-translationally modified proteins of nuclear genes KCNN2 and KCNN3. Their import into cardiac mitochondria and activation supports, at least in part, stabilization of the $\Delta\Psi_m$ by stimulating respiration in acidic conditions via mKHE [35].

An example of pathological mitochondria heterogeneity [32] is cardiac ischemia-reperfusion; here, reduced delivery of O_2 and substrates produces a triad of intertwined effects in affected myocytes: enhanced mROS generation, mCa^{2+} overloading, and ATP depletion, which together result in an altered redox state, collapse of $\Delta\Psi_m$, and mitochondrial demise by apoptosis and necrosis. Accompanying these events are lactate production, cellular acidosis, myocyte Ca^{2+} overload, failure of excitation-contraction coupling, failure of relaxation and contraction, and dysrhythmias. These topics are discussed in more detail in Chap. 8.

4.2 Endosymbiosis and Prokaryotes to Eukaryotes

Merging and or fusion of independent bacterial species may have accelerated evolutional changes. It is argued that endosymbiosis is a major horizontal gene process that led to the nucleated eukaryotes and increasing cell complexity and diversity. In contrast, autogeny suggests that cellular evolution was strictly vertical within a specific cell line. First promulgated over 50 years ago [36, 37], the theories of endosymbiotic partnerships leading to cooperation among species likely underlie the vast expansion and complexity of life during the past 1 Bya or so of evolution.

Although prokaryotes were already complex cells by the time of LUCA, there was not likely a great expansion in complexity or size in ancient Archaea or in modern Bacteria. Prokaryotes, then and now, are posited to be constrained by their lack of organelles (e.g., endoplasmic reticulum (ER), Golgi) for isolating and organizing cell function [24]. This includes an inefficient genome without a nucleus for cell fission, and also by the lack of mitochondria that provide for much more robust bioenergetics [24].

By undergoing respiration and phosphorylation across their own membranes rather than through the bacterial cell membrane, mitochondria compacted the OxPhos machinery into specialized membranes where they were also protected against changes in the extracellular milieu. By losing almost all of its genes to the genome, including respiratory components, it is argued that mitochondria could concentrate on bioenergetic efficiency and so provide more energy per gene than prokaryotes [24].

Origin of Eukaryotes It is commonly agreed from phylogenetic analyses of mitochondrial DNA that a prokaryote was the origin of the endosymbiont mitochondrion

(Sects. 4.1 and 7.4), but it is unknown and highly debated if the host cell was also a prokaryote or an eukaryote. That is, which came first, the organelle mitochondrion, or the organelle nucleus? Phylogenomic analysis suggests both were prokaryotes (Sect. 4.1) [8–12]. It is also unclear how and when some in the prokaryote pool became eukaryotes. Did the proto-mitochondrial genome have some relationship to drive formation of the nucleus, or other organelles, or the other way around? Interestingly, almost all eukaryotes examined to date contain mitochondria; the one known exception is a species of the order oxymonadida, named monocercomonoides, which is a flagellate that may have lost its original mitochondria [38].

What is clear is that the incorporation of organelles to organize genetic and metabolic functions resulted in an explosive and diverse speciation. Open questions on the origin and early evolution of eukaryotes are presented in Fig. 4.10 [7]. These include discussion of models and mechanisms of endosymbiosis that led to organelles, the timing of appearances and characterizations of the first eukaryote common ancestor (FECA) and the last eukaryote common ancestor (LECA), the evolution of distinct eukaryotic proteomes, and the elucidation of the root ancestors of Eukaryota, if it is not a distinct domain [7, 39].

Nevertheless, evidence from present day bacteria suggests that the membrane bound nucleus of eukaryotes had an endosymbiotic origin, as did mitochondria and chloroplasts [40]. It is unknown if the eukaryotic nucleus represents the first endosymbiont event but most data suggest that mitochondria were established in a prokaryote before the nucleus [40–42]. It is proposed that the first eukaryote common ancestor (FECA) originated a nucleus from two archaea-like prokaryotic cells at least 3.5 Bya, whether before or after mitochondrial endosymbiosis.

One might have been a large cell with an actin cytoskeleton, and the another a much smaller cell with a tubulin-centrin cytoskeleton [40]. The actin cytoskeleton of the larger cell is proposed to have had protrusions that enclosed the putative smaller and tubulin-centrin based flagellum-like archaea, transforming it into the first eukaryotic cell. The mobile flagellar guest cell may have resembled a gamete; so it is possible that the advent of sexual reproduction in eukaryotes also began as an endosymbiotic partnership [40].

LECA is considered to have differentiated into a cell with most all the organelles observed today, including: cell membrane, nucleus, mitochondria, Golgi apparatus, endoplasmic reticulum (ER), and an actin-tubulin cytoskeleton, among others [7, 39]. Like prokaryotes, most eukaryotes species are single-celled. Horizontal (lateral) gene transfer also likely contributed to the early species differentiation [7].

The accession of eukaryotes from prokaryotes provided first an encapsulated nuclear genome whereby about 2.2 Bya, a prokaryote could establish itself and reposition its circular DNA within its host DNA, whether it was a prokaryote or a eukaryote. Even though eukaryote defines the presence of a nucleus encompassing DNA, it also includes the linear DNA, histones, and chromosomes of the nucleus, as well as all the other organelles common to eukaryotes. All complex life on Earth today is based on the LECA and share the lineage of the first eukaryotic common ancestor (FECA).

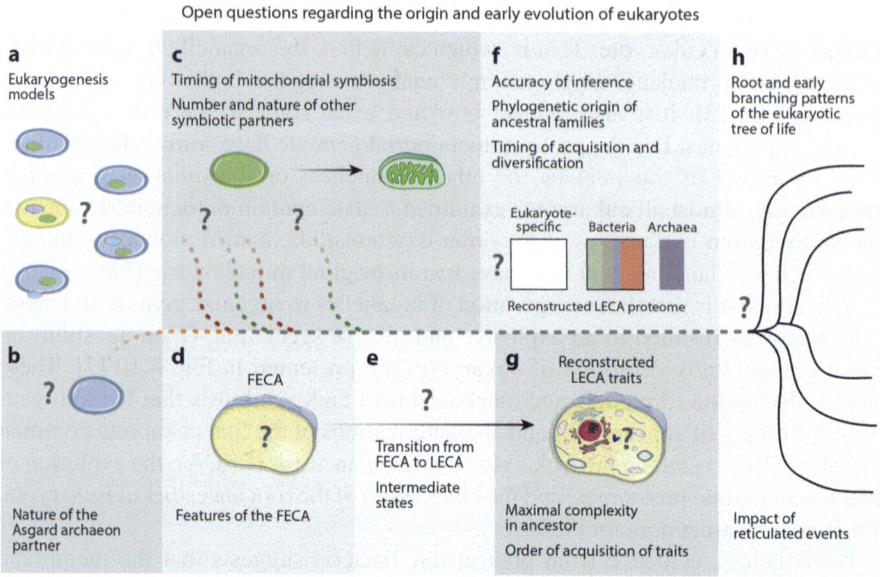

Fig. 4.10 Open questions regarding the origin and early evolution of eukaryotes: (**a**) Several not fully exclusive models of symbioses involved in eukaryogenesis have been proposed that have different metabolic, phylogenetic, or structural implications. (**b**) Although much progress has been achieved in recent years regarding the nature of the archaeal ancestor of eukaryotes, many questions remain regarding its structural, genomic, and metabolic traits. (**c**) All models include symbiosis with an alphaproteobacterial ancestor that later became the mitochondrion, yet the exact metabolic nature and the timing of acquisition of this ancestor are unknown. Similarly, the number, timing, and nature of other putative symbiotic partners contributing to the eukaryotic lineage are debated. (**d**) The features of the FECA are far from established. (**e**) The time span of the transition from FECA to LECA is unknown, and so is the nature of the intermediates. (**f**) Our knowledge of the LECA relies on the presence of gene families or traits, inferred based on the analysis of extant eukaryotes. We do not know the accuracy of these reconstructions. Phylogenetic analysis of gene families present in the LECA shows an excess of bacterial proteins and a multiplicity of phylogenetic adscriptions to diverse bacterial groups. The origin of a significant fraction of eukaryote-specific proteins is unclear. We know very little about when the different proteomic components of the LECA originated, where they were acquired, and how they diversified through duplication events. (**g**) Most complex eukaryotic traits are inferred to have been present in the LECA, but we do not know in which order they appeared. (**h**) Despite much progress, uncertainty surrounds the position of the eukaryotic root and the relative branching order of the major lineages. Importantly, we do not know the extent and potential impact of reticulated processes (horizontal gene transfer, hybridization, etc.) during the early diversification of eukaryotes. Abbreviations: *FECA* first eukaryotic common ancestor, *LECA* last eukaryotic common ancestor. (Cell images adapted from Servier Medical Art. Diagram (Fig. 1) and its legend resourced with license (https://creativecommons.org/licenses/CC-Byby/3.0/ from Gabaldon [7] (review))

Ribosomes and Translation At the time of the hypothetical LUCA, the standard genetic code (SGC) was explicit in its universality and it remains highly conserved; a given 3 base codon encodes only one amino acid in all of Earth's lifeforms (although there are minor insignificant exceptions). For the four DNA nucleobases (A,G,C,T) there are a possible $4^3 = 64$ combinations but there are only 20 amino acids utilized in lifeforms. Thus, there is a built-in redundancy in the genetic code where up to four different codons code for the same amino acid. Generally, the last base in the triplet codon is variable for a given amino acid, for example, GGG, GGC, GGA, and GGT all code for glycine. This reduces DNA base errors from having an effect to alter the amino acid encoded. Interestingly, it was found that the second position of a codon has the lowest error rate in translation of an amino acid into a sequence whereas the third base has the highest rate [43]. Although redundant, the code is not ambiguous as a given codon encodes only one amino acid.

In eukaryotes, ribosomes are made in the nucleolus, which contains clustered DNA strands that contain all the genes needed for rRNA synthesis. The rRNA, functionally a ribozyme, is highly conserved, single stranded RNA that binds in places with itself like hairpin loops and folds and twists onto itself. rRNA is noncoding (Sect. 3.3), that is, it cannot code for translating protein [44]. The rRNA is bound to proteins composed mostly of basic and aromatic amino acid residues that permit them to twist and tightly stack with the rRNA to make up the ribosome.

Each completed ribosome has two components a large subunit (LSU) and a small subunit (SSU) each composed of about 60% rRNA, 40% protein. The mRNA coding sequence is read between the subunits to translate the genetic code into a protein. The function of ribosomal proteins is to provide scaffolding for linking each amino acid to another making a chain. The function of rRNA is to align the codons on mRNA with the anticodons of tRNA and to catalyze the peptide bond. Completed ribosomes are exported from the nucleus to the cytoplasm.

During translation on the ribosome at the peptidyl transferase center (PTC) [45] each successive mRNA codon is matched to its specific anticodon, on tRNA, aligning two tRNAs together so that a peptide link can be catalyzed, connecting one amino acid to another. Unlike the prebiotic ribozyme of the past, there is no chemically-dependent link between RNA nucleobases and amino acid residues as this work is done by tRNA within the ribosome. Ribosomal proteins cannot bind at the PTC; the catalytic activity of the PTC is performed only by rRNA. At the PTC the peptide bond is formed as the aminoacyl-tRNA (A site) attaches to the terminal carboxyl group of the peptidyl (P site) tRNA. The protein enzymes formed on ribosomes carry out thousands of different catalytic activities in all cells.

Origin of Viruses A virus is a particle defined by its genetic features of genome propagation and dissemination [46]; a viron is viral DNA or RNA encapsulated by a capsid protein shell [47]. Viral genomes are characterized by virus specific genes and folding protein products not normally found in cell genomes. The viral cycle has a dynamic stage of cell infection and replication, and a stage of release of virons from the cell into the environment where they may be deconstructed or enter other host cells [48].

Although viruses can be mutualistic symbionts with living cells (like mitochondria derived from bacteria), they are usually considered parasitic based on their capability to commandeer the cell nucleus to make viral particles (virons) and viral proteins that can disengage normal cellular metabolism and defenses for the benefit of viral protein machinery [48]. This may elicit cellular antibody responses to suppress viral replication and or protein synthesis [49]. Viruses are just packets of genetic material and protein; some contain their own, or use the cell's RNA- or DNA-dependent RNA or DNA polymerases to inject their genomes into the host cell genome. However, because viruses cannot grow, produce energy, or utilize substrates to make ATP or translate proteins, which must be done by the host cell, they are not considered as lifeforms.

It is unlikely that viruses evolved before cells; it is more likely that viruses developed along with ancient cells that preceded LUCA [50] (Sects. 3.1, 3.2 and 3.3), which hypothetically represents, not the first cells, but rather the last cells with common ancestry that diversified into modern cells with nearly the same genetic and protein codes [48]. LUCA was likely early prey to viruses [51]. A cell genome today can contain 8–10% bits of ancient viral genome [52]. Most incorporated ancient viral DNA or RNA genes are not harmful to the host cell, and in some cases may be beneficial. For example, viruses can participate in horizontal gene transfer HGT) via retroviruses and some viral DNA to drive evolution [53]. In another example, envelop proteins encoded by endogenous retroviruses can protect against invasion by other viruses [52].

Viral genes can be utilized to understand the evolutionary history of individual viral genes rather than the whole viron [48]. That is, they track changes in viral genes over time, but not the composite virus itself. Ancient cells may have benefited from releasing DNA or RNA particles (viruses) from vesicles, and then taking them back in for incorporation in the host genome [54]. Without any direct evidence, it is possible that RNA and DNA viruses were derived from ancient cell RNA and cell DNA, respectively, with RNA being the ancestor of DNA (Sects. 3.2 and 3.3) [48]. One alternative is that cell RNA might have been later transformed to cell DNA by viral RNA and retrotranscribing (RT) viruses, which made the transcription of the cell genetic code more stable [48]. These and other possible scenarios are elaborated in Fig. 4.11 [48].

Common human diseases caused by RNA viruses include adenovirus (common cold, influenza), SARS, MERS, COVID-19, hepatitis C, West Nile fever, Ebola, rabies, polio, mumps, and measles. RT-RNA viruses, such as the human immunodeficiency virus (HIV), are highly infectious because of the fast replication of RNA in the cell by viral RNA-dependent RNA polymerase and because they target cells and proteins of the immune system [56]. The polymerases of the vast majority of RNA viruses lack 3′ exonuclease proofreading activity and hence are more error-prone than those of DNA viruses [57]. Thus, viruses, and especially RT viruses, mutate frequently because RNA transcription results in many errors compared with DNA transcription. Recall that the stronger deoxyribose sugar bonds of DNA resists hydrolysis compared to the hydroxyl bonds of RNA (Sects. 1.3, 3.2, 3.3 and 5.1).

After LECA Presumably with LUCA, vertical and horizontal genetic recombination led to changes in phenotype, which were adapted or rejected during evolution; eventually, this resulted in the proposed two or three domains of life we know today. Mitochondria came into existence perhaps 2–2.5 By later after LUCA Byr developing an adaptive, symbiotic relationship between a captured cell and its host that allowed foodstuffs to be oxidized more efficiently to generate ATP. Since the last eukaryote common ancestor (LECA) blossomed in animals, their cells multiplied

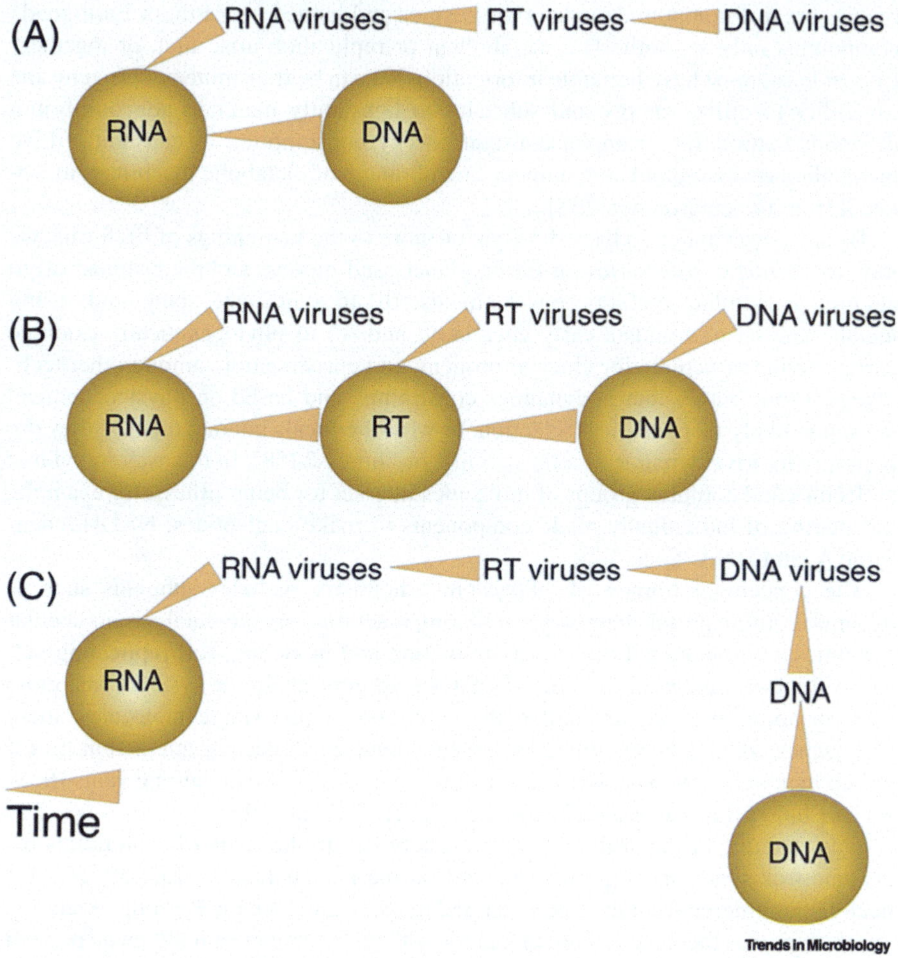

Fig. 4.11 Different scenarios for the possible evolution of different virus replicon groups: (**a**) RNA viruses evolved from RNA cells and later evolved into retrotranscribing (RT) and DNA viruses. In parallel, RNA cells evolved into DNA cells. (**b**) The evolution of RNA, RT, and DNA viruses followed the emergence of RNA, RT, and DNA cells, respectively. (**c**) RNA viruses evolved from RNA cells and later evolved into RT and DNA viruses. RNA cells evolved into DNA cells once DNA was invented by viruses (text Ref. [50]. (Cartoon (Fig. 2) and its legend resourced with license (https://creativecommons.org/licenses/by/4.0/) from Nasir et al. [48])

and differentiated to carry out different tasks. Except for one known unicellular eukaryote (Sect. 4.2), mitochondria are present in all eukaryotic cells, so they are a distinguishing characteristic of all eukaryotes.

Systems Chemistry to Probe Life's Origin As described in the Sections above, we still do not know how all these processes were put together to initiate life on Earth. Was life based on autotrophic or heterotrophic chemical origins, or both? Which came first, RNA or peptides, or did they coevolve? Was life a chemical accident that started and stopped, or was it continuous once it got started? Did life require organic compound made in the cosmos and landed on Earth, or Earth made compounds only, or both? Did metabolism or replication arise first, or together? Life, at least, must have heritable information that can be transmitted to progeny, the capability to utilize energy and substates to constantly maintain nonequilibrium thermodynamics, for example, constant energy consumption to constantly drive metabolic processes, and containment (membranes) of metabolic machinery to isolate it from the environment [58].

Because we cannot go back 3.5 Bya, or more to the beginnings of lifeforms, we look to chemical discoveries on Earth, planets and moons, and the cosmos: (a) to attempt to simulate prebiotic environments; (b) to synthesize compounds using metal catalysts to simulate early chemistry; and (c) to phylogenetically examine current living structures for clues to progeny and classification, among other techniques. From purely thermodynamics constraints, and based on physicochemical laws, it is likely that chemical selection, or why chemicals interact the way they do, is simply to drive dynamic kinetic stability, or "fitness" [58]. In this way individual molecules and complex groups of molecules interact for better fitness, for example, the sharing of individually made components to make nucleotides, NADH, coenzyme A, and DNA (Sect. 3.2).

The conceptual framework of systems chemistry includes subfields such as dynamic combinatorial chemistry, oscillating reactions, organo-catalysis, molecular dynamics, computational molecular simulation and modeling, self-replicating networks, high-throughput biochemical analysis, in vitro evolution systems, and more [58]. In addition to understanding the evolution of just single molecules, these approaches are now being utilized to process complex molecular interactions in the milieu in which they exist, like rock surfaces (Sect. 3.3), ocean thermal vents (Sect. 3.1), or lipid bilayer membranes (Sects. 1.2, 2.1, 3.1, and 7.1).

One example of the utility of systems chemistry in the study of evolution is the enzyme-free synthesis of pyrimidine and purine ribonucleotides [55, 59, 60]. Do nucleotides emerge from the free sugar and the base and then the P_i group (Sect. 1.4) attaches, or was there a common precursor with atoms that eventually made parts of both the base and sugar? If so, this would bypass the classical separation of the C-O chemistry of sugars with the C-N chemistry of nucleobases. A key finding is that the presence of P_i in the reaction buffer controls the reaction pathway to a high yield product and uniform stereochemistry. An extensive review provides many more examples of the use of the integrative approaches of systems chemistry to unravel the secrets of prebiotic chemistry [58].

4.3 Electrons and Ions as Transporters (Current) of Energy

In the next chapter we leave the evolution of bioenergetics and turn to the makeup and function of a present day bioenergetic product of evolution, the cardiac mitochondrion. But first, we compare electricity in a physical circuit with bioelectricity in a solution.

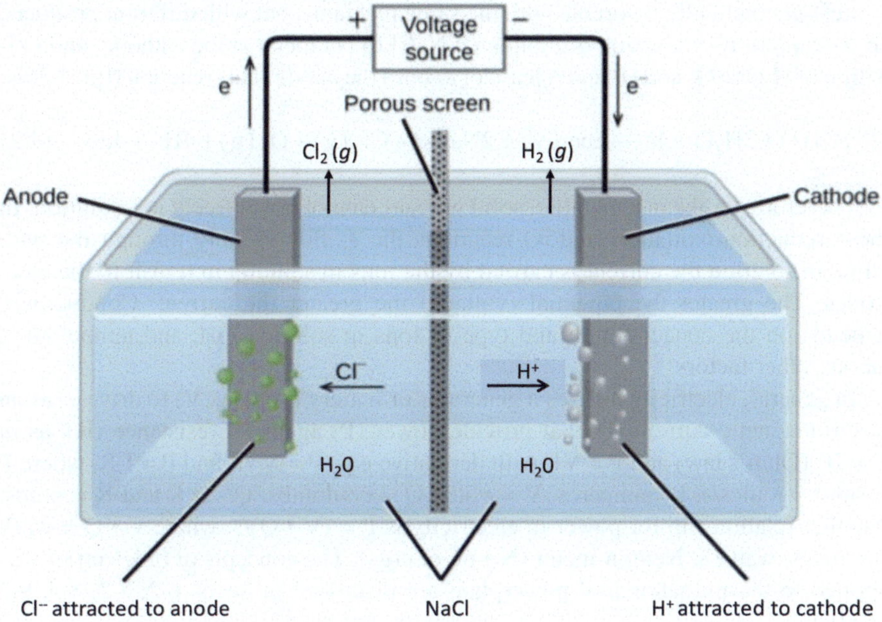

Fig. 4.12 Process of electrolysis of H_2O and NaCl: In a solution of NaCl in H_2O at pH 7, application of a voltage to produce an e- current causes H_2O to be reduced to H_2 gas at the cathode and Cl^- to be oxidized to Cl_2 gas at the anode; another product is OH^-. (Revised scheme resourced and available at: https://courses.lumenlearning.com/chemistryformajors/chapter/electrolysis/)

In an electrical circuit with an applied voltage across a copper wire (conductor) the 29th *e*- of $_{29}Cu^{64}$ is freed to physically move down the wire but only at a speed of about 1 mm/s (drift velocity). However, the electromagnetic wave rippling through the electron (*e*-) cloud can travel up to 60–90% of the speed of light (signal velocity). The protons or neutrons of copper do not move down the wire.

When solid NaCl salt is dissolved into H_2O the Na and Cl ions dissociate into hydrated Na^+ and Cl^- ions where Na^+ is attracted to the negatively charged O part of H_2O and Cl^- is attracted to the positively charged H_2 part of H_2O. Figure 4.12 shows how an electric current applied to this electrolyte solution at pH 7, causes H^+ cations to migrate to the cathode and Cl^- anions to migrate to the anode creating an ion current. Electrons collected on the cathode are used to reduce H_2O (actually, H_3O^+) to H_2 gas and Cl^- is oxidized to Cl_2 gas with *e*- released onto the anode. A diaphragm is required to prevent the Cl_2 gas from reacting with NaOH that accumulates at the cathode. The electrodes are commonly made of graphite because its valence electrons move about freely and because it is highly resistant to chemical erosion. The overall reaction (Eq. 4.1) is:

$$2\mathrm{Na}^+(\mathrm{aq}) + 2\mathrm{Cl}^-(\mathrm{aq}) + 2\mathrm{H}_2\mathrm{O} + 4e- + \mathrm{energy} \rightleftharpoons 2\mathrm{Na}^+(\mathrm{aq}) + 2\mathrm{OH}^-(\mathrm{aq}) + 4\mathrm{e}- + \mathrm{H}_2(\mathrm{g}) + \mathrm{Cl}_2(\mathrm{g}) \quad (4.1)$$

Less energetically favorable with the same reactants, but with different products, are *e*- capture for reduction of molten (*l*) NaCl to Na metal at the cathode, and oxidation of H_2O to O_2 and H^+ and release of *e*- onto the anode. This reaction (Eq. 4.2) is:

$$2\text{NaCl}(\text{l}) + 2\text{H}_2\text{O} + 4e- + \text{energy} \rightarrow 2\text{Na}(\text{s}) + \text{Cl}_2(\text{g}) + \text{O}_2(\text{g}) + 4\text{H}^+ + 4e- \quad (4.2)$$

Therefore, unlike in a metallic, solid wire, *e*- cannot travel freely in a solution. In these reduction-oxidation (redox) reactions the *e*- flow is only through the wire while in solution the current is carried by the ions in solution in transit to the electrodes. The greater the potential (voltage) the greater the current. Conductivity depends on the concentration and type of ions in solution, pH, and temperature, among other factors.

In general, electricity utilizes a generator or battery (voltage, V) to drive e- as an electromagnetic current (I), that provide power (P) against a resistance (R), as in $V = IR$ (Ohm's law) and $P = VI$, with derivative eqs. $P = V^2/R$ and $P = I^2R$, where P (watts) = joules/s, I = amperes, V = volts, Q = coulombs, Q/s = I, and R = ohms. Another relationship for power in electricity is $P = (V \bullet Q)/s$, where $V \bullet Q$ = work (in joules, watt • s, Neuton-meter (N • m, or ergs). The concepts of bioelectricity as applied to the mitochondrial powerplant are discussed in Sects. 6.2, 7.2, 8.4, 9.2 and 9.4.

References

1. Klamt S, Grammel H, Straube R, Ghosh R, Gilles ED. Modeling the electron transport chain of purple non-sulfur bacteria. Mol Syst Biol. 2008;4:156. Epub 20080115. https://doi.org/10.1038/msb4100191. PubMed PMID: 18197174; PMCID: PMC2238716.
2. Alberts B, Johnson A, Lewis J, Raff M, Roberts K, Walter P. The evolution of electron-transport chains. In: Molecular biology of the cell. 4th ed. New York: Garland Science; 2002. Available from: https://www.ncbi.nlm.nih.gov/books/NBK26849/.
3. Weiss MC, Sousa FL, Mrnjavac N, Neukirchen S, Roettger M, Nelson-Sathi S, Martin WF. The physiology and habitat of the last universal common ancestor. Nat Microbiol. 2016;1(9):16116. Epub 20160725. https://doi.org/10.1038/nmicrobiol.2016.116. PubMed PMID: 27562259.
4. Weiss MC, Preiner M, Xavier JC, Zimorski V, Martin WF. The last universal common ancestor between ancient Earth chemistry and the onset of genetics. PLoS Genet. 2018;14(8):e1007518. Epub 20180816. https://doi.org/10.1371/journal.pgen.1007518. PubMed PMID: 30114187; PMCID: PMC6095482.
5. Cavalier-Smith T. Origin of mitochondria by intracellular enslavement of a photosynthetic purple bacterium. Proc Biol Sci. 2006;273(1596):1943–52. https://doi.org/10.1098/rspb.2006.3531. PubMed PMID: 16822756; PMCID: PMC1634775.
6. Andersson SG, Kurland CG. Origins of mitochondria and hydrogenosomes. Curr Opin Microbiol. 1999;2(5):535–41. https://doi.org/10.1016/s1369-5274(99)00013-2. PubMed PMID: 10508728.
7. Gabaldon T. Origin and early evolution of the eukaryotic cell. Annu Rev Microbiol. 2021;75:631–47. Epub 20210803. https://doi.org/10.1146/annurev-micro-090817-062213. PubMed PMID: 34343017.

8. Gray MW, Burger G, Lang BF. Mitochondrial evolution. Science. 1999;283(5407):1476–81. https://doi.org/10.1126/science.283.5407.1476. PubMed PMID: 10066161.
9. Kailing F, Lieberman J, Wang J, Turner JL, Goldman AD. Evolution of cellular organization along the first branches of the tree of life. J Mol Evol. 2024. Epub 20240717. https://doi.org/10.1007/s00239-024-10188-7. PubMed PMID: 39020132.
10. Eme L, Spang A, Lombard J, Stairs CW, Ettema TJG. Archaea and the origin of eukaryotes. Nat Rev Microbiol. 2017;15(12):711–23. https://doi.org/10.1038/nrmicro.2017.133. PubMed PMID: 29123225.
11. Zaremba-Niedzwiedzka K, Caceres EF, Saw JH, Backstrom D, Juzokaite L, Vancaester E, Seitz KW, Anantharaman K, Starnawski P, Kjeldsen KU, Stott MB, Nunoura T, Banfield JF, Schramm A, Baker BJ, Spang A, Ettema TJ. Asgard archaea illuminate the origin of eukaryotic cellular complexity. Nature. 2017;541(7637):353–8. Epub 20170111. https://doi.org/10.1038/nature21031. PubMed PMID: 28077874.
12. Fournier GP, Poole AM. A briefly argued case that asgard Archaea are part of the eukaryote tree. Front Microbiol. 2018;9:1896. Epub 20180815. https://doi.org/10.3389/fmicb.2018.01896. PubMed PMID: 30158917; PMCID: PMC6104171.
13. Williams TA, Cox CJ, Foster PG, Szollosi GJ, Embley TM. Phylogenomics provides robust support for a two-domains tree of life. Nat Ecol Evol. 2020;4(1):138–47. Epub 20191209. https://doi.org/10.1038/s41559-019-1040-x. PubMed PMID: 31819234; PMCID: PMC6942926.
14. Woese CR, Fox GE. Phylogenetic structure of the prokaryotic domain: the primary kingdoms. Proc Natl Acad Sci USA. 1977;74(11):5088–90. https://doi.org/10.1073/pnas.74.11.5088. PubMed PMID: 270744; PMCID: PMC432104.
15. Milshteyn D, Cooper G, Deamer D. Chemiosmotic energy for primitive cellular life: proton gradients are generated across lipid membranes by redox reactions coupled to meteoritic quinones. Sci Rep. 2019;9(1):12447. Epub 20190828. https://doi.org/10.1038/s41598-019-48328-5. PubMed PMID: 31462644; PMCID: PMC6713726.
16. Neupane P, Bhuju S, Thapa N, Bhattarai HK. ATP synthase: structure, function and Inhibition. Biomol Concepts. 2019;10(1):1–10. Epub 20190307. https://doi.org/10.1515/bmc-2019-0001. PubMed PMID: 30888962.
17. Jasielec JJ, Filipek R, Dolowy K, Lewenstam A. Precipitation of inorganic salts in mitochondrial matrix. Membranes (Basel). 2020;10(5). Epub 20200427. https://doi.org/10.3390/membranes10050081. PubMed PMID: 32349446; PMCID: PMC7281443.
18. Boguszewska K, Szewczuk M, Kazmierczak-Baranska J, Karwowski BT. The similarities between human mitochondria and bacteria in the context of structure, genome, and base excision repair system. Molecules. 2020;25(12). Epub 20200621. https://doi.org/10.3390/molecules25122857. PubMed PMID: 32575813; PMCID: PMC7356350.
19. Zeth K. Structure and evolution of mitochondrial outer membrane proteins of beta-barrel topology. Biochim Biophys Acta. 2010;1797(6–7):1292–9. Epub 20100505. https://doi.org/10.1016/j.bbabio.2010.04.019. PubMed PMID: 20450883.
20. Wiedemann N, Truscott KN, Pfannschmidt S, Guiard B, Meisinger C, Pfanner N. Biogenesis of the protein import channel Tom40 of the mitochondrial outer membrane: intermembrane space components are involved in an early stage of the assembly pathway. J Biol Chem. 2004;279(18):18188–94. Epub 20040220. https://doi.org/10.1074/jbc.M400050200. PubMed PMID: 14978039.
21. Adams KL, Palmer JD. Evolution of mitochondrial gene content: gene loss and transfer to the nucleus. Mol Phylogenet Evol. 2003;29(3):380–95. https://doi.org/10.1016/s1055-7903(03)00194-5. PubMed PMID: 14615181.
22. Yamauchi A. Rate of gene transfer from mitochondria to nucleus: effects of cytoplasmic inheritance system and intensity of intracellular competition. Genetics. 2005;171(3):1387–96. Epub 20050803. https://doi.org/10.1534/genetics.104.036350. PubMed PMID: 16079242; PMCID: PMC1456839.

23. Bock R. Witnessing genome evolution: experimental reconstruction of endosymbiotic and horizontal gene transfer. Annu Rev Genet. 2017;51:1–22. Epub 20170828. https://doi.org/10.1146/annurev-genet-120215-035329. PubMed PMID: 28846455.
24. Lane N. Bioenergetic constraints on the evolution of complex life. Cold Spring Harb Perspect Biol. 2014;6(5):a015982. Epub 20140501. https://doi.org/10.1101/cshperspect.a015982. PubMed PMID: 24789818; PMCID: PMC3996473.
25. Munoz-Gomez SA. The energetic costs of cellular complexity in evolution. Trends Microbiol. 2024;32(8):746–55. Epub 20240201. https://doi.org/10.1016/j.tim.2024.01.003. PubMed PMID: 38307786.
26. Lynch M, Marinov GK. The bioenergetic costs of a gene. Proc Natl Acad Sci USA. 2015;112(51):15690–5. Epub 20151102. https://doi.org/10.1073/pnas.1514974112. PubMed PMID: 26575626; PMCID: PMC4697398.
27. Hoppel CL, Tandler B, Fujioka H, Riva A. Dynamic organization of mitochondria in human heart and in myocardial disease. Int J Biochem Cell Biol. 2009;41(10):1949–56. Epub 20090514. https://doi.org/10.1016/j.biocel.2009.05.004. PubMed PMID: 19446651; PMCID: PMC3221317.
28. Wei W, Chinnery PF. Inheritance of mitochondrial DNA in humans: implications for rare and common diseases. J Intern Med. 2020;287(6):634–44. Epub 20200318. https://doi.org/10.1111/joim.13047. PubMed PMID: 32187761; PMCID: PMC8641369.
29. Parakatselaki ME, Ladoukakis ED. mtDNA heteroplasmy: origin, detection, significance, and evolutionary consequences. Life (Basel). 2021;11(7). Epub 20210629. https://doi.org/10.3390/life11070633. PubMed PMID: 34209862; PMCID: PMC8307225.
30. Keogh MJ, Chinnery PF. Mitochondrial DNA mutations in neurodegeneration. Biochim Biophys Acta. 2015;1847(11):1401–11. Epub 20150523. https://doi.org/10.1016/j.bbabio.2015.05.015. PubMed PMID: 26014345.
31. Kuznetsov AV, Margreiter R. Heterogeneity of mitochondria and mitochondrial function within cells as another level of mitochondrial complexity. Int J Mol Sci. 2009;10(4):1911–29. Epub 20090424. https://doi.org/10.3390/ijms10041911. PubMed PMID: 19468346; PMCID: PMC2680654.
32. Chen L, Zhou M, Li H, Liu D, Liao P, Zong Y, Zhang C, Zou W, Gao J. Mitochondrial heterogeneity in diseases. Signal Transduct Target Ther. 2023;8(1):311. Epub 20230823. https://doi.org/10.1038/s41392-023-01546-w. PubMed PMID: 37607925; PMCID: PMC10444818.
33. Stowe DF, Gadicherla AK, Zhou Y, Aldakkak M, Cheng Q, Kwok WM, Jiang MT, Heisner JS, Yang M, Camara AK. Protection against cardiac injury by small Ca^{2+}-sensitive K^+ channels identified in guinea pig cardiac inner mitochondrial membrane. Biochim Biophys Acta. 2013;1828(2):427–42. Epub 20120908. https://doi.org/10.1016/j.bbamem.2012.08.031. PubMed PMID: 22982251; PMCID: PMC3534888.
34. Yang M, Camara AKS, Aldakkak M, Kwok WM, Stowe DF. Identity and function of a cardiac mitochondrial small conductance Ca^{2+}-activated K^+ channel splice variant. Biochim Biophys Acta Bioenerg. 2017;1858(6):442–58. Epub 20170322. https://doi.org/10.1016/j.bbabio.2017.03.005. PubMed PMID: 28342809; PMCID: PMC5749404.
35. Malas KM, Lambert DS, Heisner JS, Camara AKS, Stowe DF. Time and charge/pH-dependent activation of K^+ channel-mediated K^+ influx and K^+/H^+ exchange in guinea pig heart isolated mitochondria; role in bioenergetic stability. Biochim Biophys Acta Bioenerg. 2022;1863(8):148908. Epub 20220809. https://doi.org/10.1016/j.bbabio.2022.148908. PubMed PMID: 35961396.
36. Margulis L. Origin of eukaryotic cells: evidence and research implications for a theory of the origin and evolution of microbial, plant, and animal cells on the precambrian earth. Illustrated ed. Yale University Press; 1970. 349 p. ISBN 0300013531, 9780300013535.
37. Sagan L. On the origin of mitosing cells. J Theor Biol. 1967;14(3):255–74. https://doi.org/10.1016/0022-5193(67)90079-3. PubMed PMID: 11541392.
38. Karnkowska A, Vacek V, Zubacova Z, Treitli SC, Petrzelkova R, Eme L, Novak L, Zarsky V, Barlow LD, Herman EK, Soukal P, Hroudova M, Dolezal P, Stairs CW, Roger AJ, Elias

M, Dacks JB, Vlcek C, Hampl V. A eukaryote without a mitochondrial organelle. Curr Biol. 2016;26(10):1274–84. Epub 20160512. https://doi.org/10.1016/j.cub.2016.03.053. PubMed PMID: 27185558.

39. Poole AM, Gribaldo S. Eukaryotic origins: how and when was the mitochondrion acquired? Cold Spring Harb Perspect Biol. 2014;6(12):a015990. Epub 20140718. https://doi.org/10.1101/cshperspect.a015990. PubMed PMID: 25038049; PMCID: PMC4292153.
40. Baluska F, Lyons S. Archaeal origins of eukaryotic cell and nucleus. Biosystems. 2021;203:104375. Epub 20210204. https://doi.org/10.1016/j.biosystems.2021.104375. PubMed PMID: 33549602.
41. Baum B, Spang A. On the origin of the nucleus: a hypothesis. Microbiol Mol Biol Rev. 2023;87(4):e0018621. Epub 20231129. https://doi.org/10.1128/mmbr.00186-21. PubMed PMID: 38018971; PMCID: PMC10732040.
42. Ettema TJ. Evolution: mitochondria in the second act. Nature. 2016;531(7592):39–40. Epub 20160203. https://doi.org/10.1038/nature16876. PubMed PMID: 26840482.
43. Woese CR, Olsen GJ, Ibba M, Soll D. Aminoacyl-tRNA synthetases, the genetic code, and the evolutionary process. Microbiol Mol Biol Rev. 2000;64(1):202–36. https://doi.org/10.1128/MMBR.64.1.202-236.2000. PubMed PMID: 10704480; PMCID: PMC98992.
44. Lodish H, Berk A, Zipursky SL, Matsudaira P, Baltimore D, Darnell J. The three roles of RNA in protein synthesis. In: Molecular cell biology. 4th ed. W. H. Freeman; 2002.
45. Jose MV, Bobadilla JR, Zamudio GS, de Farias ST. Symmetrical distributions of aminoacyl-tRNA synthetases during the evolution of the genetic code. Theory Biosci. 2023;142(3):211–9. Epub 20230705. https://doi.org/10.1007/s12064-023-00394-0. PubMed PMID: 37402895; PMCID: PMC10423125.
46. Nasir A, Sun FJ, Kim KM, Caetano-Anolles G. Untangling the origin of viruses and their impact on cellular evolution. Ann N Y Acad Sci. 2015;1341:61–74. Epub 20150310. https://doi.org/10.1111/nyas.12735. PubMed PMID: 25758413.
47. Raoult D, Forterre P. Redefining viruses: lessons from Mimivirus. Nat Rev Microbiol. 2008;6(4):315–9. Epub 20080303. https://doi.org/10.1038/nrmicro1858. PubMed PMID: 18311164.
48. Nasir A, Romero-Severson E, Claverie JM. Investigating the concept and origin of viruses. Trends Microbiol. 2020;28(12):959–67. https://doi.org/10.1016/j.tim.2020.08.003. PubMed PMID: 33158732; PMCID: PMC7609044.
49. Murin CD, Wilson IA, Ward AB. Antibody responses to viral infections: a structural perspective across three different enveloped viruses. Nat Microbiol. 2019;4(5):734–47. Epub 20190318. https://doi.org/10.1038/s41564-019-0392-y. PubMed PMID: 30886356; PMCID: PMC6818971.
50. Forterre P, Prangishvili D. The origin of viruses. Res Microbiol. 2009;160(7):466–72. Epub 20090730. https://doi.org/10.1016/j.resmic.2009.07.008. PubMed PMID: 19647075.
51. Krupovic M, Dolja VV, Koonin EV. The LUCA and its complex virome. Nat Rev Microbiol. 2020;18(11):661–70. Epub 20200714. https://doi.org/10.1038/s41579-020-0408-x. PubMed PMID: 32665595.
52. Frank JA, Singh M, Cullen HB, Kirou RA, Benkaddour-Boumzaouad M, Cortes JL, Garcia Perez J, Coyne CB, Feschotte C. Evolution and antiviral activity of a human protein of retroviral origin. Science. 2022;378(6618):422–8. Epub 20221027. https://doi.org/10.1126/science.abq7871. PubMed PMID: 36302021; PMCID: PMC10542854.
53. Irwin NAT, Pittis AA, Richards TA, Keeling PJ. Systematic evaluation of horizontal gene transfer between eukaryotes and viruses. Nat Microbiol. 2022;7(2):327–36. Epub 20211231. https://doi.org/10.1038/s41564-021-01026-3. PubMed PMID: 34972821.
54. Gill S, Catchpole R, Forterre P. Extracellular membrane vesicles in the three domains of life and beyond. FEMS Microbiol Rev. 2019;43(3):273–303. https://doi.org/10.1093/femsre/fuy042. PubMed PMID: 30476045; PMCID: PMC6524685.
55. Fernandez-Garcia C, Coggins AJ, Powner MW. A chemist's perspective on the role of phosphorus at the origins of life. Life (Basel). 2017;7(3). Epub 20170713. https://doi.org/10.3390/life7030031. PubMed PMID: 28703763; PMCID: PMC5617956.

56. Duffy S. Why are RNA virus mutation rates so damn high? PLoS Biol. 2018;16(8):e3000003. Epub 20180813. https://doi.org/10.1371/journal.pbio.3000003. PubMed PMID: 30102691; PMCID: PMC6107253.
57. Sanjuan R, Domingo-Calap P. Mechanisms of viral mutation. Cell Mol Life Sci. 2016;73(23):4433–48. Epub 20160708. https://doi.org/10.1007/s00018-016-2299-6. PubMed PMID: 27392606; PMCID: PMC5075021.
58. Ruiz-Mirazo K, Briones C, de la Escosura A. Prebiotic systems chemistry: new perspectives for the origins of life. Chem Rev. 2014;114(1):285–366. Epub 20131031. https://doi.org/10.1021/cr2004844. PubMed PMID: 24171674.
59. Powner MW, Sutherland JD. Prebiotic chemistry: a new modus operandi. Philos Trans R Soc Lond Ser B Biol Sci. 2011;366(1580):2870–7. https://doi.org/10.1098/rstb.2011.0134. PubMed PMID: 21930577; PMCID: PMC3158916.
60. Xu J, Green NJ, Gibard C, Krishnamurthy R, Sutherland JD. Prebiotic phosphorylation of 2-thiouridine provides either nucleotides or DNA building blocks via photoreduction. Nat Chem. 2019;11(5):457–62. Epub 20190401. https://doi.org/10.1038/s41557-019-0225-x. PubMed PMID: 30936523; PMCID: PMC6597365.

Part II
Bioenergetics in Modern High-Energy Mitochondria

Overview of Part II: Through chemical evolution, the atoms created during and after the Big Bang interacted with each other depending on their bond energies, the presence of light and heat, and the environment and circumstances in which the reactions proceeded. Chemiosmosis is postulated as the evolutionary mechanism by which respiration and phosphorylation occur; however, it remains unclear exactly how protons can be "pumped" through a membrane. Nevertheless, once photosynthesis and oxidative phosphorylation were adopted by Nature, there existed the machinery to convert light or substrates to a chemical energy source for building molecules and macromolecules. Jumping ahead to the present time, the high-energy cardiac mitochondria of mammalian species are capable of a five to sixfold increase in respiration and ATP production from rest to strenuous exercise. Discussed in detail in Part II are glycolysis, the tricarboxylic acid cycle, the proposed electron transport system driving proton pump system (respiration), and the proton uptake, ATP synthase system (phosphorylation). Although electrons are tightly coupled to their charge carriers during electron transport, O_2 molecules can attract electrons if forward electron transfer is slowed, or if reverse electron transfer occurs, so that the superoxide radical is formed and quickly converted to hydrogen peroxide by superoxide dismutases. Proton uptake into the inner membrane space outside of uptake by ATP synthase uncouples oxidative phosphorylation and produces heat. During acute stress, mitochondrial Ca^{2+} loading can overwhelm the Ca^{2+} buffering system and induce Ca^{2+} and K^+ entry to enhance respiration by proton exchange. Mitochondria communicate with nuclear and sarcoplasmic factors to regulate mitochondrial density, size, fitness, function, and demise.

Chapter 5
Mitochondrial Function and Bioenergetics

Abstract Up to the present time, electron transport enzymes and ATP synthase in the mitochondrial supercomplex, ribosomal machinery, genome function, and overall cellular function have become increasingly complex as cells became diversified and specialized into organs. Not all our genes are known to be necessary for life, but many of the other genes likely provide redundancy or duplicity for ensuring fitness for survival, synergy, repair, replacement, and adaptations for stress or injury, among other actions. The number of nuclear genes regulating mitochondrial function is now very high. The basic chemiosmotic mechanism is probably highly conserved but many added mitochondrial enzymes now facilitate the processes of glycolysis, carbon turnover in the tricarboxylic acid cycle, the electron transport system, and ATP synthesis. The details of these reactions in modern cardiac mitochondria are described here and utilize equations to understand the transfer of protons and electrons with metabolism from substrates and O_2 to water and carbon dioxide. Photosynthetic bacteria, algae, and plants utilize light for chemiosmosis to make ATP and sugars as substrates for all non-photosynthesizing animals. In glycolysis (fermentation) sugars are broken down to reduce a universally established cofactor, the nicotinamide-adenine-dinucleotide complex, and to make substrate-level ATP; further metabolism by chemiosmosis and oxidative phosphorylation produces much more ATP. Other universal and important cofactors that transfer electrons and protons in a continuous cycle are nicotinamide-adenine-dinucleotide-phosphate and the flavoenzymes. Mobile electron carriers (ubiquinones) within the respiratory complexes I, III, and IV promote proton pumping and sequentially begin with a low affinity for electrons with the highest affinity for electrons during the final reduction of O_2 to water. Proton pumping provides for high mitochondrial redox, pH, and membrane charge potentials. ATP consumption has a tendency to lower these potentials during re-phosphorylation at complex V because of proton reentry, which then stimulates respiration (oxidation) to restore the potentials to basal levels. Because of the variability of ATP demand, as in exercise vs. rest, cardiac mitochondria are well adapted to rapidly reset the level of bioenergetic activity. Bioenergetics can be studied in isolated hearts and in isolated mitochondria using ion-selective fluorescence dyes, ADP, substrates, activators, and inhibitors.

D. F. Stowe, *Evolution of Bioenergetics from Elements to Life*,
https://doi.org/10.1007/978-3-031-76668-8_5

Keywords Anaplerosis · Antioxidants · Bioenergetics · Biosynthesis · Cardiac mitochondria · Cataplerosis · Chemiosmosis · Coenzymes · Electron transfer system · Electron tunneling · Energy transfer cofactors · Eukaryotes · Evolutionary adaptions · Fermentation · Glycolysis · Inner mitochondrial membrane · Inner mitochondrial membrane space · Metal catalysts · Metal cofactors · Mitochondrial fission · Mitochondrial fusion · Mitochondrial measurements · Mitochondrial membrane potential · Mitochondrial supercomplex · Mobile electron carriers · Natural selection · Oxidative phosphorylation · Pentose phosphate pathway · Post-translational modifications · Prokaryotes · Proton chemical gradient · Proton motive force · Purine pathway · Reduction potential · Quinones · Redox homeostasis · Redox pairs · Redox salvage pathways · Ribosomes · Ribozymes · Semi-ubiquinone radical · Substrate level phosphorylation · Tricarboxylic acid cycle · Ubiquinones

5.1 Electron and Proton (H^+) Transport in Glycolysis and the Tricarboxylic Acid Cycle

Chapters 1, 2, 3, and 4 trace the origins of elements, atoms, ions, molecules, and macromolecules that eventually led to membranes, glycolysis, the TCA cycle, chemiosmosis, and high-energy bonds in living cells. In evolution, traits of a given species are selected that might offer an advantage to the prodigy of that species, either by DNA mutations or epigenetics (if it occurs in mammals), or long-term changes in the environment could exert genetic pressure for one species to evolve into another species. Some of these processes became packaged for better efficiency during the evolution of mitochondria, but that was not the end of the evolution of bioenergetics.

This chapter and Chaps. 6, 7, 8 and 9 are not about chemical evolution and the origins of chemiosmosis and lifeforms but about bioenergetics and function in cardiac mitochondria, a present-day product of evolution. These Sections take the reader to our current understanding of cellular bioenergetics since the structure and function of the mitochondrion and its cell have become much more complex over the last Byr, or so. Many adaptions have been made in mitochondria since their beginning. In addition to maintaining the basic machinery of OxPhos, initially using metal catalysts and co-factors, multiple complex, convergent, and divergent pathways have developed that regulate mitochondria and their interactions with all cellular activities. Over time mitochondria have imported more than a thousand proteins encoded by the nuclear genome to carry out these processes.

Evolutionary Adaptions Many adaptions involve selection of genes encoding enzymes that make reactions more favorable or efficient, often accompanied by post-translational modifications to the enzymes giving them greater utility. If 10–30% of an animal species genes are deemed essential for life, what is the utility of much or the remaining genes? It is probable that some genes may no longer encode a functional protein, but many of the genes may have led to proteins and enzymes that are only occasionally useful, as in a change in environment, a disease process, an injury, or to override a deleterious gene. The primordial ribozyme became encased by 50 or so enzymes that support translation in the ribosome (Fig. 3.30). For mitochondria, the complexity of ETS enzymes, ATP synthase, and others likely convey selective advantages for fitness, e.g., responsiveness, repair, replacement, fission, fusion) due to adaptions that provide for redundancy, versatility, synergy, and others. Many of these complex actions and interactions of mitochondria are described next.

The bioenergetic machinery is similar in all eukaryotic cells of a species. Adult mammalian muscle and nerve cells do not normally divide but they are the greatest consumers of energy because of their very high energy needs for sensing, coordination, and locomotion. Throughout evolution thousands of enzymes were developed to aid in biochemical reactions, while redundant and protective pathways were adapted to backup primary pathways and to protect against stress and or to repair damage to cell constituents and tissues. This chapter (Sects. 5.1, 5.2 and 5.3) first covers the bioenergetic machinery of present-day mitochondria in cardiomyocytes. The principles apply to cells in any organ and across any mammalian or avian species.

NAD^+/NADH As introduced in Sects. 3.2 and 3.3, reduced nicotinamide adenine dinucleotide (NADH) (Fig. 3.24) is a universal cofactor required for energy transfer from reduced substrates (nutrients) to make ATP [1–3]. NAD^+, in its reduced form NADH, acts as a cofactor in multiple reduction-oxidation (redox) reactions related to energy production, glycolysis, the tricarboxylic acid (TCA) cycle, OxPhos, fatty acid oxidation, and other reactions. NAD^+ is kept in equilibrium in part by biosynthesis and consumption, but mostly by recycling [1, 3–7], as detailed in Fig. 5.1 [4]. The role of NAD^+/NADH in maintaining intracellular redox homeostasis is a critical factor in all lifeforms. Figure 5.2 [4] shows how the NAD^+/NADH and $NADP^+$/NADPH redox pairs are at the center of all redox reactions [4].

As an *e-* carrier pair, NAD^+/NADH is not consumed but is continually recycled in its important redox function mode. NAD^+ is synthesized in the cytosol and in organelles by linking adenosine monophosphate (adenosine + ribose, + phosphate, i.e., AMP) to nicotinamide mononucleotide (nicotinamide + ribose + phosphate, NMN) (Fig. 3.24) [3]. The precursor, niacin (nicotinic acid and nicotinamide, vitamin B_3), is mostly obtained in the diet but it is also synthesized in plants and animals from the amino acid tryptophan. As a coenzyme, NAD^+ oxidizes reduced reactants (RH_2), like glucose, and reduces oxidized NAD^+ to NADH ($RH_2 + NAD^+ \rightarrow NADH + H^+ + R$). Because NAD^+ participates in so many reactions it is efficiently recycled via these salvage pathways from nicotinamide (NAM), a byproduct of NAD^+ consuming reactions. Figure 5.3 [4] furnishes a composite description of recycling of the NAD^+ redox pool, the pathway for NAD^+ biosynthesis, and the degradation pathways for NAD^+.

NAD^+ functions as a link between metabolism and signaling in many physiological processes. The main source of NAD^+ in this role is called the salvage pathway because the catabolic product NAM from the above NAD^+-consumed enzymes is

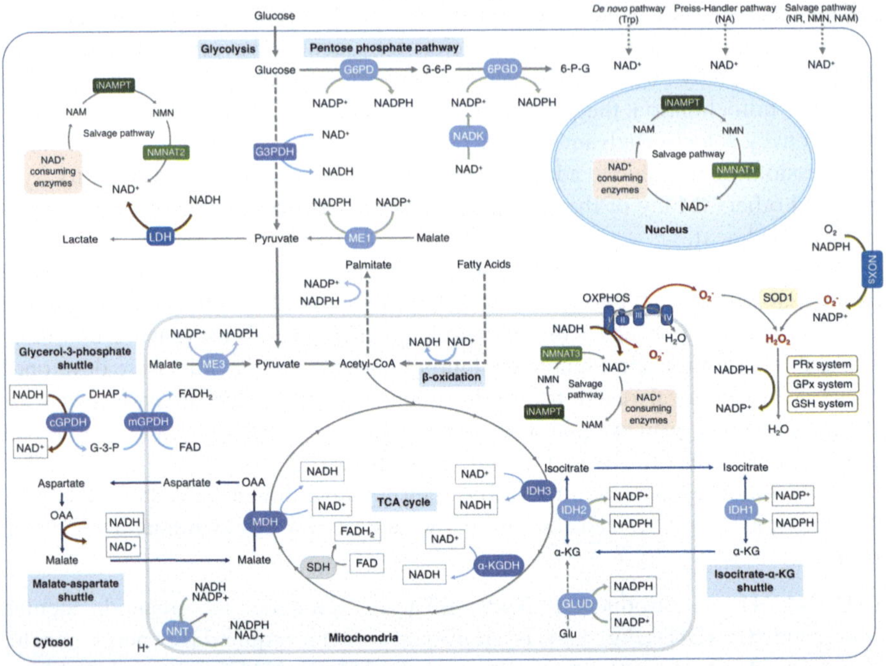

Fig. 5.1 Subcellular equilibrium of NAD+: NAD+ homeostasis is maintained by biosynthesis, consumption, and recycling in different subcellular compartments including cytosol, nucleus, and mitochondria. NAD+ precursors including Trp, NA, NR, NMN and NAM, are metabolized into NAD+ via Preiss-Handler, de novo, and salvage pathways, respectively. NAD+ can receive hydride to yield the reduced form NADH in the metabolic processes including glycolysis, FAO, and the TCA cycle. NADH provides an electron pair to drive mitochondrial OxPhos for generating ATP and converting lactic acid to pyruvate. Cytosolic and mitochondrial NADH is exchanged through malate-aspartate and glycerol-3-phosphate shuttles, while cytosolic and mitochondrial NADPH is exchanged by the isocitrate-α-KG shuttle. NAD+ can also be phosphorylated into NADP+ by NAD+ kinases including nicotinamide nucleotide transhydrogenase (NNT) and NAD kinases (NADKs). Cytosolic NADP+ is reduced into NADPH by G6PD and 6PGD in the pentose phosphate pathway, and by ME1 in the conversion of malate to pyruvate. Mitochondrial NADPH is produced by IDH2, GLUD, NNT and ME3. The NADPH is required for activating NOXs and synthesizing palmitate. Abbreviations: *FAO* fatty acid oxidation, *α-KGDH* alpha-ketoglutarate dehydrogenase, *GLUD* glutamate dehydrogenase, *NNT* nicotinamide nucleotide transhydrogenase, *G3PDH* glyceraldehyde 3-phosphate dehydrogenase, *6PGD* 6-phosphogluconate dehydrogenase, *G6PD* glucose-6-phosphate dehydrogenase, *GPx* glutathione peroxidases, *IDH1/2* isocitrate dehydrogenase 1 and 2, *MDH* malate dehydrogenase, *ME1/3* malic enzyme, *NADK* NAD+ kinase, *NOXs* NADPH oxidases, *OXPHOS* oxidative phosphorylation, *PPP* pentose phosphate pathway, *PRx* peroxiredoxin, *SDH* succinate dehydrogenase, *SOD1–3* superoxide dismutase type 1–3, *TCA cycle* tricarboxylic acid cycle, *GSH* Glutathione, *LDH* Lactate dehydrogenase. (Cartoon with schemes (Fig. 2) and its modified legend resourced with license (https://creativecommons.org/licenses/by/4.0/) from Xie et al. [4] (review))

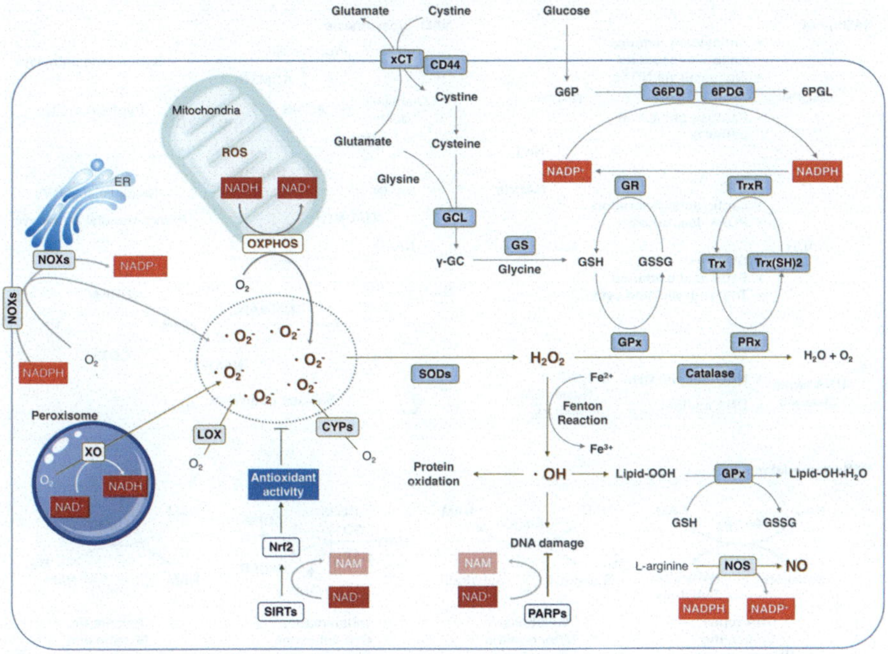

Fig. 5.2 NAD^+ metabolism controls redox homeostasis: ROS could be produced either from metabolic reactions in mitochondria, such as OxPhos, or from a range of cytosolic enzymes, including NOXs, XO, LOX, CYPs, all of which need the NADH/NADPH serving as electron donor. To maintain redox homeostasis, both enzymatic and non-enzymatic antioxidant system components exhibit their effects in coordination with each other to contract with the ROS. GSH, the most abundant of non-enzymatic antioxidants, is synthesized from glutamate, cysteine, and glycine catalyzed by two consecutive cytosolic enzymes, GCL and GS. Importantly, NADPH serves as the reductive power for ROS-detoxifying enzymes including glutathione reductases (GR) and thioredoxin reductases (TrxR) to maintain the reduced forms of GSH and Trx (SH)2 in response to ROS produced from mitochondria or NOXs. Abbreviations: *6PGD* 6-phosphogluconate dehydrogenase, *CYPs* Cytochromes P450, *G6PD* glucose-6-phosphate dehydrogenase, *GCL*, *GR* glutathione reductases, *GS*, *LOX*, *NAD* nicotinamide adenine dinucleotide, *NOXs, NADPH oxidases*, *NADPH* nicotinamide adenine dinucleotide phosphate, *OXPHOS* oxidative phosphorylation, *PRx* peroxiredoxin, *GPx* glutathione peroxidases, *SOD1/2* superoxide dismutase 1 and 2, *Trx* thioredoxin, *TrxR* thioredoxin reductases, *XO* xanthine oxidase. (Cartoon with schemes (Fig. 3) and its legend resourced with license (https://creativecommons.org/licenses/by/4.0/) from Xie et al. [4])

recycled [2–4] to reconstitute NAD^+ as depicted in Fig. 5.4 [3]. NAD^+ is 600- to 1100-fold more abundant in cells than is NADH because it participates in multiple molecular, non-redox processes, including DNA repair, cell signaling, post-translational modifications, mitochondrial metabolism, inflammatory responses, senescence, and apoptosis [3, 4, 6–11]. The non-redox role of NAD^+ in aging and metabolic diseases is a subject of much current research [12].

In these non-redox reactions, NAD^+ is consumed as a substrate by different groups of enzymes, including sirtuins, which remove chemical groups and poly(ADP-ribosyl) polymerases (PARPs), which mediate DNA repair [13]. NAD^+

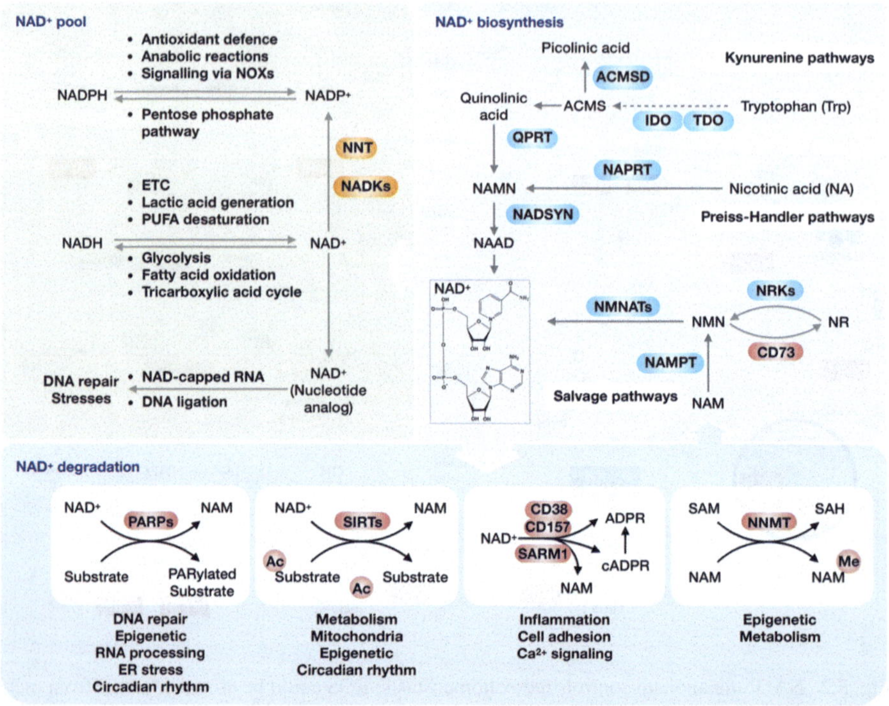

Fig. 5.3 Overview of the NAD+ metabolism and its physiological function: Mammalian cells can synthesize NAD+ de novo from tryptophan by the kynurenine pathway or from NA by the Preiss-Handler pathway, while most NAD+ is recycled via salvage pathways from nicotinamide (NAM), a by-production of NAD+-consuming reactions. NAD+ can be reduced into NADH in metabolic processes, including glycolysis, fatty acid oxidation and the TCA cycle. NADH, in turn, drives generation of ATP via OxPhos, production of lactic acid from pyruvate, and desaturation of poly unsaturated fatty acids (PUFAs). NADPH plays an essential role in antioxidant defense and regulates cellular signaling via NADPH oxidases (NOXs). Moreover, NAD+ is found to decorate various RNAs in different organisms as nucleotide analog and serves as an alternative adenylation donor for DNA ligation in NHEJ repair. NAD+ also acts as a co-substrate for a wide variety of enzymes, including PARPs, sirtuins, CD38/CD157 and SARM1and so impacts metabolism, genomic stability, gene expression, inflammation, circadian rhythm, and stress resistance. Using NAD+ as a co-substrate, both PARPs and sirtuins regulate their target molecules, generating NAM as a by-product. The CD38/CD157 and SARM1 also catalyze NAD+ to NAM, producing ADPR and cADPR. Abbreviations: *IDOs* indoleamine 2,3-dioxygenase, *QA* quinolinic acid, *NAMN* nicotinate mononucleotide, *QPRT* quinolinate phosphoribosyl-transferase, *NAPRT* nicotinic acid phosphoribosyl-transferase, *NMNATs* nicotinamide mononucleotide adenylyl transferases, *NADSYN* NAD synthase, *NR* nicotinamide riboside, *Trp* tryptophan, *NADKs* NAD+ kinases, *PARPs* poly (ADP-ribose) polymerases, *NNT* nicotinamide nucleotide transhydrogenase, *TDO* tryptophan 2,3-dioxygenase, *SARM1* sterile alpha and TIR motif containing 1, *NNMT* Nicotinamide N-methyltransferase, *NMN* nicotinamide mononucleotide, *PUFAs* polyunsaturated fatty acids, *NAM* nicotinamide. ICartoon with schemes (Fig. 1) and its legend resourced with license (https://creativecommons.org/licenses/by/4.0/) from Xie et al. [4] (review))

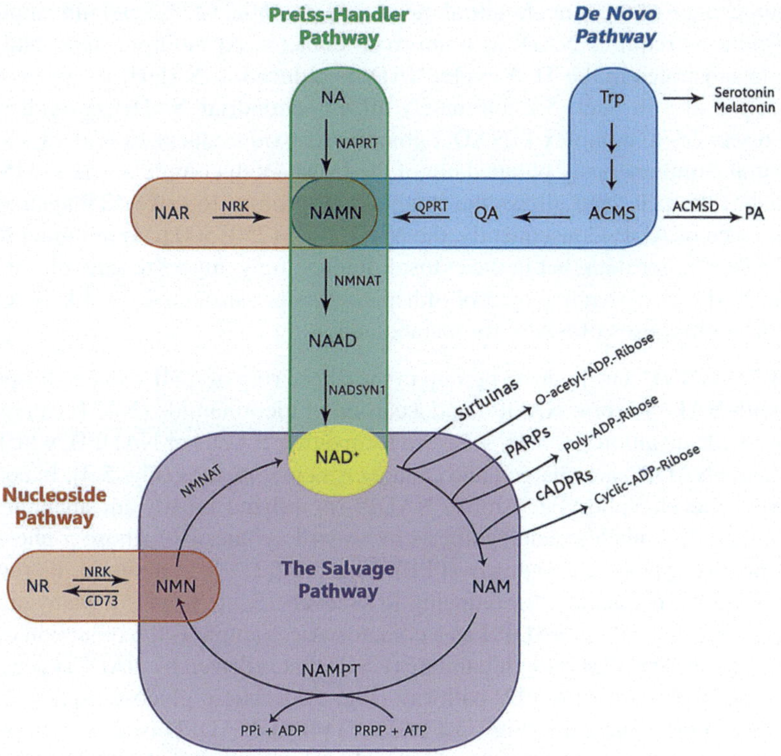

Fig. 5.4 Simplified version of NAD+ biosynthetic pathways: NAD+ can be synthesized from different dietary precursors by the de novo pathway (from Trp), the Preiss–Handler pathway (from NA), and the nucleoside pathway (from nicotinic NAR or NR). However, the main source of NAD+ is the salvage pathway where the catabolic product (NAM), from NAD+-consumed enzymes (sirtuins, PARPs and cADPRSs), is recycled to reconstitute NAD+. Refer to source for abbreviations. (Cartoon (Fig. 3) and its legend resourced with license (https://creativecommons.org/licenses/by/4.0/) from Navas and Carnero [3] (review))

is consumed during transcriptional regulation and metabolism by transferring long chains and branches of ADP-ribosyl polymers to their protein substrates, and by the cyclic ADP-ribose synthases (cADPRS, e.g., CD38 and CD157), a family of ecto-enzymes that use NAD+ as a donator of ADP-ribose, with release of NAM as the main recycling reaction product [13].

In the redox role of NAD+, the slightly more electronegative N (position 1), shown by the purple colored N+ of Fig. 3.24 in the 6-member ring connected to ribose, is "+" charged (missing a valence *e*- for stability) and so functions as a sink for 1*e*- from the hydride ion, while the hydride ion and the other *e*- bind to the number 4 C of the ring [14]. When NADH reduces another molecule, it is re-oxidized to NAD+; this is essential for stripping H atoms and *e*- from additional substrates in a continuous cycle. Both a hydride ion (H^-) and a proton (H^+) are removed from a substrate by NAD+ with the proton released into solution.

Maintenance of the mitochondrial membrane potential ($\Delta\Psi_m$) and mitochondrial ATP synthesis requires NAD^+ as a universal cofactor. At multiple steps, and with multiple substrates in the TCA cycle, NAD^+ is reduced to NADH. As discussed in this section and in Sect. 5.2, ultimately all mitochondrial NADH is oxidized by donating its 2*e*- to complex I (NADH ubiquinone oxidoreductase) of the ETS with the proton simultaneously pumped out of the IMM. With complexes III and IV, this establishes the $\Delta\Psi_m$ and allows the re-uptake of protons to drive ATP synthesis in the presence of ADP. Consequently, the NAD^+ pool (250–500 μM) in mitochondria is 3–5 times larger than that in the cytosol. Interestingly, myocyte sarcoplasm has a smaller NAD^+ pool than the cytosol of hepatocytes or astrocytes; this likely reflects their differential uses of NAD^+ for metabolism.

$NADP^+$/NADPH The hydroxyl group in the ribose ring of AMP can be phosphorylated with NAD^+ to form $NADP^+$, and a carbon of nicotinamide (NAM) can readily accept a hydride anion ($H^- = H^+ + 2e$-) to form either NADH or NADPH, which like NAD^+, and $NADP^+$, are charged and cannot cross membranes (Fig. 5.3). In contrast to NAD^+, the phosphorylated form, $NADP^+$, is utilized mostly for anabolic processes, including nucleic acid synthesis by way of synthesis of ribose 5-phosphate in the pentose phosphate pathway (PPP) (Fig. 5.1) [3, 9]. In addition, its reduced form, NADPH, is essential for reducing ROS levels (Sect. 8.4) [15]. Many antioxidant proteins depend on NADPH as a cofactor. An example is the reduction of glutathione disulfide (GSSG) to glutathione (GSH) that is driven by NADPH generated during the first step of the PPP pathway (Fig. 5.1), where glucose-6-phosphate is converted to 6-P-gluconate (Figs. 5.1 and 5.2) [4, 9]. NADPH is also an important cofactor in the synthesis of fatty acids, which utilize acetyl-CoA as a precursor for membrane synthesis in proliferating cells [15].

FMN, FAD Flavoenzymes (Fig. 5.5) [16], which contain flavin mononucleotide (FMN, flavin + ribose + phosphate + adenosine) or flavin adenine dinucleotide (FAD, flavin + ribose + 2 phosphates + adenosine) as prosthetic groups, catalyze many of the one and two *e*- oxidation/reduction reactions critical to aerobic respiration. Flavoproteins are mainly located in mitochondria because of their redox power. Examples are FMN in complex I and FAD in complex II of the ETS. FMN and FAD carry out redox reactions similarly to NAD^+, one difference being that two protons are accepted by FMN or FAD instead of one hydride ion by NAD^+ (but 2 e- are accepted by both redox systems) and another is that FAD has a more positive reduction potential than NAD^+ and so is a stronger oxidizing agent. In eukaryotes, the precursor riboflavin (vitamin B_2) must be obtained in the diet, but all plants and most microorganisms are capable of synthesizing riboflavin from two major precursors: ribulose 5-phosphate (Ribu5P; pentose phosphate pathway) and guanosine triphosphate (GTP; purine pathway) [17, 18].

The flavins of FMN and FAD are three ring structures (isoalloxazines) (Fig. 5.5) tightly bound to their enzymes. Each prosthetic group first donates a single *e*- and a proton to form the semiquinone radical intermediates, $FMN^\bullet$ and $FAD^\bullet$, and then donates another *e*- and a proton to fully reduce the cofactors to $FMNH_2$ and $FADH_2$.

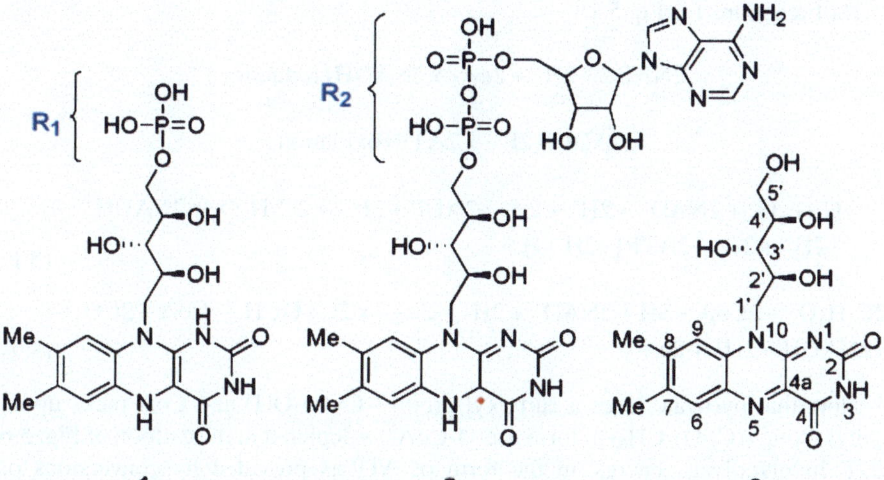

Fig. 5.5 Structures of FMN and FAD: FMN (1) and FAD (2) differ from riboflavin (3) in their ribityl side chains, which contain a phosphate residue (R1) or an ADP moiety (R2) in ester linkage with its terminal hydroxyl group, respectively. The isoalloxazine ring of flavins can exist in any one of three oxidation states. FMN is shown in the two-electron-reduced hydroquinone state, FAD as a one-electron-reduced neutral semiquinone (red dot on FAD ring), and riboflavin in its oxidized form. (Structures (Scheme 1) and legend resourced with permission (copyright Am. Chem. Soc., 2024) from Mansoorabadi et al. [16] (perspective))

FAD-dependent proteins function in many metabolic pathways including the ETS, DNA repair, nucleotide biosynthesis, β-oxidation of fatty acids, amino acid catabolism, and synthesis of other cofactors such as co-enzyme A (CoA) (Figs. 1.16 and 3.8), co-enzyme Q_{10} (CoQ) (Sects 5.3 and 6.1), and heme groups [16]. Like NADH and NADPH, $FMNH_2$ and $FADH_2$ donate their *e*- and protons to other substrates, leaving FMN and FAD to complete the redox cycle. Details of the three metabolic processes of substrate oxidation, i.e., glycolysis, TCA cycle, and electron and proton transfer, are discussed next.

Glycolysis The flow (transfer) of *e*- in redox reactions is responsible for all of the work done by living organisms [19]. The source of *e*- are reduced food stuffs, e.g., glucose, amino acids, or free fatty acids (FFA) from dietary sugars, proteins, and fats, which through glycolysis in the cytosol, or β-oxidation in the mitochondrial matrix (Fig. 5.1), forms acetyl-CoA [20]. Pyruvate is carried into the matrix from the cytosol by the pyruvate mitochondrial carrier (PMC) [21]. During glycolysis alone, net oxidation of glucose to pyruvic acid (Eq. 5.1), and on to acetyl-CoA (Eq. 5.2), each acetyl-CoA transfers 1 *e*- and 1 H^+ to 2 NAD^+ to form 2 NADH. Note that in most subsequent balanced reactions, *e*- and or H^+ are not cancelled out but left in to show their transfers during redox reactions, i.e., from reactants to products:

$$C_6H_{12}O_6 \rightarrow 2C_3H_4O_3 + 2H^+ + 2e - + (2H_2O)(\text{oxidation})$$

Half reactions for Eq. 5.1:

$$2\text{NAD}^+ + 2\text{H}^+ + 2e- \rightarrow 2\text{NADH}(\text{reduction})$$

$$2\text{ADP} + 2\text{P}_\text{i} \rightarrow 2\text{ATP}(\text{oxidation})$$

$$\text{C}_6\text{H}_{12}\text{O}_6 + 2\text{NAD}^+ + 2\text{H}^+ + 2e - + 2\text{ADP} + 2\text{P}_\text{i} \rightarrow 2\text{C}_3\text{H}_4\text{O}_3 + 2\text{NADH} + 2\text{H}^+ + 2e - + 2\text{ATP}(+2\text{H}_2\text{O}) \quad (5.1)$$

$$2\text{C}_3\text{H}_4\text{O}_3 + 2\text{CoA} - \text{SH} + 2\text{NAD}^+ + 2\text{H}^+ + 2e- \rightarrow 2\text{C=OCH}_3 - \text{CoA} + 2\text{CO}_2 + 2\text{NADH} + 2\text{H}^+ + 2e- \quad (5.2)$$

Note that pyruvate loses a carboxyl group ((C=O)-OH) and CoA picks up an acetyl group ((C=O)-CH_3) to form acetyl-CoA, as depicted in the cartoon of Fig. 5.6 [22]. In glycolysis, energy in the form of ATP is provided by conversions of 1,3-bisphosphoglycerate to 3-phosphoglycerate (by phosphoglycerate kinase) and phosphoenolpyruvate to pyruvate (by pyruvate kinase), and in the TCA cycle (Fig. 5.7) [23] by the reaction of succinyl-CoA to succinate and CoA-SH (by succinyl coenzyme A synthetase) [20]. In H_2O with a reactant of lower or higher energy, the reversible reaction is (Eq. 5.3):

$$\text{ATP}^{4-} + \text{H}_2\text{O} \rightleftharpoons \text{ADP}^{3-} + \text{HPO}_4^{\ 2-} + \text{H}^+ \quad (5.3)$$

The removal of one P_i by hydrolysis from high-energy ATP yields a Gibbs free energy of $\Delta G^{o\prime} = -57$ KJ/mol (14 kcal/mol) in living organisms [24], but the

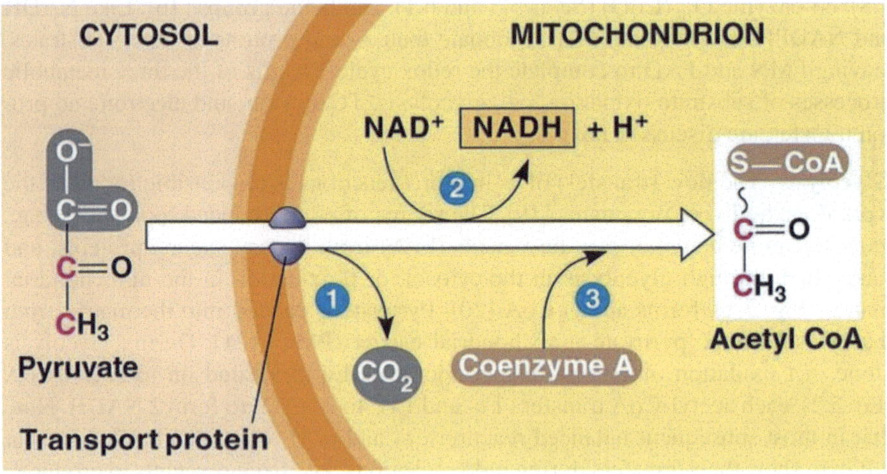

Fig. 5.6 Transfer of carboxyl group from pyruvate to co-enzyme A. (Cartoon resourced with license (https://creativecommons.org/licenses/by-nc-sa/4.0) from LIbreTexts and credited to Pearson Education, Inc., publishing as Pearson Benjamin Cummings [22])

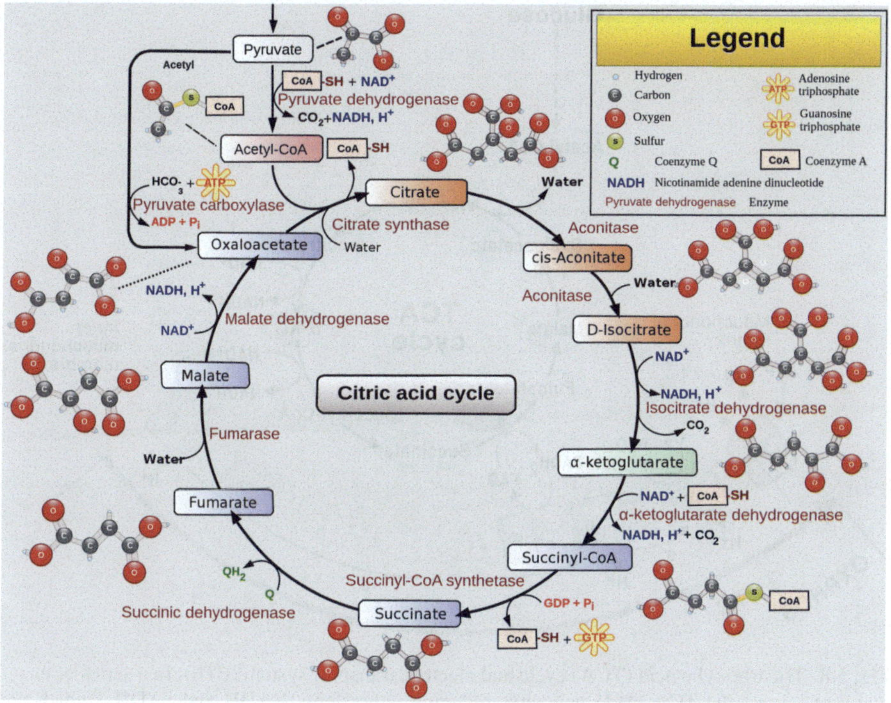

Fig. 5.7 Reactants of tricarboxylic acid (TCA) cycle: The TCA cycle, a.k.a. citric acid cycle, describes most of the reactants, chemical structures, enzymes, co-factors, and sites absorbed and released energy. Note 2-C acetyl-CoA, 4-C oxaloacetate, and 6-C citrate; also note loss of 1-C at α-ketoglutarate and loss of 1-C at succinyl-CoA. ((https://creativecommons.org/licenses/by-sa/4.0/) Diagram with structures resourced from Wikipedia Commons: https://upload.wikimedia.org/wikipedia/commons/thumb/0/0b/Citric_acid_cycle_with_aconitate_2.svg/2560px-Citric_acid_cycle_with_aconitate_2.svg.png)

negative free energy actually comes from the phosphorylation of the reactant by ATP since the reactant would have bonds that have lower energy when not phosphorylated; this energy is utilized to complete a reaction, e.g., in the first step of glycolysis, ATP phosphorylates (energizes) glucose to make glucose 6-P_i + ADP and the energy acquired by the glucose bond with P_i is used to make fructose 6-P_i.

TCA Cycle Figure 5.7 depicts the reactants, chemical structures, enzymes, cofactors, and released energy of the tricarboxylic acid (TCA) cycle. Note how the 6-C citrate is oxidized to the 4-C oxaloacetate (OAA), and then reduced back to citrate with the 2-Cs of acetyl-CoA. Therefore, a functioning TCA cycle requires a continuous pool of acetyl-CoA and a supply of TCA cycle intermediates to synthesize oxaloacetate (Fig. 5.8) [25]. Specifically, cytosolic acetyl CoA moves from the cytosol to the matrix via the citrate shuttle where citrate is oxidized to oxaloacetate, and with acetyl-CoA, to citrate by citrate synthase. During the TCA cycle, the 6-C substrate (citrate) loses 2-C in two steps and after four more steps the 4-C product

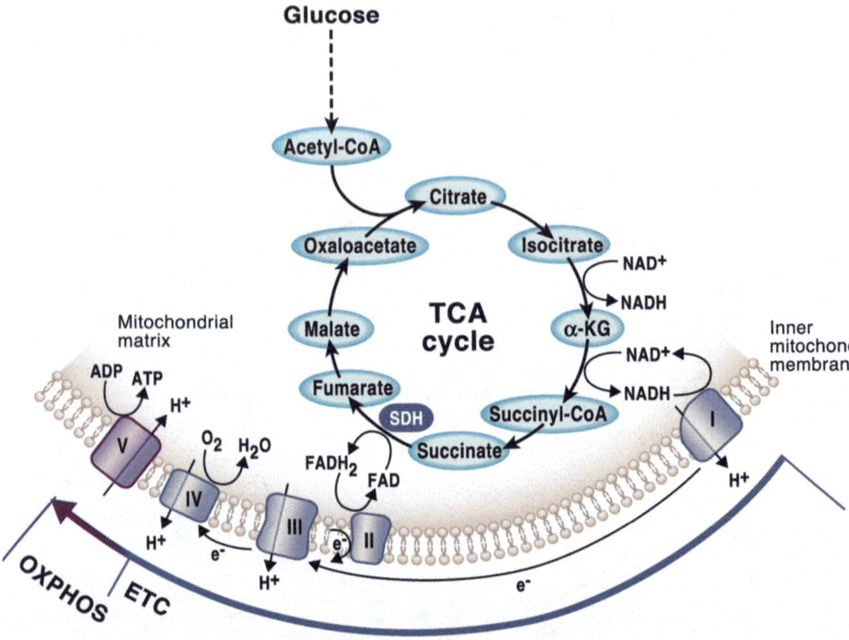

Fig. 5.8 Tricarboxylic acid (TCA) cycle and electron transport system (ETS): In a series of enzymatic reactions, the TCA cycle generates reducing equivalents NADH and $FADH_2$, which are required to transfer electrons to the mitochondrial respiratory chain, also known as the ETS. As electrons are funneled through respiratory complexes in the IMM, a functional ETS generates a mitochondrial membrane potential ($\Delta\Psi_m$) that is used to produce ATP. This process requires the presence of O_2 and is known as oxidative phosphorylation (OxPhos). Mitochondrial complex I and II in the ETS replenish NAD^+ and FAD, respectively, allowing the oxidative TCA cycle to function. Succinate dehydrogenase (SDH) is the only enzyme that participates in both the TCA cycle and the ETS. (Diagram (Fig. 3) and its legend resourced with license (https://creativecommons.org/licenses/by/4.0/) from Martinez-Reyes and Chandel [23] (review))

(oxaloacetate) again combines with acetyl-CoA to continue the cycle. The net reaction (Eq. 5.4) for two cycles (see Fig. 5.7) is:

$$\begin{aligned}&2C{=}O-CH_3-CoA+6NAD^++12e-+2FAD+2ADP+2P_i\\&+4H_2O\rightarrow 4CO_2+6NADH+6H^++12e-+2FADH_2+2ATP+2CoA-SH\end{aligned}\tag{5.4}$$

where FAD is reduced to $FADH_2$. The ATP (Eqs. 5.1 and 5.4) made anaerobically during glycolysis and TCA cycling is called substrate level phosphorylation, meaning ATP not made at complex V. Ultimately, only two net moles of ATP are made by these steps. NADH and $FADH_2$ are the two *e*- carriers that supply all *e*- for the electron transfer system (ETS) [4]. TCA intermediates are removed (cataplerosis) to supply substrates (e.g., α-ketoglutarate (αKG) → glutamate) for biosynthetic pathways (amino acids and nucleotides) as shown in Fig. 5.9a [25]. In contrast, amino acids and fatty acids can also be consumed (anaplerosis) by the TCA cycle to

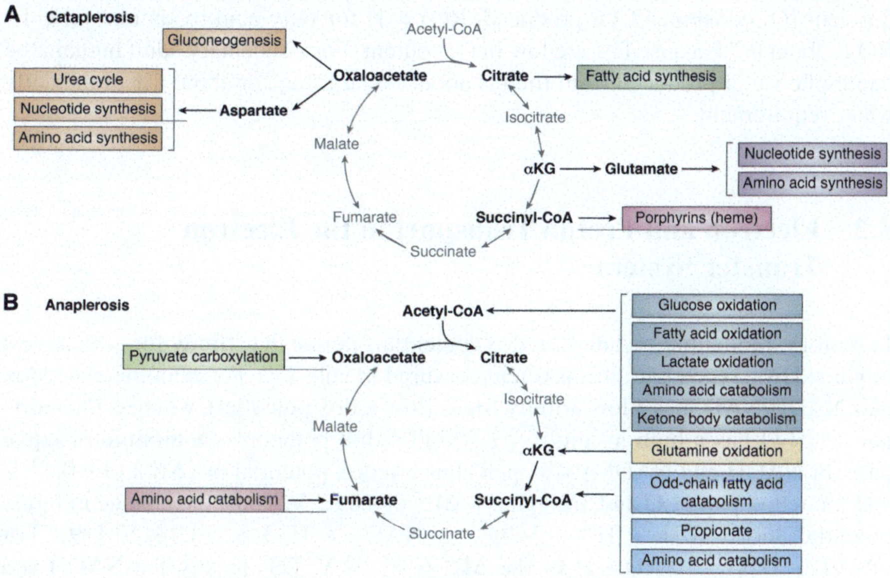

Fig. 5.9 Outputs and inputs into the tricarboxylic acid (TCA) cycle: A functioning TCA cycle requires a continuous pool of acetyl-CoA and supply of TCA cycle intermediates that can be used to synthesize oxaloacetate. (**a**) Removal of TCA cycle intermediates ("cataplerosis") occurs at multiple steps of the cycle to supply precursors for biosynthetic processes or feed into other metabolic pathways. (**b**) Replacement of TCA cycle intermediates ("anaplerosis") is required to support continuous production of oxaloacetate. Sources of anaplerosis are shown in this panel. Several pathways either produce acetyl-CoA directly, or produce pyruvate, an indirect source of acetyl-CoA through the activity of the pyruvate dehydrogenase (PDH) complex. (Diagram (Fig. 3) and its legend resourced with license (https://creativecommons.org/licenses/by/4.0/) from Arnold and Finley [25] (review))

maintain TCA intermediates for metabolism as shown in Fig. 5.9b [25]. For example, glutamate/glutamine and aspartate enter the TCA cycle at the levels of αKG and OAA, respectively [26]. As the *e*- are passed along the ETS with increasing oxidation potentials, they ultimately reduce molecular O_2 to H_2O. NADH and $FADH_2$ are very strong reducing agents. The stoichiometries for high energy NADH entering at complex I, and $FADH_2$ entering at complex II (Eq. 5.5), or from the substrate glucose (Eq. 5.6) through to its complete oxidation to H_2O at complex IV are given as:

$$\begin{aligned}&10\text{NADH}+20e-+10\text{H}^{+}+2\text{FADH}_{2}+4e-+4\text{H}^{+}+6\text{O}_{2}\rightarrow 10\text{NAD}^{+}\\&+20e-+10\text{H}^{+}+2\text{FAD}+4e-+4\text{H}^{+}+12\text{H}_{2}\text{O}\end{aligned}\tag{5.5}$$

$$\begin{aligned}&\text{C}_{6}\text{H}_{12}\text{O}_{6}+6\text{O}_{2}+24\text{NAD}^{+}\left(\text{or equivalent}\right)+24e-\rightarrow 6\text{CO}_{2}\\&+6\text{H}_{2}\text{O}+24\text{NADH}+24e-\end{aligned}\tag{5.6}$$

Simply put, oxidation of one mole of glucose utilizes 6 moles of O_2 and produces 6 moles of CO_2 and H_2O, plus 24 reducing equivalents. Therefore, the respiration quotient (O_2 consumed/CO_2 produced, RQ) = 1; for fatty acid oxidation only, the RQ is about 0.7 because fats are low in O_2 content. For a sedentary adult human, the metabolic water produced from food is about 300 mL/day, or about 8% of the daily water requirement.

5.2 Electron and Proton Transport in the Electron Transfer System

In biology, reduction/oxidation (redox) potentials define the affinity for *e*- to be lost or gained by a redox pair; this is often measured in volts (V). For example, the redox pair NADH/NAD^+ has a low affinity for *e*- (low redox potential), whereas the redox pair O_2/H_2O has a high affinity for *e*- (high redox potential). A mixture of equal parts of NADH and NAD^+ (redox pair) has a redox potential of ($\Delta E^{o\prime}$) of −0.32 V and the redox pair H_2O and 1/2O_2 has a $\Delta E^{o\prime}$ of +0.82 V. Thus, the change in redox potential for NADH + H^+ + 1/2O_2 → NAD^+ + H_2O is +1.14 V [19]. For $FADH_2$ + 1/2O_2 → FAD + H_2O, the $\Delta E^{o\prime}$ is +1.04 V. This means that NADH and $FADH_2$ are strong *e*- donors and that O_2 is a strong *e*- acceptor. In terms of change in $\Delta G^{o\prime}$, these values are −220 and −200 KJ/mol, respectively. These energies are much too large if released all at once as the short burst of energy would be lost quickly as heat.

Thus, within the ETS, reducing equivalents are transported from molecule to molecule (*e*- carriers) in the direction of higher reduction potentials at each step from NADH or $FADH_2$ to H_2O [19]. The energy (Eq. 5.6) released stepwise with transfer of these 24 *e*- (12 *e*- pairs) is tied to pumping H^+ through the IMM at complexes I, III and IV; This is the basis of chemiosmosis described earlier in Sects. 1.3, 2.2, 3.3 and 4.1. Redox potentials (ΔE) across complexes I, III, and IV, the proton motive force (*pmf*, also written as Δp_m), and the net driving force (ΔG) are depicted in Fig. 5.10 [27].

Assessing Redox State The status of the overall redox state can be monitored in isolated hearts by measuring the tissue autofluorescence of NADH/NAD^+ and or $FADH_2$/FAD pairs [28–32] as shown in Fig. 5.11 [29]. The results show that opening of a cardiac K^+ channel by a drug (DCEB) before global ischemia enhances the redox state during ischemia and normalizes it during reperfusion. In another example (Fig. 5.12), brief exposure to the general anesthetic sevoflurane ($C_4H_3F_7O$) (anesthetic preconditioning, APC), or brief ischemia (ischemic preconditioning, IPC) before 30 min global ischemia, improves the NADH redox state (NADH > NAD^+) during reperfusion after ischemia. The higher the NADH, or the lower the FAD (higher $FADH_2$), the more reducing equivalents are available for respiration to proceed. On reperfusion the delivery of new reducing equivalents is incompletely met by *e*- acceptance by O_2 in the ETS.

NADH and FAD can also be measured in isolated mitochondria (Fig. 5.13) to examine the redox state during simultaneous changes in the IMM pH gradient

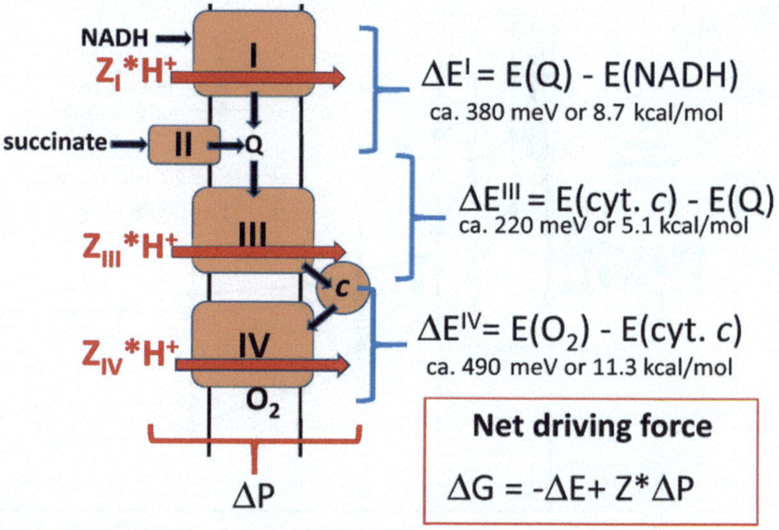

Fig. 5.10 Scheme of *e-* and H^+ transfer with ΔE in the mitochondrial respiratory chain: Complexes I, III, and IV are depicted in the IMM, together with complex II, cytochrome *c* and ubiquinone (Q), NADH and succinate and the acceptor, O_2. The combination of N,N,N′,N′-tetramethyl-phenylene-diamine (TMPD) and ascorbate (not shown for clarity) donates *e-* directly to cytochrome *c* and thus engages complex IV only. Black arrows depict the H^+ and *e-* transfers in the chain, eventually reducing O_2 to H_2O. Red arrows symbolize associated proton motive events where Z_n is the effective proton motive H^+/e- stoichiometry, 2 for complex I and complex IV, and 1 for complex III (see "Methods" in text). Roman numerals are the common names of the complexes. ΔEs are typical average redox potential differences across each of the complexes I, III, IV. The net free energy change (ΔG) for any combination of complexes is the arithmetic sum of the relevant redox potential difference and the proton motive force (Δp_m) multiplied by the H^+/e- stoichiometry (*Z*), where ΔE has a negative sign and is larger than Δp_m for the case where redox chemistry drives H^+ translocation. (Diagram (Fig. 1) and its legend resourced with license (https://creativecommons.org/licenses/by/4.0/) from Wikstrom and Springett [27])

(ΔpH_m), the IMM potential gradient ($\Delta\Psi_m$), m[Ca^{2+}], and mO_2 consumption (mitochondrial respiration) under experimental conditions such as alternative substrates or addition of ADP (state 3 respiration) [33, 34], simulated ischemia, altered ionic conditions, or the presence of respiratory enzyme blockers [35–37]. The plots of Fig. 5.13 demonstrate the effects of the anesthetic isoflurane ($C_3H_2ClF_5O$) on NADH in the presence of complex I and II inhibitors during state 4 (post ADP phosphorylation). The redox state of downstream anti-oxidant systems is ultimately dependent on mitochondrial viability with maintenance of the reduced states of NADH and $FADH_2$ [38]. The exergonic transfer of *e-* is intrinsically coupled to the endergonic pumping of protons across the IMM. As the *e-* are passed along from one donor to another, the associated proteins alter their shape to push protons out of the matrix creating a proton motive force (*pmf* or Δp_m) as depicted in Fig. 5.10.

Electron Carriers Beginning with NADH and $FADH_2$, the four types of *e-* carriers are (1) FMN with 2 Cu atoms that transfers one *e-* at a time, (2) iron sulfur (either

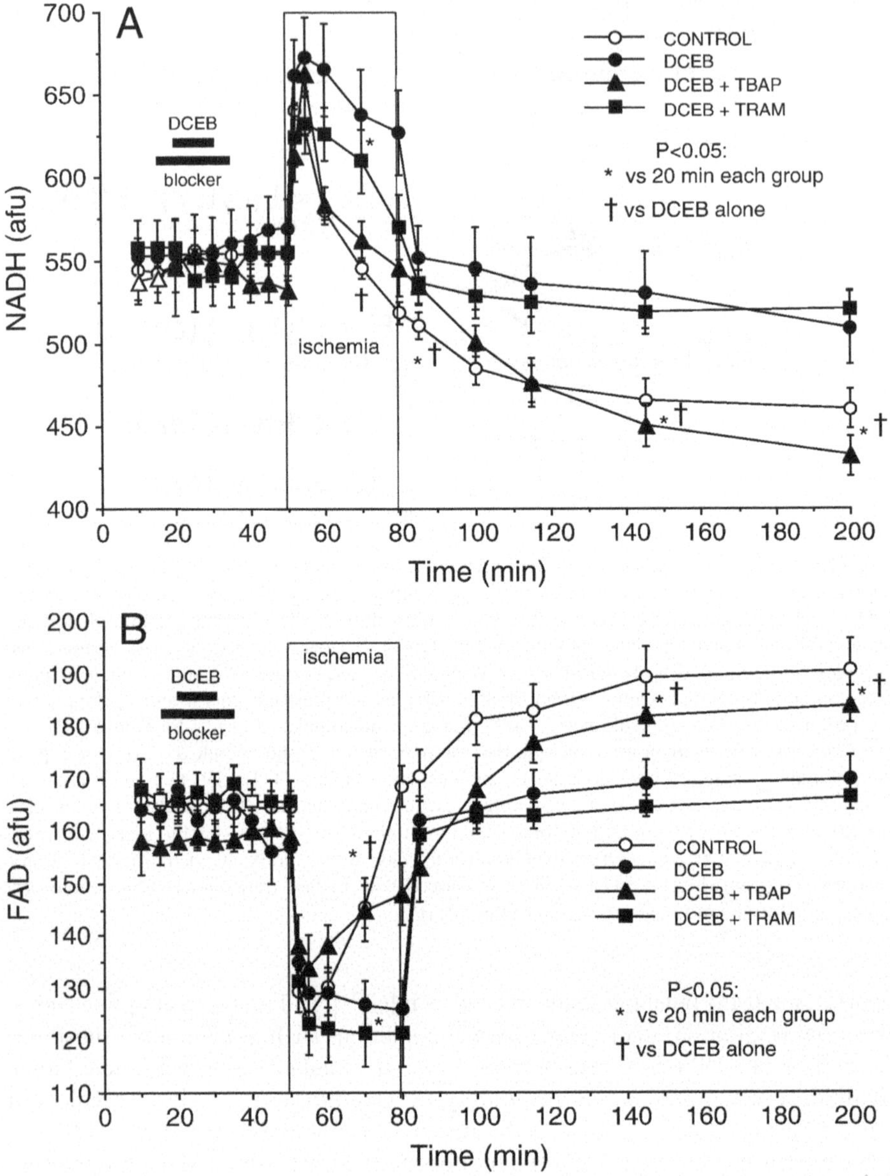

Fig. 5.11 NADH and FAD measurements in isolated hearts: Improved redox state assessed by (**a**) NADH and (**b**) FAD autofluorescence after preconditioning with SK_{Ca} channel activator DCEB. Note the inverse changes in NADH and FAD during ischemia and reperfusion and the more normalized responses in the DCEB and DCEB+TRAM groups. A chemical superoxide ($O_2^{\bullet-}$) scavenger, TBAP, reversed the protective effects DCEB, whereas TRAM (triarylmethane), an antagonist of intermediate conductance K_{Ca} channels did not. (Plot (Fig. 3) and legend resourced with permission from copyright owner of Stowe et al. [29])

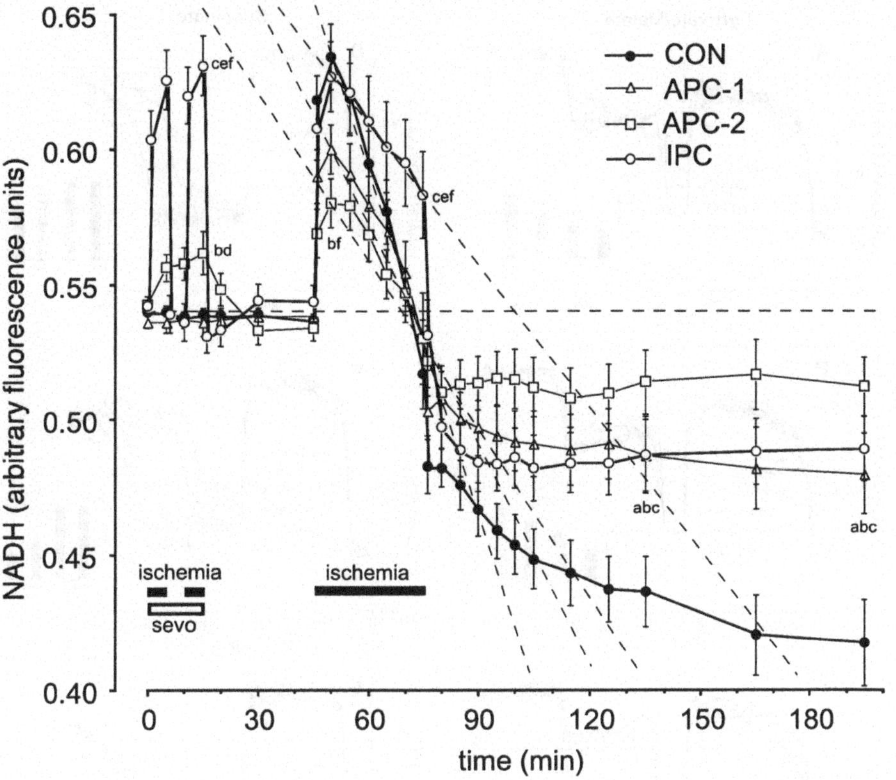

Fig. 5.12 Time course of NADH during cardiac ischemia in isolated hearts: Shown are NADH levels during exposure to anesthetic preconditioning (APC) of 0.5 (APC-1) or 1.3 mM sevoflurane (sevo) (APC-2) for 15 min or two 5-min periods of ischemic preconditioning (IPC) as well as before, during, and after 30 min of global ischemia compared with nonpreconditioned ischemic control (CON) hearts; n = 10 for each group. Preconditioning stimuli in IPC and APC-2, but not APC-1, caused a completely reversible increase in NADH. Note the lower increase in NADH during early ischemia after APC and the slower decline in NADH after IPC or APC during prolonged ischemia. Throughout reperfusion, the deviation in NADH fluorescence from baseline levels was markedly lower in IPC and APC conditioned hearts than in CON hearts. Statistical symbols for $P < 0.05$ are: aAPC-1 vs. CON, bAPC-2 vs. CON, cIPC vs. CON, dAPC-1 vs. APC-2, eAPC-1 vs. IPC, and fAPC-2 vs. IPC. Slopes (dotted lines) show rates of decline in NADH from 5 to 30 min ischemia for each group. (Plot (Fig. 1) and its legend resourced with permission from copyright owner of Riess et al. [28])

2Fe-2S or 4Fe-4S) centers that transfer one *e*- at a time, (3) cytochromes in which Fe, placed at the center of a porphyrin ring (Fig. 2.5), changes its valence from Fe^{3+} to Fe^{2+} when it accepts one *e*- at a time, and (4) ubiquinone (Fig. 4.2) that also has a heme carrier and picks up one H^+ for every *e*- it accepts [19]. The affinity for the *e*- carriers for *e*- is lowest for carrier 1 and highest for carrier 4 (increasing redox potentials) so the transfer of *e*- generally follows this routing [19]. There are excellent reviews with details and diagrams on the mechanisms of the ETS [19, 39–42].

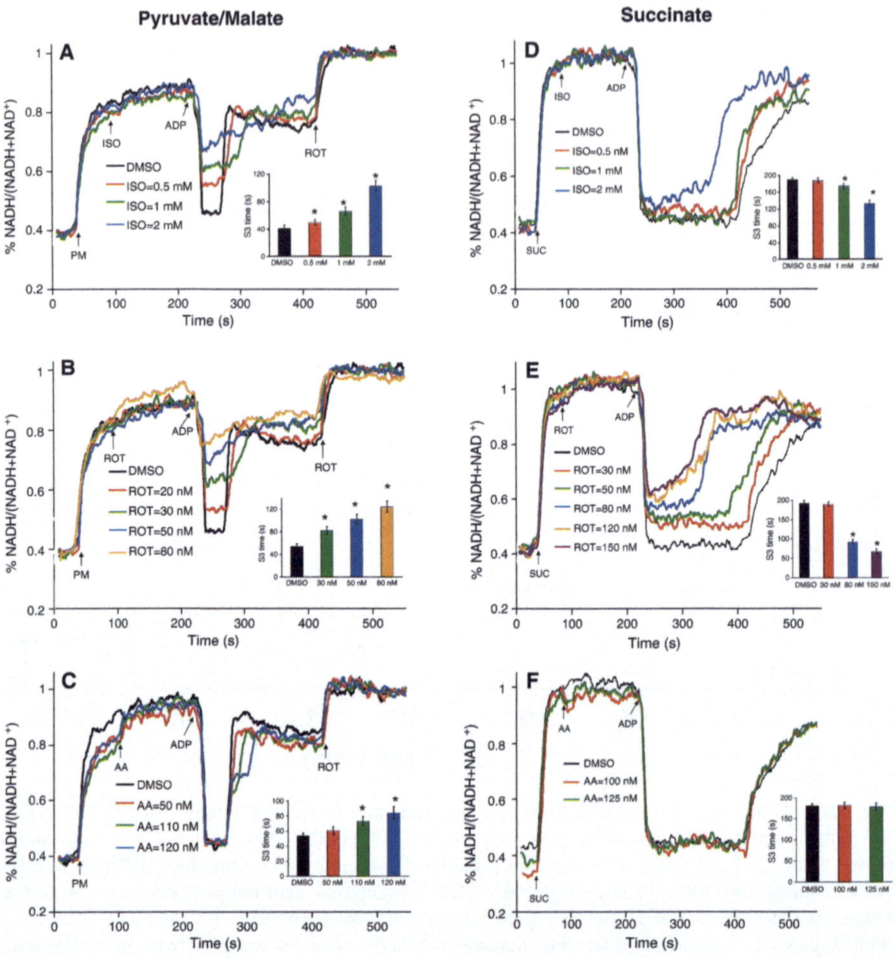

Fig. 5.13 NADH measurements in isolated cardiac mitochondria: Normalized mitochondrial redox state (NADH/NADH + NAD$^+$) measured during states 2, 3, and 4 respiration with different substrates and either the volatile anesthetic isoflurane (ISO) or a blocker of an electron transfer system (ETS) complex. State 2 was induced with 10 mM pyruvate malate (PM) (**a–c**) or succinate (SUC) (**d–f**) and state 3 with 250 μM ADP; state 4 is after phosphorylation of all ADP to ATP. ISO, complex I blocker rotenone (ROT), or complex III blocker antimycin A (AA) was added after PM or SUC and before ADP. ADP induced a transient decrease in the reduced state of NADH. The insets summarize effects of ISO compared to ROT and AA on duration of ADP-induced state 3 NADH oxidation (S3 time, s). Note similarities and differences of responses to each drug. NADH transients were normalized on a scale of 0–1, i.e., between fully oxidized to fully reduced states by adding the mitochondrial uncoupler FCCP or the complex I inhibitor ROT, respectively. Data represent mean ± SD of 3 replicates. (Plots (Fig. 2) and its legend resourced with permission from copyright owner of Agarwal et al. [33])

The enzymes of the respiratory complexes each have different redox potentials for donating and accepting *e*-. The mobile *e*- carriers, cytochrome *c* and ubiquinone, diffuse rapidly in the IMM so that each enzyme complex donates and receives one *e*- about once every 5–20 ms. [19] Importantly, each carrier can only interact with its adjacent carrier to conduct an orderly transfer of *e*-. If an *e*- carrier with a low redox potential were to collide with a *e*- carrier with a high redox potential, a short circuit would occur as *e*- can jump across a gap of up to 2 nm; this is called *e*- tunneling. Short circuits are diminished by insulation provided by carrying the *e*- deep inside the hydrophobic lining of proteins of the respiratory chain [19].

Proton Pumping An essential component of chemiosmosis, proton pumping occurs from the matrix to the intermembrane space (IMS) and establishes a $\Delta\Psi_m$ with an inward negative voltage gradient of approximately −180 mV as well as a mitochondrial H^+ chemical gradient (ΔpH_m) (alkaline inside) of about 0.5–0.7 pH units [36] in the presence of oxidizable substrate and O_2. These two gradients make up the total proton motive force (*pmf*), again also described as Δp_m or $\Delta\mu H^+$, as depicted by proton flux among the respiratory complexes in Fig. 5.14 [43, 44]. H^+ movement across the IMM affects both the $\Delta\Psi_m$ and ΔpH_m voltage and chemical gradients, while transport of non-proton cations, e.g., Ca^{2+} and K^+ uptake, and adenine nucleotide translocase (ANT) activity, rely only on $\Delta\Psi_{m.}$ and their chemical gradients. Non-electrogenic H^+ exchangers, and the P_i-H^+ phosphate co-transporter (P_iC), rely only on the ΔpH_m voltage and the H^+ chemical gradients [36, 44, 45]. These components are depicted in the cartoon of Fig. 5.15 [44].

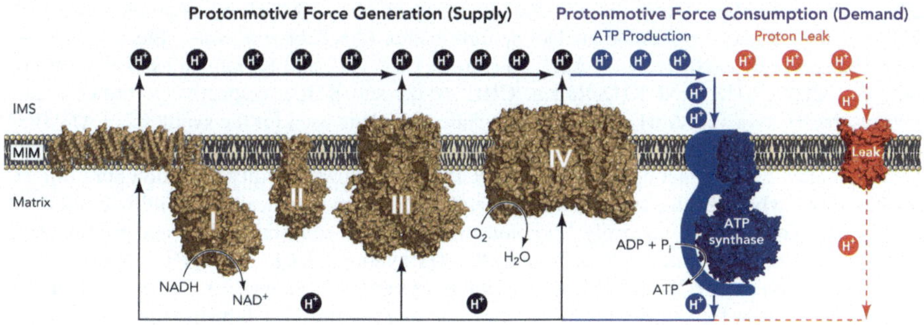

Fig. 5.14 Generation and consumption of the mitochondrial proton motive force (*pmf*): Electrons harvested from oxidizable substrates are passed through the respiratory chain in an exergonic process that drives proton (H^+) pumping by respiratory complexes I, III, and IV. The resulting electrochemical H^+ gradient across the IMM can be dissipated in two ways: (1) through the F_oF_1 ATP synthase, where relieving the *pmf* drives ADP phosphorylation, and (2) via H^+ leak pathways that do not generate ATP but regulate physiological processes including non-shivering thermogenesis, possibly glucose-stimulated insulin secretion, and protection from oxidative damage. One H^+ leak pathway is structurally represented by adenine nucleotide translocase (ANT), which can mediate both basal and inducible H^+ flux. See citation for origin of structures for complexes I to V, and ANT. (Cartoon (Fig. 1) and its abridged, modified legend resourced with permission from copyright owner of Divakaruni and Brand [43] (review))

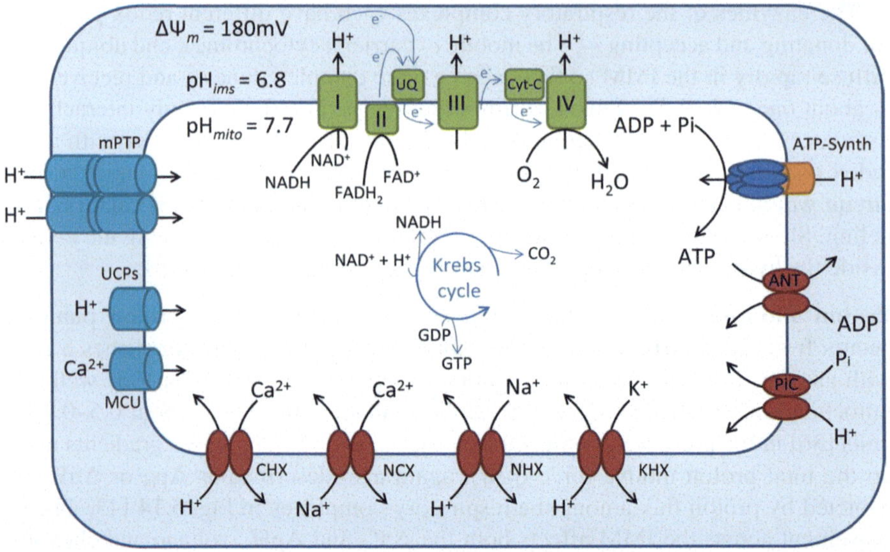

Fig. 5.15 Determinants of the mitochondrial pH gradient: Protons are pumped from the matrix into the inner membrane space (IMS) by respiratory chain complexes I, III, and IV (green boxes) as *e*- transfer from reduced substrates in the matrix to O_2. Pumping of electrically charged protons generates a $\Delta\Psi_m$ of −180 mV and a pH gradient ($\Delta pH_m = pH_m - pH_e$) of approximately 0.9 pH units as the matrix becomes more alkaline than the IMS. The proton circuit is in thermodynamic equilibrium and changes in $\Delta\Psi_m$ cause opposing changes in ΔpH_m by altering the energy required for pumping of protons by respiratory chain complexes. $\Delta\Psi_m$ and ΔpH_m add up to generate a Δp_m (*pmf*) used by ATP synthase (blue-orange barrel) to generate ATP from ADP and P_i in the matrix. $\Delta\Psi_m$ drives Ca^{2+} uptake across the mitochondrial Ca^{2+} uniporter (MCU; blue cylinder) and ADP/ATP exchange across the adenine nucleotide translocator (ANT; brown ovals). Electroneutral H^+ ion exchangers rely exclusively on ΔpH_m to extrude Ca^{2+}, Na^+, and K^+ ions in exchange for protons, i.e., CHX, NHX, and KHX (a.k.a. CHE, NHE, and KHE), respectively (brown ovals), whereas the P_iC relies on ΔpH_m to import inorganic phosphate used for the synthesis of ATP (P_iC; brown ovals). The coupling of H^+ and ion fluxes implies that changes in Na^+, K^+, Ca^{2+}, and P_i gradients can alter ΔpH_m. Uncoupling proteins (UCPs) and the permeability transition pore (mPTP) opening (blue cylinders) dissipate both ΔpH_m and $\Delta\Psi_m$ to generate heat and to initiate cell death by $\Delta\Psi_m$ depolarization, respectively. Variations in pH_m reflect equilibrium between proton pumping by the respiratory chain; Δp dissipation by the ATP synthase, UCPs, and mPTP; ΔpH_m dissipation by KHX, NHX, CHX, and PiC, and adaptive responses to changes in cytosolic pH and in $\Delta\Psi_m$. Krebs cycle = TCA cycle; mPTP = "mitochondrial permeability transition pore". (Diagram (Fig. 1) and its modified legend resourced with license (https://creativecommons.org/licenses/BY-NC-SA/4.0/) from Santo-Domingo and Demaurex [44] (perspective))

In essence, proton pumping and/or *e*- transfer via complex I (NADH ubiquinone oxidoreductase, a.k.a., NADH dehydrogenase), complex II (succinate dehydrogenase), complex III (cytochrome *c* reductase), and complex IV (cytochrome *c* oxidase) are complex processes in which an *e*- goes through the linked system of electron carrier proteins that stepwise decreases its reduction potential and causes a release in energy as heat and the energy-requiring pumping of H^+ from the matrix to the IMS (Fig. 5.14). This results in an electrochemical charge ("−") gradient in which the matrix voltage is normally more negative than the inner membrane space (IMS) voltage, and the matrix [H^+] is less than in the IMS [H^+] [38, 41, 42, 46].

Oxidative Phosphorylation Mitochondrial respiration (proton pumping with O_2 consumption) and phosphorylation (proton influx at complex V to drive ATP production from ADP + P_i) are the two separate processes of oxidative phosphorylation (OxPhos) as they can individually occur without the other, but they are mostly linked (coupled) to utilize foodstuffs to produce energy for protein synthesis, metabolism, bioelectricity, and muscle contraction and relaxation, among others [47]. When respiration (oxidation) is coupled to uptake of protons by complex V (ATP synthase), the energy derived from photosynthesis or oxidation of foodstuff is converted to ATP (phosphorylation) as a near universal source of energy for metabolism in all living things. One glucose molecule can form only 2 net molecules of ATP through the TCA cycle while about 26–28 molecules of ATP are generated via H^+ uptake at complex V when protons are pumped and the associated *e*- are passed on to O_2. Due to the complexity and limitations in experimental data, an understanding of OxPhos and cell metabolism has been aided by computational analysis [48–51].

5.3 Fast but Stepwise Release of Energy by the Respiratory Complexes

The respirasome is a tightly integrated supercomplex of the enzyme complexes I, III, and IV that together have the vital roles of stepwise transfer of *e*- from the *e*-donors NADH, FMH_2, and $FADH_2$ to their final *e*- acceptor O_2; this is accompanied by outward H^+ pumping across the IMM by complexes I, III, and IV to create both the transmembrane $\Delta\Psi_m$ and ΔpH gradients [52–54]. Aggregation of the respirasome complexes is dependent on cardiolipin present in the IMM. For example, cardiolipin serves as a charge neutralizer for the lysine residues near the presumed interaction domain between complex III (cyt bc_1 site) and complex IV, thereby stabilizing the supercomplex formation [55]. Available data indicate that by stabilizing the complexes together, this facilitates efficient *e*- transfer [56] with minimal opportunity for free radicals to form, particular at complex I [57].

Respiratory Complexes I, III, and IV Each respiratory complex is composed of independently translated protein subunits (Fig. 5.14). For example, a homology of complex I subunits with those in other protein complexes reveal that complex I originated from the unification of pre-evolved subcomplexes in archaeans, each with distinct functions that can be combined in different ways, referred to as evolutionary modules [58, 59]. Present day chloroplasts and cyanobacteria contain a complex I-like enzyme having only 11 subunits [60]. This enzyme lacks the N-module which harbors the NADH binding site and the flavin and Fe-S cluster prosthetic groups. This compact 11-subunit complex I is proposed to be ancestral to all present-day complex I enzymes [60].

Mammalian complex I contains 45 subunits, comprising the classic 14 core subunits that house the catalytic machinery, which have been conserved from bacteria to humans. In addition there is a mammalian-specific cohort of 31 supernumerary subunits, the function of which may be to furnish enzyme stability for the conserved core subunits [61, 62]. A more detailed description of the function of each linked complex is given next to highlight the tight and fast respiratory process when "reducing equivalents" are called upon to provide energy.

At *complex I*, NADH dehydrogenase first transfers *e*- to FMN, next to 7 or 8 Fe-S clusters one *e*- at a time, and then to coenzyme Q_{10} (i.e., ubiquinone, CoQ) [57]. It is unclear if the number of protons pumped per 2*e*- at complex I is 3 or 4 [39, 47] (Eq. 5.7):

$$\begin{aligned}&\mathrm{NADH}+\mathrm{H}^{+}+\mathrm{CoQ}+2e-+4\mathrm{H}^{+}(\mathrm{matrix})\rightarrow\mathrm{NAD}^{+}\\&+\mathrm{CoQH}_{2}+2e-+4\mathrm{H}^{+}(\mathrm{IMS})\end{aligned}\tag{5.7}$$

At *complex II*, succinate dehydrogenase, succinate (an intermediate from the TCA cycle) is oxidized to fumarate and its *e*- are transferred to three 2Fe-2S clusters in complex II and then to CoQ, which has a similar action to that in complex I. FAD is part of complex II and acts as a catalyst so $FADH_2$ is not actually a product in this reaction. The energy produced at complex II ($\Delta G^{o\prime} = -5.8$ KJ/mol) is too low to allow proton transfer across the IMM [47]. Because complex II does not pump protons, many do not consider this part of the respiratory complex (Eq. 5.8):

$$\begin{aligned}&\mathrm{succinate}+(\mathrm{FAD})+2\mathrm{H}^{+}+2e-\rightarrow\mathrm{fumarate}+2\mathrm{H}^{+}(\mathrm{matrix})\\&+(\mathrm{FADH}_{2})+2e-,\mathrm{and}\ \mathrm{FADH}_{2}+\mathrm{CoQ}\rightarrow\mathrm{FAD}^{+}+\mathrm{CoQH}_{2}\end{aligned}\tag{5.8}$$

At *complex III*, ubiquinol-cytochrome *c* reductase, $CoQH_2$ (ubiquinol), from complex II, transfers one of its *e*- to Rieske subunits (two Fe-S clusters) and then to cytochrome *c,* at site 1, and the other *e*- to cytochrome *b* and then to ubiquinone (CoQ), at site 2, forming a transitional radical, semi-ubiquinone ($CoQH^{\bullet-}$, a.k.a., SQ or $UQ^{\bullet}$) as depicted on a structural model of complex III in Fig. 5.16 [39]. The heme groups alternate between Fe^{2+} and Fe^{3+} during *e*- transfer. Ubiquinol is oxidized to ubiquinone. These actions furnish the energy to pump two H^+ from the matrix to the IMS. This is called step one of the Q cycle. Step 2 is a partial repeat of step 1; another $CoQH_2$ transfers its two *e*- to sites 1 and 2 and pumps two H^+ as before. The difference is that the *e*- transferred to cytochrome *b* is used to reduce the semi-ubiquinone radical $CoQH^{\bullet-}$ to ubiquinone, thus completing the full cycle back to $CoQH_2$ to again transfer its electrons (Eq. 5.9):

$$\begin{aligned}&2\mathrm{CoQH}_{2}(\mathrm{site}\,1)+\mathrm{CoQ}(\mathrm{site}\,2)+2\,\mathrm{cytochrome}\,c(\mathrm{oxidized})+2\mathrm{H}^{+}(\mathrm{matrix})\\&+2e-\rightarrow2\mathrm{CoQ}(\mathrm{site}\,1)+\mathrm{CoQH}_{2}(\mathrm{site}\,2)+2\,\mathrm{cytochrome}\,c(\mathrm{reduced})\\&+4\mathrm{H}^{+}(\mathrm{IMS})+2e-\end{aligned}\tag{5.9}$$

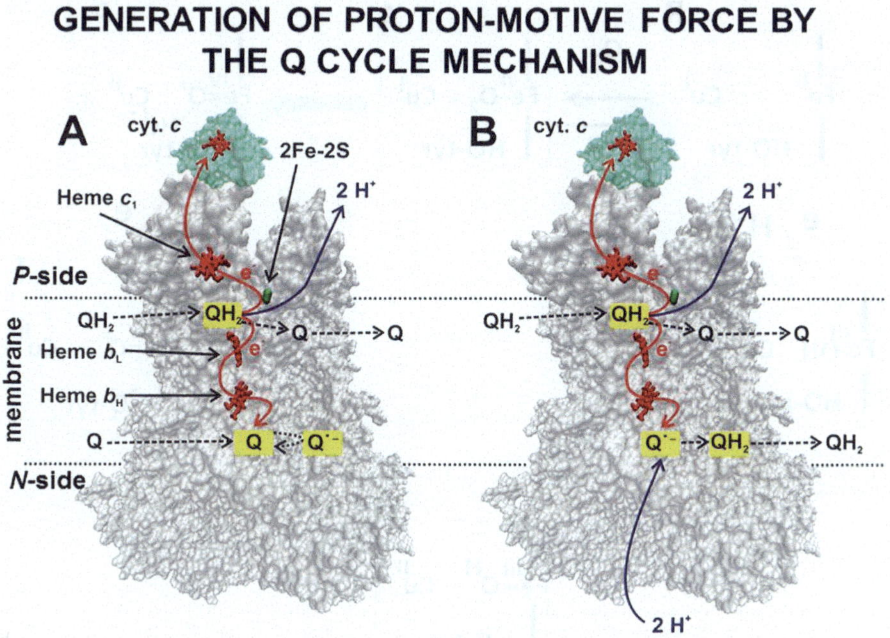

Fig. 5.16 Q cycle of complex III: The Q cycle mechanism is described as the sum of two turnovers (**a**, **b**), each starting with oxidation of hydroquinone (QH_2) near the P-side of the membrane. The low- and high-potential hemes *b* are called heme *b*L and heme *b*H, respectively. $Q^{•-}$, semiquinone; 2Fe-2S, the Rieske nonheme Fe center primarily accepts an electron from QH_2 and delivers it to heme c_1. (Diagram (Fig. 4) and its legend resourced with permission of owner (Copyright 2024, American Chemical Society) of Wikstrom et al. [39])

At *complex IV*, cytochrome *c* oxidase (COX), four reduced cytochrome *c* molecules are oxidized by COX with the 4 *e*- transferred one at a time through cytochrome heme proteins *a*, a_3, and Cu_B groups with each *e*- trapped by O_2 [39, 42], facilitating the rapid reduction of O_2 to H_2O as diagramed in Fig. 5.17 [39]. One molecule of COX can bind, activate, and reduce up to 250 molecules of O_2 per second [63]. As often expressed, O_2 is a very strong oxidizing agent. This occurs because molecular O_2 has two unpaired *e*- with parallel spin states in separate orbitals; this causes a high electronegativity so that O_2 strongly attracts *e*- (paramagnetism). This disposes O_2 to its reduction to H_2O by a univalent pathway that stepwise but rapidly, produces $O_2^{•-}$ (superoxide), then with two protons and a 2nd *e*-, H_2O_2, with a 3rd *e*-, OH^- (and $OH^•$), and finally with two more protons and the 4th *e*-, two molecules of H_2O [19, 64]. The initial pickup of one *e*- by molecular O_2 to $O_2^{•-}$ triggers the recruitment of the next 3 *e*- needed to make $2H_2O$. Figure 5.18 [41] shows Lewis structures of various ROS, the stepwise reduction of O_2 to H_2O during aerobic respiration, and the redox potential for reduction of each intermediate. Babcock [63] provides a summary on how these discoveries were made. The overall complex IV reaction is (Eq. 5.10):

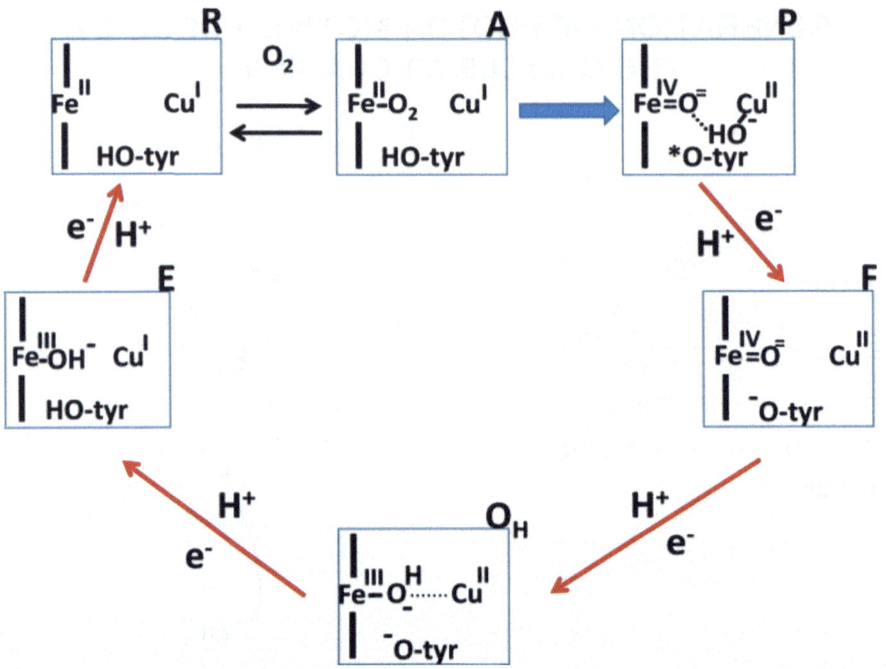

Fig. 5.17 Catalytic intermediates of the binuclear center of complex IV: The binuclear center is shown as a rectangle consisting of heme a_3 (Fe), the nearby Cu_B (Cu), and the conserved tyrosine (tyr-OH, tyrosine; tyr-O^-, tyrosinate; tyr-$O^\bullet$, a neutral tyrosine radical, is covalently linked to one of the three histidine ligands of Cu_B. The three histidine ligands of Cu_B and the proximal histidine of heme a_3 are not shown. The arbitrary name of each intermediate state is found outside the right corner of each rectangle. Red arrows specify reactions that each involve *e*- transfer to the site from the low spin heme a, net transfer of one substrate H_+ from the N-side of the membrane to complete O_2-reduction chemistry, and uptake (from the N-side) and release (to the P-side) of another proton (i.e. H^+ pumping). Hence, red arrows indicate partial redox reactions that are reversible at high Δp_m. The blue arrow depicts the irreversible partial reaction step where the O=O bond of O_2 is broken in a reaction not linked to energy transduction. Note that the classical "ferrous-oxy" intermediate A (Ref. [26]) is more accurately described as a ferric-superoxide state of heme a_3 (Ref. [5]). (Cartoon (Fig. 4) and its modified legend resourced with license (https://creativecommons.org/licenses/BY-NC-SA/4.0/) from Wikstrom and Springett [39])

$$4\,\text{cytochrome}\,c\left(\text{reduced}\right)+O_2+4H^+\left(\text{matrix}\right)+4e- \\ \rightarrow 4\,\text{cytochrome}\,c\left(\text{oxidized}\right)+4H^+\left(\text{IMS}\right)+4e-+2H_2O \tag{5.10}$$

The inorganic chemical reaction of making H_2O from activated diatomic H_2 gas in the presence of diatomic O_2 is highly exothermic and requires little activating energy. The energies of the gas reactants are much higher than the product H_2O; thus, the reduction of O_2, and the oxidation of H_2, can result in a rapid explosion (combustion) of light, heat, and sound. To effectively utilize H_2 as a commercial energy source, catalysts can slow down the gas reactions [65] whereas fuel cells

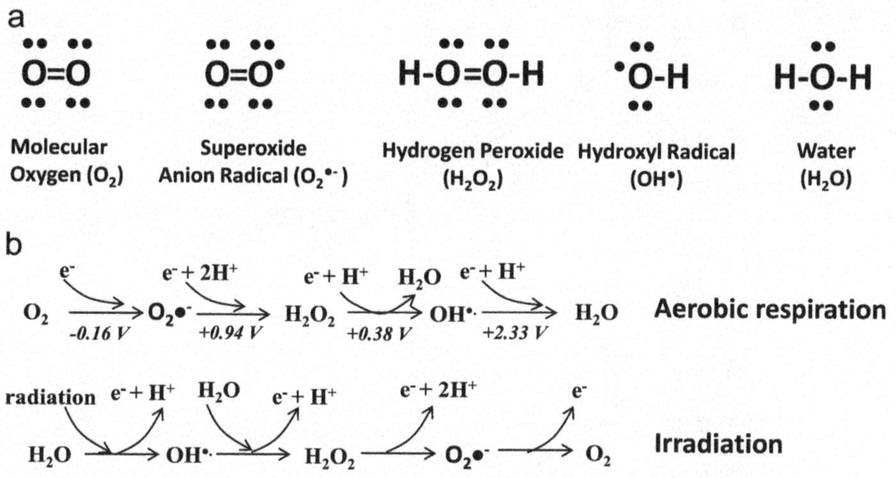

Fig. 5.18 Reduction of O_2 to H_2O and its free radical intermediates: (**a**) Lewis structures for molecular oxygen (O_2) and its singlet electron derivatives superoxide ($O_2^{\bullet-}$), hydrogen peroxide (H_2O_2), and hydroxyl radical ($OH^{\bullet-}$). (**b**) Step wise reduction of O_2 to H_2O during aerobic respiration. The standard redox potential for reduction of each intermediate is also shown. (Reproduced and modified from Ref. [23] in Mailloux). Irradiation-mediated cleavage of H_2O which produces $OH^{\bullet-}$ accounts for the damaging effects of radiation therapy. (Structures (Fig. 1) and its legend resourced with license (https://creativecommons.org/licenses/BY-NC-SA/4.0/) from Mailloux [41] (review))

generally send an electric current to first reduce either atmospheric O_2 to $O_2^{\bullet-}$ or oxidize H_2 to protons to control the release of energy by the fuel cells and so to make electricity by the controlled release of *e-* [66]. Thus, in organic or biologic chemistry, the utility of a reactant (H_2) or reducing equivalents (H^+ as in NADH) to reduce O_2 and release energy must be conducted stepwise at higher reduction potentials at each step. However, the step transfer of 4 *e-* to O_2, as described during oxidation of reduced cytochrome c, is similar, just controlled.

Blocking Reduction of O_2 to H_2O The strong electronegativities of O and N (Fig. 1.10 and Sect. 1.2) can be demonstrated by the mitochondrial poisons carbon monoxide (CO) [67, 68] and cyanide (CN^-) [69, 70], both of which inhibit the complex IV mediated reduction of O_2 to H_2O. The O in CO attracts *e-* from Fe^{2+} at the a_3 heme prosthetic ring of reduced cytochrome *c* oxidase to form carbonyl iron ($5CO + Fe^{2+} \rightarrow Fe(CO)_5$,), and the N from CN^- attracts *e-* from Fe^{2+} at the same site and at oxidized Cu_B to form cyanidoferrate ($6CN^- + Fe^{2+} \rightarrow [Fe(CN)_6]^{4-}$) and copper cyanide ($4CN^- + Cu^+ \rightarrow [Cu(CN)_4]^{3-}$), respectively. Nitric oxide ($NO^{\bullet}$), produced naturally, also competes reversibly with O_2 [71] at the reduced binuclear center a_3/Cu_B of complex IV [71] as assessed in isolated mitochondria, where Fig. 5.19 [72] shows experiments in which respiration is inhibited by adding increasing concentrations of $NO^{\bullet}$ under state 3 (phosphorylating) and 4 (non-phosphorylating) conditions [72]. $NO^{\bullet}$ generation, its reactants, and its effects are discussed in detail in Sects. 1.4, 8.4 and 9.3.

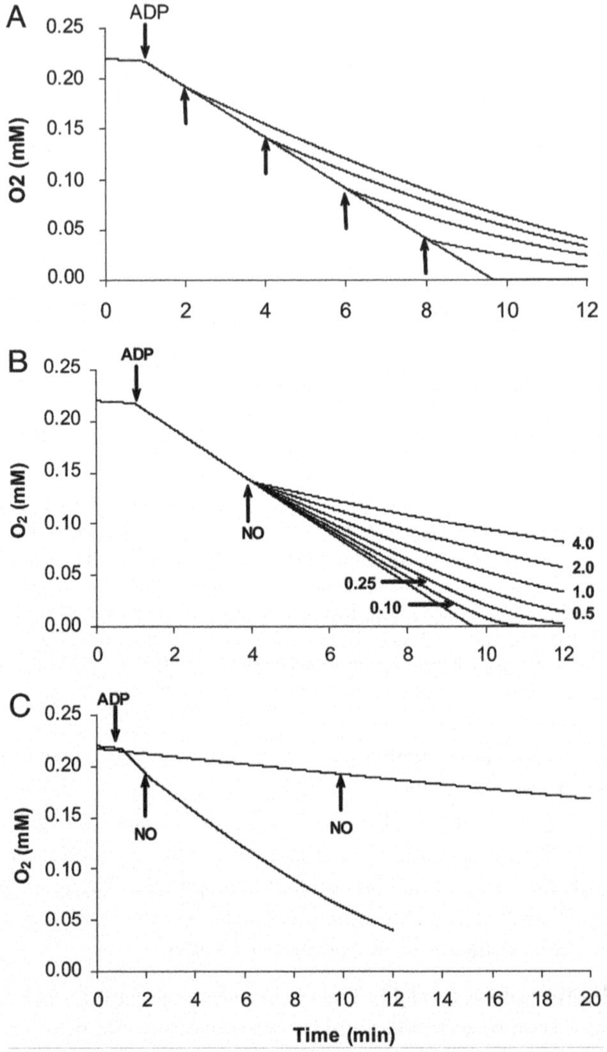

Fig. 5.19 Simulation of typical experiments with isolated mitochondria on inhibiting respiration by NO$^•$: (**a**) NO$^•$ was added at 1 μM as indicated by the upward arrows. (**b**) NO$^•$ was added at 4.0, 2.0, 1.0, 0.5, 0.25, or 0.10 μM at 4 min, when O_2 concentration was 142 μM. (**c**) NO$^•$ at 1 μM was added to respiration with high turnover (state 3-like) and with low turnover (state 4-like). Model 1 was used with the following parameters: COX initial concentration was 0.014 μM. At time 0, it was assumed that all COX was in the form Fe_{a3}^{2+}–Cu_B^{+}, which is equivalent to 0.1 mg of mitochondrial protein per ml; O_2 initial concentration was 220 μM; $k_{NO\bullet on} = 4 \times 10^7\ M^{-1}\ s^{-1}$; $k_{NO\bullet off} = 0.13\ s^{-1}$; $kO_2app = 1.4 \times 10^8\ M^{-1}\ s^{-1}$; $k_{IV} = 3\ s^{-1}$ (to simulate state 4 respiration); and $k_{IV} = 30\ s^{-1}$ (to simulate state 3 respiration). Transition between state 4 and state 3 is indicated by the addition of ADP. (Plot (Fig. 1) and its modified legend resourced with permission of owner (Copyright 2024, National Academy of Sciences) of Antunes et al. [72])

Measuring Mitochondrial Bioenergetics and Function Our understanding of how mitochondria function stems from experimentation in part from the use of fluorescence spectrophotometry for extra- and intra-mitochondrial markers of NADH and FAD redox states (Sect. 5.3), $\Delta\Psi_m$, mCa^{2+}, pH_m, $O_2^{\bullet-}$ and H_2O_2, and other probes for O_2 consumption (respiration), K^+ (mitochondrial volume), specific proteins (western blots), immunohistochemistry, electron microscopy, and is added by drugs that are known to alter specific aspects of mitochondrial function. Mitochondrial imaging is more frequently used to examine morphology, e.g., distribution, arrangement, volume and shape, and mitochondrial functional characteristics, e.g., $\Delta\Psi_m$, O_2 consumption, redox state, mROS, and Ca^{2+} in cells of a variety of organs and tissues. In addition to confocal microscopy, which permits both spatial and temporal analysis of mitochondrial morphology and function (Fig. 4.8), newer methods include fluorescence resonance energy transfer (FRET) for imaging protein-protein interactions and conformational changes, and stimulated emission depletion (STED) microscopy that allows for higher lateral and axial resolution or organelles, with long term imaging capability in living cells [73]. Other imaging techniques include transmission and scanning electron microscopy (TEM, SEM) and cryo-electron microscopy [74].

The utilization of isolated mitochondria to understand their functions removes information about the important interactions of mitochondria with their host cells or tissue (Sects. 7.2, 7.3 and 7.4). However, isolating mitochondria allows for examination of the effects of known drugs that block *e*- transport at specific sites within the mitochondrial ETS and helps to determine the effects of other substances or drugs that may interfere with *e*- transport in a similar manner. Many specific inhibitors of respiration are known [32, 33, 38, 75], as summarized in Fig. 5.20 [38].

TABLE 2. DRUGS COMMONLY USED TO ASSESS DEDUCTIVELY THE SITES OF ELECTRON LEAK AND SUPEROXIDE GENERATION WITHIN THE RESPIRATORY CHAIN

Drug	*Mechanism*	*Effect*
Complex I sites	*Inhibitors of NADH oxidation*	
Rotenone (ROT)	Inhibits transfer of e^- from Q-binding site to complex I	$+O_2^{\bullet-}$
Amobarbital (AMO)	Blocks transfer of e^- from Q_{Nf} to Q_{Ns} distal to N2 Fe/S center	$+O_2^{\bullet-}$
Piericidin A (PCA)	Inhibits e^- by competing with Q_B for binding site in complex I	$+O_2^{\bullet-}$
Complex II sites	*Inhibitors of succinate dehydrogenase*	
Malonate (MAL)	Competitive inhibitor of complex II	$+O_2^{\bullet-}$
Methylmalonate (MML)	Competitive inhibitor of complex II	$+O_2^{\bullet-}$
3-Nitroproprionate (3NP)	Irreversible inhibitor of complex II	$+O_2^{\bullet-}$
Atpenins (APT)	Noncompetitive inhibitor of complex II	$+O_2^{\bullet-}$
Complex III sites	*Inhibitors of quinone (Q) cycle*	
Stigmatellin (STG)	Prevents transfer of 1st e^- from QH_2 to Fe/S protein (ISP)	$\pm O_2^{\bullet-a}$
Myxothiazol (MYX)	Prevents binding of QH_2 at Q_o site	$+O_2^{\bullet-}$
Antimycin A (ANA)	Prevents transfer of 2nd e^- to Q_i site	$+O_2^{\bullet-}$
Complex IV sites	*Inhibitors of O_2 binding*	
K^+ cyanide (KCN)	Allosterically inhibits cytochrome oxidase	$\pm O_2^{\bullet-b}$
Na^+ azide (NaN_3)	Allosterically inhibits cytochrome oxidase	$\pm O_2^{\bullet-}$
Carbon monoxide (CO)	Competes with binding of O_2 to cytochrome oxidase	$\pm O_2^{\bullet-}$
Nitric oxide ($NO^\bullet$)	Competes with binding of O_2 to cytochrome oxidase	$\pm O_2^{\bullet-}$
	General oxidants	
Menadione (vitK_3)	Stimulates redox cycling by one e^- reductive enzymes	$+O_2^{\bullet-}$
Quinones, doxorubicin	Reacts with Fe^{3+} compounds to produce $Fe^{2+} + e^-$	$+O_2^{\bullet-}$
H_2O_2	Reacts with Fe compounds to produce $Fe(IV{=}O)^\bullet$ and $Fe^{+3} + {}^\bullet OH + {}^- OH$ Fenton effect (minor)	$+{}^\bullet OH$

Note: Effects of some drugs on $O_2^{\bullet-}$ generation[a,b] depend on the substrate used and whether other drugs that act on the respiratory system are present.

+, increase; −, decrease.

Fig. 5.20 Drugs commonly used to block the electron transport system (ETS): Blocking electron transfer causes electron leak that leads to release of superoxide. (Description (Table 2) resourced with permission from copyright owner of Stowe and Camara [38] (review))

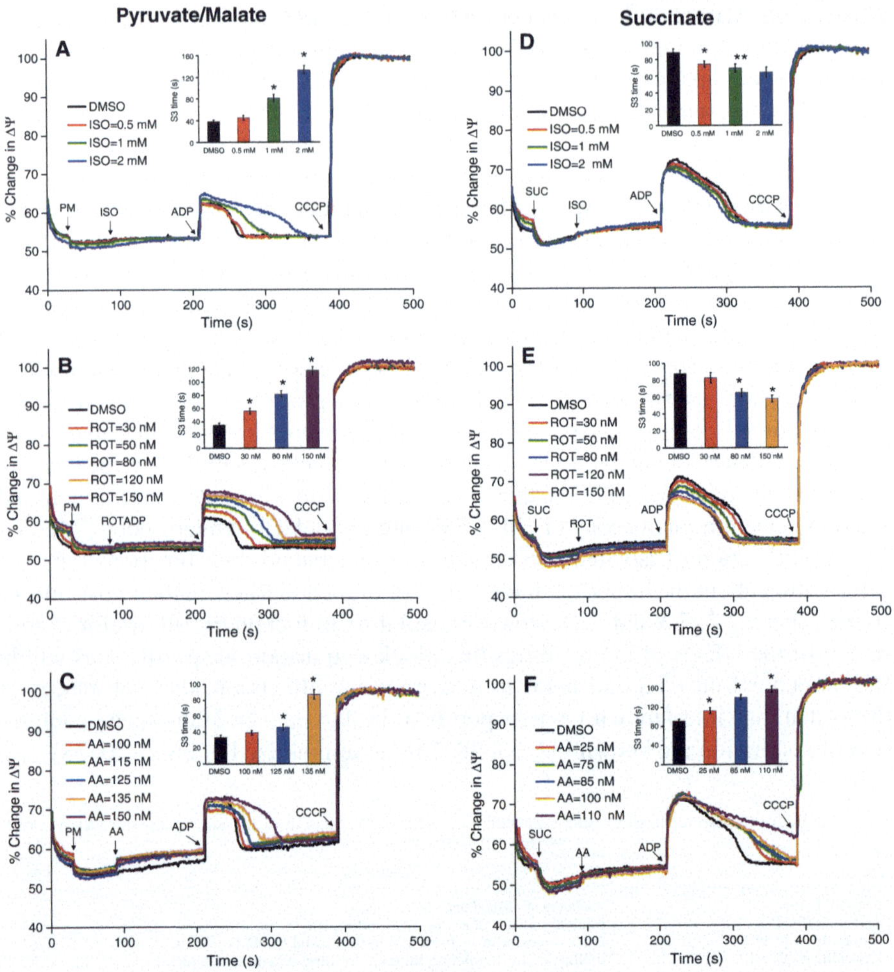

Fig. 5.21 Measurement of mitochondrial membrane potential ($\Delta\Psi_m$) in isolated mitochondria: $\Delta\Psi_m$ was measured during states 2, 3, and 4 respiration with different substrates and after adding either the volatile anesthetic isoflurane (ISO) or blockers of ETS complexes. ISO, rotenone (ROT), or antimycin A (AA) was given after pyruvate malate (PM) (**a–c**) or succinate (SUC) (**d–f**), and before ADP, which induced a transient depolarization. The insets summarize effects of ISO compared to ROT and AA on duration of ADP-induced state 3 decreases in $\Delta\Psi_m$ (S3 time, s). Note similarities and differences of responses to each drug. $\Delta\Psi_m$ transients were normalized with respect to maximal (100%) depolarization obtained with added mitochondrial respiratory uncoupler carbonyl cyanide chlorophenylhydrazone (CCCP). Data represent mean ± SD of 3 replicates. (Plots (Fig. 3) and its legend resourced with permission from copyright owner of Agarwal et al. [33])

$\Delta\Psi_m$ is typically measured in isolated mitochondria by a fluorescent probe, e.g., TMRM (tetramethyl-rhodamine methyl ester perchlorate). As shown in typical experiments (Figs. 5.21 and 5.22) [34], the effects of a drug, e.g., isoflurane, on $\Delta\Psi_m$ can be compared with the effects of inhibitors of complex I and III under different substrate conditions. As with mitochondrial assessment of NADH (Fig. 5.13) and

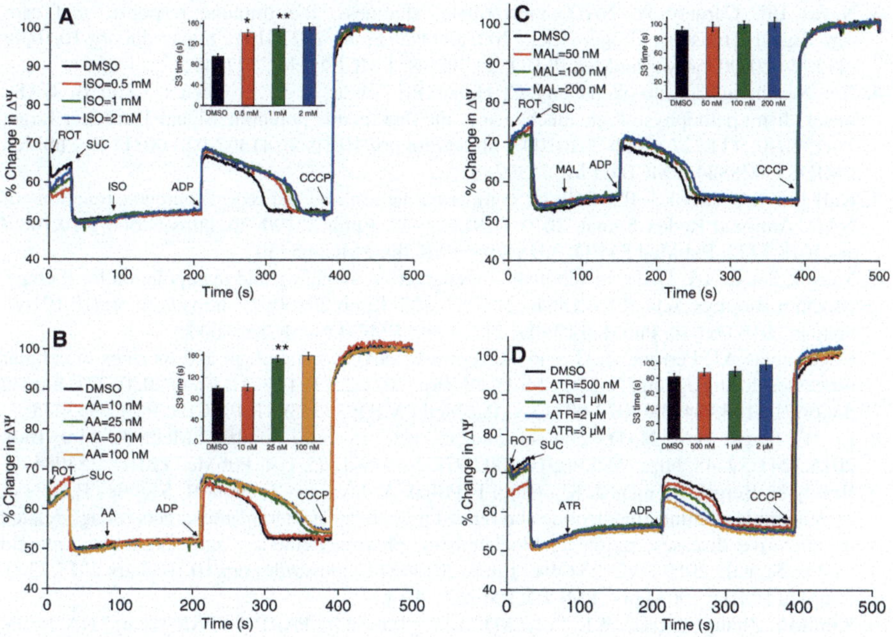

Fig. 5.22 Normalized mitochondrial membrane potential ($\Delta\Psi_m$) during states 2, 3, and 4 respiration: Isolated mitochondria were treated with succinate (SUC) + rotenone (ROT) and either isoflurane (ISO) or a blocker of the ETS. ROT (1 μM) was given before other drugs to maximally block complex I. ISO (**a**), complex III blocker antimycin A (AA) (**b**), complex II blocker malonate (MAL) (**c**), or complex V blocker atractyloside (ATR) (**d**) was added after SUC (**a–d**) and before adding ADP, which induced a transient depolarization. The insets summarize effects of ISO compared to AA, MAL, and ATR on duration of state 3 ADP-induced decreases in $\Delta\Psi_m$ (S3 time, s). Note similarities and differences of responses to each drug. $\Delta\Psi_m$ transients were normalized with respect to maximal (100%) depolarization obtained with added respiratory uncoupler carbonyl cyanide chlorophenylhydrazone (CCCP). Data represent mean ± SD of 3 replicates. (Plots (Fig. 4) and its legend resourced with permission from copyright owner of Agarwal et al. [33])

respiration (O_2 consumption) (Sect. 6.4), assessment of changes in $\Delta\Psi_m$ (Figs. 5.21 and 5.22) furnishes valuable information on mitochondrial responses to new drugs by testing them against known respiratory inhibitors or activators.

References

1. Dunn JD, Alvarez LA, Zhang X, Soldati T. Reactive oxygen species and mitochondria: a nexus of cellular homeostasis. Redox Biol. 2015;6:472–85. Epub 20150910. https://doi.org/10.1016/j.redox.2015.09.005. PubMed PMID: 26432659; PMCID: PMC4596921.
2. Stein LR, Imai S. The dynamic regulation of NAD metabolism in mitochondria. Trends Endocrinol Metab. 2012;23(9):420–8. Epub 20120721. https://doi.org/10.1016/j.tem.2012.06.005. PubMed PMID: 22819213; PMCID: PMC3683958.

3. Navas LE, Carnero A. NAD^+ metabolism, stemness, the immune response, and cancer. Signal Transduct Target Ther. 2021;6(1):2. Epub 20210101. https://doi.org/10.1038/s41392-020-00354-w. PubMed PMID: 33384409; PMCID: PMC7775471.
4. Xie N, Zhang L, Gao W, Huang C, Huber PE, Zhou X, Li C, Shen G, Zou B. NAD^+ metabolism: pathophysiologic mechanisms and therapeutic potential. Signal Transduct Target Ther. 2020;5(1):227. Epub 20201007. https://doi.org/10.1038/s41392-020-00311-7. PubMed PMID: 33028824; PMCID: PMC7539288.
5. Kulkarni CA, Brookes PS. Cellular compartmentation and the redox/nonredox functions of NAD. Antioxid Redox Signal. 2019;31(9):623–42. Epub 20190326. https://doi.org/10.1089/ars.2018.7722. PubMed PMID: 30784294; PMCID: PMC6657305.
6. Yang Y, Sauve AA. NAD^+ metabolism: bioenergetics, signaling and manipulation for therapy. Biochim Biophys Acta. 2016;1864(12):1787–800. Epub 20160629. https://doi.org/10.1016/j.bbapap.2016.06.014. PubMed PMID: 27374990; PMCID: PMC5521000.
7. Covarrubias AJ, Perrone R, Grozio A, Verdin E. NAD^+ metabolism and its roles in cellular processes during ageing. Nat Rev Mol Cell Biol. 2021;22(2):119–41. Epub 20201222. https://doi.org/10.1038/s41580-020-00313-x. PubMed PMID: 33353981; PMCID: PMC7963035.
8. Li W, Sauve AA. NAD^+ content and its role in mitochondria. Methods Mol Biol. 2015;1241:39–48. https://doi.org/10.1007/978-1-4939-1875-1_4. PubMed PMID: 25308486.
9. Braidy N, Berg J, Clement J, Khorshidi F, Poljak A, Jayasena T, Grant R, Sachdev P. Role of nicotinamide adenine dinucleotide and related precursors as therapeutic targets for age-related degenerative diseases: rationale, biochemistry, pharmacokinetics, and outcomes. Antioxid Redox Signal. 2019;30(2):251–94. Epub 20180511. https://doi.org/10.1089/ars.2017.7269. PubMed PMID: 29634344; PMCID: PMC6277084.
10. Kim MY, Zhang T, Kraus WL. Poly(ADP-ribosyl)ation by PARP-1: 'PAR-laying' NAD^+ into a nuclear signal. Genes Dev. 2005;19(17):1951–67. https://doi.org/10.1101/gad.1331805. PubMed PMID: 16140981.
11. Anderson KA, Madsen AS, Olsen CA, Hirschey MD. Metabolic control by sirtuins and other enzymes that sense NAD^+, NADH, or their ratio. Biochim Biophys Acta Bioenerg. 2017;1858(12):991–8. Epub 20170922. https://doi.org/10.1016/j.bbabio.2017.09.005. PubMed PMID: 28947253; PMCID: PMC5648639.
12. Poljsak B, Kovac V, Spalj S, Milisav I. The central role of the NAD^+ molecule in the development of aging and the prevention of chronic age-related diseases: strategies for NAD^+ modulation. Int J Mol Sci. 2023;24(3). Epub 20230203. https://doi.org/10.3390/ijms24032959. PubMed PMID: 36769283; PMCID: PMC9917998.
13. Langelier MF, Zandarashvili L, Aguiar PM, Black BE, Pascal JM. NAD^+ analog reveals PARP-1 substrate-blocking mechanism and allosteric communication from catalytic center to DNA-binding domains. Nat Commun. 2018;9(1):844. Epub 20180227. https://doi.org/10.1038/s41467-018-03234-8. PubMed PMID: 29487285; PMCID: PMC5829251.
14. Htet Y, Tennyson AG. NAD^+ as a hydride donor and reductant. J Am Chem Soc. 2016;138(49):15833–6. Epub 20161130. https://doi.org/10.1021/jacs.6b10451. PubMed PMID: 27960347.
15. Xiao W, Wang RS, Handy DE, Loscalzo J. NAD(H) and NADP(H) redox couples and cellular energy metabolism. Antioxid Redox Signal. 2018;28(3):251–72. Epub 20170728. https://doi.org/10.1089/ars.2017.7216. PubMed PMID: 28648096; PMCID: PMC5737637.
16. Mansoorabadi SO, Thibodeaux CJ, Liu HW. The diverse roles of flavin coenzymes—nature's most versatile thespians. J Org Chem. 2007;72(17):6329–42. Epub 20070620. https://doi.org/10.1021/jo0703092. PubMed PMID: 17580897; PMCID: PMC2519020.
17. Averianova LA, Balabanova LA, Son OM, Podvolotskaya AB, Tekutyeva LA. Production of vitamin B_2 (Riboflavin) by microorganisms: an overview. Front Bioeng Biotechnol. 2020;8:570828. Epub 20201112. https://doi.org/10.3389/fbioe.2020.570828. PubMed PMID: 33304888; PMCID: PMC7693651.
18. Abbas CA, Sibirny AA. Genetic control of biosynthesis and transport of riboflavin and flavin nucleotides and construction of robust biotechnological producers. Microbiol Mol Biol Rev.

2011;75(2):321–60. https://doi.org/10.1128/MMBR.00030-10. PubMed PMID: 21646432; PMCID: PMC3122625.

19. Alberts B, Johnson A, Lewis J, Raff M, Roberts K, Walter P. Electron-transport chains and their proton pumps. In: Molecular biology of the cell. 4th ed. New York: Garland Science; 2002. Available from: https://www.ncbi.nlm.nih.gov/books/NBK26904/.
20. Bonora M, Patergnani S, Rimessi A, De Marchi E, Suski JM, Bononi A, Giorgi C, Marchi S, Missiroli S, Poletti F, Wieckowski MR, Pinton P. ATP synthesis and storage. Purinergic Signal. 2012;8(3):343–57. Epub 20120412. https://doi.org/10.1007/s11302-012-9305-8. PubMed PMID: 22528680; PMCID: PMC3360099.
21. McCommis KS, Finck BN. Mitochondrial pyruvate transport: a historical perspective and future research directions. Biochem J. 2015;466(3):443–54. https://doi.org/10.1042/BJ20141171. PubMed PMID: 25748677; PMCID: PMC4464838.
22. LibreTexts. Introductory and general biology. 12th ed. Raven BIology; 2023. Available from: https://bio.libretexts.org/Bookshelves/Introductory_and_General_Biology/Map%3A_Raven_Biology_12th_Edition/03%3A_The_Chemical_Building_Blocks_of_Life.
23. Martinez-Reyes I, Chandel NS. Mitochondrial TCA cycle metabolites control physiology and disease. Nat Commun. 2020;11(1):102. Epub 20200103. https://doi.org/10.1038/s41467-019-13668-3. PubMed PMID: 31900386; PMCID: PMC6941980.
24. OpenStat C. ATP-denosine triphosphate. 2023. In: LibreTexts Biology [Internet]. Available from: https://bio.libretexts.org/Bookshelves/Microbiology/Microbiology_(Boundless)/02%3A_Chemistry/2.05%3A_Organic_Compounds/2.5.06%3A_ATP-_Adenosine_Triphosphate.
25. Arnold PK, Finley LWS. Regulation and function of the mammalian tricarboxylic acid cycle. J Biol Chem. 2023;299(2):102838. Epub 20221226. https://doi.org/10.1016/j.jbc.2022.102838. PubMed PMID: 36581208; PMCID: PMC9871338.
26. Hill BG, Benavides GA, Lancaster JR Jr, Ballinger S, Dell'Italia L, Jianhua Z, Darley-Usmar VM. Integration of cellular bioenergetics with mitochondrial quality control and autophagy. Biol Chem. 2012;393(12):1485–512. https://doi.org/10.1515/hsz-2012-0198. PubMed PMID: 23092819; PMCID: PMC3594552.
27. Wikstrom M, Springett R. Thermodynamic efficiency, reversibility, and degree of coupling in energy conservation by the mitochondrial respiratory chain. Commun Biol 2020;3(1):451. Epub 20200818. https://doi.org/10.1038/s42003-020-01192-w. PubMed PMID: 32811895; PMCID: PMC7434914.
28. Riess ML, Camara AK, Chen Q, Novalija E, Rhodes SS, Stowe DF. Altered NADH and improved function by anesthetic and ischemic preconditioning in guinea pig intact hearts. Am J Physiol Heart Circ Physiol. 2002;283(1):H53–60. https://doi.org/10.1152/ajpheart.01057.2001. PubMed PMID: 12063274.
29. Stowe DF, Gadicherla AK, Zhou Y, Aldakkak M, Cheng Q, Kwok WM, Jiang MT, Heisner JS, Yang M, Camara AK. Protection against cardiac injury by small Ca^{2+}-sensitive K^+ channels identified in guinea pig cardiac inner mitochondrial membrane. Biochim Biophys Acta. 2013;1828(2):427–42. Epub 20120908. https://doi.org/10.1016/j.bbamem.2012.08.031. PubMed PMID: 22982251; PMCID: PMC3534888.
30. Stowe DF, Aldakkak M, Camara AK, Riess ML, Heinen A, Varadarajan SG, Jiang MT. Cardiac mitochondrial preconditioning by Big Ca^{2+}-sensitive K^+ channel opening requires superoxide radical generation. Am J Physiol Heart Circ Physiol. 2006;290(1):H434–40. Epub 20050826. https://doi.org/10.1152/ajpheart.00763.2005. PubMed PMID: 16126810.
31. Aldakkak M, Stowe DF, Heisner JS, Spence M, Camara AK. Enhanced Na^+/H^+ exchange during ischemia and reperfusion impairs mitochondrial bioenergetics and myocardial function. J Cardiovasc Pharmacol. 2008;52(3):236–44. https://doi.org/10.1097/FJC.0b013e3181831337. PubMed PMID: 18806604; PMCID: PMC2659015.
32. Aldakkak M, Stowe DF, Chen Q, Lesnefsky EJ, Camara AK. Inhibited mitochondrial respiration by amobarbital during cardiac ischaemia improves redox state and reduces matrix Ca^{2+} overload and ROS release. Cardiovasc Res. 2008;77(2):406–15. https://doi.org/10.1016/j.cardiores.2007.08.008. PubMed PMID: 17900548

33. Agarwal B, Dash RK, Stowe DF, Bosnjak ZJ, Camara AK. Isoflurane modulates cardiac mitochondrial bioenergetics by selectively attenuating respiratory complexes. Biochim Biophys Acta. 2014;1837(3):354–65. Epub 20131217. https://doi.org/10.1016/j.bbabio.2013.11.006. PubMed PMID: 24355434; PMCID: PMC4084852.
34. Agarwal B, Camara AK, Stowe DF, Bosnjak ZJ, Dash RK. Enhanced charge-independent mitochondrial free Ca^{2+} and attenuated ADP-induced NADH oxidation by isoflurane: Implications for cardioprotection. Biochim Biophys Acta. 2012;1817(3):453–65. Epub 20111202. https://doi.org/10.1016/j.bbabio.2011.11.011. PubMed PMID: 22155157; PMCID: PMC3269543.
35. Blomeyer CA, Bazil JN, Stowe DF, Pradhan RK, Dash RK, Camara AK. Dynamic buffering of mitochondrial Ca^{2+} during Ca^{2+} uptake and Na^{+}-induced Ca^{2+} release. J Bioenerg Biomembr. 2013;45(3):189–202. Epub 20121207. https://doi.org/10.1007/s10863-012-9483-7. PubMed PMID: 23225099; PMCID: PMC4065551.
36. Haumann J, Camara AKS, Gadicherla AK, Navarro CD, Boelens AD, Blomeyer CA, Dash RK, Boswell MR, Kwok WM, Stowe DF. Slow Ca^{2+} efflux by Ca^{2+}/H^{+} exchange in cardiac mitochondria is modulated by Ca^{2+} re-uptake via MCU, extra-mitochondrial pH, and H^{+} pumping by F_OF_1-ATPase. Front Physiol. 2018;9:1914. Epub 20190204. https://doi.org/10.3389/fphys.2018.01914. PubMed PMID: 30804812; PMCID: PMC6378946.
37. Boelens AD, Pradhan RK, Blomeyer CA, Camara AK, Dash RK, Stowe DF. Extra-matrix Mg^{2+} limits Ca^{2+} uptake and modulates Ca^{2+} uptake-independent respiration and redox state in cardiac isolated mitochondria. J Bioenerg Biomembr. 2013;45(3):203–18. Epub 20130303. https://doi.org/10.1007/s10863-013-9500-5. PubMed PMID: 23456198; PMCID: PMC3670670.
38. Stowe DF, Camara AK. Mitochondrial reactive oxygen species production in excitable cells: modulators of mitochondrial and cell function. Antioxid Redox Signal. 2009;11(6):1373–414. https://doi.org/10.1089/ars.2008.2331. PubMed PMID: 19187004; PMCID: PMC2842133.
39. Wikstrom M, Sharma V, Kaila VR, Hosler JP, Hummer G. New perspectives on proton pumping in cellular respiration. Chem Rev. 2015;115(5):2196–221. Epub 20150219. https://doi.org/10.1021/cr500448t. PubMed PMID: 25694135.
40. Hinkle PC, McCarty RE. How cells make ATP. Sci Am. 1978;238(3):104–17, 21–3. https://doi.org/10.1038/scientificamerican0378-104. PubMed PMID: 635521.
41. Mailloux RJ. Teaching the fundamentals of electron transfer reactions in mitochondria and the production and detection of reactive oxygen species. Redox Biol. 2015;4:381–98. Epub 20150207. https://doi.org/10.1016/j.redox.2015.02.001. PubMed PMID: 25744690; PMCID: PMC4348434.
42. Ramirez BE, Malmstrom BG, Winkler JR, Gray HB. The currents of life: the terminal electron-transfer complex of respiration. Proc Natl Acad Sci USA. 1995;92(26):11949–51. https://doi.org/10.1073/pnas.92.26.11949. PubMed PMID: 8618820; PMCID: PMC40272.
43. Divakaruni AS, Brand MD. The regulation and physiology of mitochondrial proton leak. Physiology (Bethesda). 2011;26(3):192–205. https://doi.org/10.1152/physiol.00046.2010. PubMed PMID: 21670165.
44. Santo-Domingo J, Demaurex N. Perspectives on: SGP symposium on mitochondrial physiology and medicine: the renaissance of mitochondrial pH. J Gen Physiol. 2012;139(6):415–23. https://doi.org/10.1085/jgp.201110767. PubMed PMID: 22641636; PMCID: PMC3362525.
45. Malas KM, Lambert DS, Heisner JS, Camara AKS, Stowe DF. Time and charge/pH-dependent activation of K^{+} channel-mediated K^{+} influx and K^{+}/H^{+} exchange in guinea pig heart isolated mitochondria; role in bioenergetic stability. Biochim Biophys Acta Bioenerg. 2022;1863(8):148908. Epub 20220809. https://doi.org/10.1016/j.bbabio.2022.148908. PubMed PMID: 35961396.
46. Ahmad M, Wolberg A, Kahwaji CI. Biochemistry, electron transport chain. Treasure Island: StatPearls Publishing; 2023. Available from: https://www.ncbi.nlm.nih.gov/books/NBK526105/.
47. Mazat JP, Ransac S, Heiske M, Devin A, Rigoulet M. Mitochondrial energetic metabolism-some general principles. IUBMB Life. 2013;65(3):171–9. https://doi.org/10.1002/iub.1138. PubMed PMID: 23441039.

48. Randall EB, Hock M, Lopez R, Marzban B, Marshall C, Beard DA. Quantitative analysis of mitochondrial ATP synthesis. Math Biosci. 2021;340:108646. Epub 20210617. https://doi.org/10.1016/j.mbs.2021.108646. PubMed PMID: 34146585; PMCID: PMC8434986.
49. Wu F, Yang F, Vinnakota KC, Beard DA. Computer modeling of mitochondrial tricarboxylic acid cycle, oxidative phosphorylation, metabolite transport, and electrophysiology. J Biol Chem. 2007;282(34):24525–37. Epub 20070625. https://doi.org/10.1074/jbc.M701024200. PubMed PMID: 17591785.
50. Beard DA. A biophysical model of the mitochondrial respiratory system and oxidative phosphorylation. PLoS Comput Biol. 2005;1(4):e36. Epub 20050909. https://doi.org/10.1371/journal.pcbi.0010036. PubMed PMID: 16163394; PMCID: PMC1201326.
51. Cortassa S, Aon MA, Marban E, Winslow RL, O'Rourke B. An integrated model of cardiac mitochondrial energy metabolism and calcium dynamics. Biophys J. 2003;84(4):2734–55. https://doi.org/10.1016/S0006-3495(03)75079-6. PubMed PMID: 12668482; PMCID: PMC1201507.
52. Guo R, Gu J, Wu M, Yang M. Amazing structure of respirasome: unveiling the secrets of cell respiration. Protein Cell. 2016;7(12):854–65. Epub 20161014. https://doi.org/10.1007/s13238-016-0329-7. PubMed PMID: 27743346; PMCID: PMC5205662.
53. Gu J, Wu M, Guo R, Yan K, Lei J, Gao N, Yang M. The architecture of the mammalian respirasome. Nature. 2016;537(7622):639–43. https://doi.org/10.1038/nature19359. PubMed PMID: 27654917.
54. Wu M, Gu J, Zong S, Guo R, Liu T, Yang M. Research journey of respirasome. Protein Cell. 2020;11(5):318–38. Epub 20200109. https://doi.org/10.1007/s13238-019-00681-x. PubMed PMID: 31919741; PMCID: PMC7196574.
55. Wenz T, Hielscher R, Hellwig P, Schagger H, Richers S, Hunte C. Role of phospholipids in respiratory cytochrome bc_1 complex catalysis and supercomplex formation. Biochim Biophys Acta. 2009;1787(6):609–16. Epub 20090228. https://doi.org/10.1016/j.bbabio.2009.02.012. PubMed PMID: 19254687.
56. Lapuente-Brun E, Moreno-Loshuertos R, Acin-Perez R, Latorre-Pellicer A, Colas C, Balsa E, Perales-Clemente E, Quiros PM, Calvo E, Rodriguez-Hernandez MA, Navas P, Cruz R, Carracedo A, Lopez-Otin C, Perez-Martos A, Fernandez-Silva P, Fernandez-Vizarra E, Enriquez JA. Supercomplex assembly determines electron flux in the mitochondrial electron transport chain. Science. 2013;340(6140):1567–70. https://doi.org/10.1126/science.1230381. PubMed PMID: 23812712.
57. Sharma LK, Lu J, Bai Y. Mitochondrial respiratory complex I: structure, function and implication in human diseases. Curr Med Chem. 2009;16(10):1266–77. https://doi.org/10.2174/092986709787846578. PubMed PMID: 19355884; PMCID: PMC4706149.
58. Friedrich T, Scheide D. The respiratory complex I of bacteria, archaea and eukarya and its module common with membrane-bound multisubunit hydrogenases. FEBS Lett. 2000;479(1–2):1–5. https://doi.org/10.1016/s0014-5793(00)01867-6. PubMed PMID: 10940377.
59. Efremov RG, Sazanov LA. The coupling mechanism of respiratory complex I—a structural and evolutionary perspective. Biochim Biophys Acta. 2012;1817(10):1785–95. Epub 20120222. https://doi.org/10.1016/j.bbabio.2012.02.015. PubMed PMID: 22386882.
60. Moparthi VK, Hagerhall C. The evolution of respiratory chain complex I from a smaller last common ancestor consisting of 11 protein subunits. J Mol Evol. 2011;72(5–6):484–97. Epub 20110520. https://doi.org/10.1007/s00239-011-9447-2. PubMed PMID: 21597881; PMCID: PMC3144371.
61. Zhu J, Vinothkumar KR, Hirst J. Structure of mammalian respiratory complex I. Nature. 2016;536(7616):354–8. Epub 20160810. https://doi.org/10.1038/nature19095. PubMed PMID: 27509854; PMCID: PMC5027920.
62. Hirst J. Why does mitochondrial complex I have so many subunits? Biochem J. 2011;437(2):e1–3. https://doi.org/10.1042/BJ20110918. PubMed PMID: 21711245.

63. Babcock GT. How oxygen is activated and reduced in respiration. Proc Natl Acad Sci USA. 1999;96(23):12971–3. https://doi.org/10.1073/pnas.96.23.12971. PubMed PMID: 10557256; PMCID: PMC33932.
64. Fridovich I. Oxygen: how do we stand it? Med Princ Pract. 2013;22(2):131–7. Epub 20120628. https://doi.org/10.1159/000339212. PubMed PMID: 22759590; PMCID: PMC5685332.
65. Heiden ZM, Rauchfuss TB. Homogeneous catalytic reduction of dioxygen using transfer hydrogenation catalysts. J Am Chem Soc. 2007;129(46):14303–10. Epub 20071025. https://doi.org/10.1021/ja073774p. PubMed PMID: 17958423.
66. Shao M, Chang Q, Dodelet JP, Chenitz R. Recent advances in electrocatalysts for oxygen reduction reaction. Chem Rev. 2016;116(6):3594–657. Epub 20160217. https://doi.org/10.1021/acs.chemrev.5b00462. PubMed PMID: 26886420.
67. Alonso JR, Cardellach F, Lopez S, Casademont J, Miro O. Carbon monoxide specifically inhibits cytochrome c oxidase of human mitochondrial respiratory chain. Pharmacol Toxicol. 2003;93(3):142–6. https://doi.org/10.1034/j.1600-0773.2003.930306.x. PubMed PMID: 12969439.
68. Brzezinski P, Malmstrom BG. The reduction of cytochrome c oxidase by carbon monoxide. FEBS Lett. 1985;187(1):111–4. https://doi.org/10.1016/0014-5793(85)81224-2. PubMed PMID: 2991001.
69. Parashar A, Venkatachalam A, Gideon DA, Manoj KM. Cyanide does more to inhibit heme enzymes, than merely serving as an active-site ligand. Biochem Biophys Res Commun. 2014;455(3–4):190–3. Epub 20141106. https://doi.org/10.1016/j.bbrc.2014.10.137. PubMed PMID: 25449264.
70. Lawson-Smith P, Jansen EC, Hyldegaard O. Cyanide intoxication as part of smoke inhalation—a review on diagnosis and treatment from the emergency perspective. Scand J Trauma Resusc Emerg Med. 2011;19:14. Epub 20110303. https://doi.org/10.1186/1757-7241-19-14. PubMed PMID: 21371322; PMCID: PMC3058018.
71. Pannala VR, Camara AK, Dash RK. Modeling the detailed kinetics of mitochondrial cytochrome c oxidase: catalytic mechanism and nitric oxide inhibition. J Appl Physiol (1985). 2016;121(5):1196–207. Epub 20160915. https://doi.org/10.1152/japplphysiol.00524.2016. PubMed PMID: 27633738; PMCID: PMC5142251.
72. Antunes F, Boveris A, Cadenas E. On the mechanism and biology of cytochrome oxidase inhibition by nitric oxide. Proc Natl Acad Sci USA. 2004;101(48):16774–9. Epub 20041116. https://doi.org/10.1073/pnas.0405368101. PubMed PMID: 15546991; PMCID: PMC534717.
73. Wu Z, Xu X, Xi P. Stimulated emission depletion microscopy for biological imaging in four dimensions: a review. Microsc Res Tech. 2021;84(9):1947–58. Epub 20210313. https://doi.org/10.1002/jemt.23750. PubMed PMID: 33713513.
74. Strubbe-Rivera JO, Schrad JR, Pavlov EV, Conway JF, Parent KN, Bazil JN. The mitochondrial permeability transition phenomenon elucidated by cryo-EM reveals the genuine impact of calcium overload on mitochondrial structure and function. Sci Rep. 2021;11(1):1037. Epub 20210113. https://doi.org/10.1038/s41598-020-80398-8. PubMed PMID: 33441863; PMCID: PMC7806632.
75. Chen Q, Vazquez EJ, Moghaddas S, Hoppel CL, Lesnefsky EJ. Production of reactive oxygen species by mitochondria: central role of complex III. J Biol Chem. 2003;278(38):36027–31. Epub 20030702. https://doi.org/10.1074/jbc.M304854200. PubMed PMID: 12840017.

Chapter 6
Process of Oxidative Phosphorylation and Free Radical Generation

Abstract Before phosphorylation of ADP to ATP occurs at complex V, secondary to matrix proton uptake, respiration must supply the protons that are pumped out across the inner mitochondrial membrane to generate the proton motive force. As this is associated with electron transfer from one mobile electron carrier to another, it is essential that an electron is captured first by the next charge carrier before highly electronegative atoms such as O_2 can capture the electrons that could result in an excess of superoxide radicals. Normally, this is kept to a minimum by several factors that keep forward electron transfer at a fast pace and inhibit reverse electron transfer. This is aided by tight distance coupling of the mobile carriers, the presence of hydrophobic amino acids in the respiratory complexes that keep water away from the electrons, and by the presence of mitochondrial and cytosolic superoxide dismutases, which scavenge for superoxide ions and convert them to hydrogen peroxide, a gas that can transverse membranes. Meanwhile, the reuptake of protons drives ATP synthesis, and the fewer protons entering the matrix for the ATP generated, the higher is the coupling ratio (ATP generated/protons taken up), a marker of mitochondrial efficiency. Proton leak is the matrix entry of H^+ ions by any mechanism, for example via cation exchangers, other than via ATP synthase. Proton leak stimulates respiration because it decreases the energy of two components of the proton motive force, the voltage ($\Delta\Psi_m$), and proton (ΔpH_m) gradients. Low levels of electron leak can lead to formation of hydrogen peroxide, adapted to use as a gaseous modulator of mitochondrial and cell function. Glycolytic and tricarboxylic acid cycle enzyme activities are regulated allosterically by substrates, metabolic intermediates, metal and organic cofactors, ATP/ADP ratios, and others. Respiration is modulated by the proton motive force and by factors that modulate it, like protons, but also by Ca^{2+} and K^+ ions. The individual respiratory complexes can also be inhibited or stimulated; for example, complex IV is inhibited by nitric oxide or ATP. An understanding of these mechanisms typically requires dynamic measurements of multiple mitochondrial factors in situ or in vivo, such as membrane potential, pH gradient, redox state, respiration (O_2 consumption), ATP/ADP ratios, enzyme activities, cation concentrations, and reactive oxygen species.

D. F. Stowe, *Evolution of Bioenergetics from Elements to Life*,
https://doi.org/10.1007/978-3-031-76668-8_6

Keywords Assessing free radicals · Bioelectricity · Bioenergetics · Cardioprotection · Chemiosmosis · Electron leak · Electronegativity · Electron transport system · Fermentation · Forward electron transfer · Gibbs free energy · Glycolysis · Hydrogen peroxide · Inner mitochondrial membrane · Mitochondrial coupling · Mitochondrial ion exchangers · Mitochondrial pH gradient · Mitochondrial K^+ channels · Mitochondrial K^+-H^+ exchanger · Mitochondrial stoichiometry · Mobile electron carriers · Nitric oxide · O_2 consumption · Oxidative phosphorylation · Quinone · Peroxynitrite · Proton circuit · Proton leak · Proton motive force · Reactive oxygen species · Respiratory control index · Respirasome · Reverse electron transfer · Semi-ubiquinone radical · Superoxide dismutases · Superoxide radical · Tricarboxylic acid cycle · Ubiquinol · Ubiquinone

6.1 Free Radical $O_2^{\bullet-}$ Generation and ROS Scavenging in Cardiac Mitochondria

In glycolysis (Eq. 5.1), a net of $4H^+$ and 4*e*- are essentially removed from glucose to form pyruvic acid with the $4H^+$ and 4*e*- transferred together from $2NAD^+$ to 2NADH. Similarly, protons and *e*- are transferred together during the TCA cycle. However, this process differs in the ETS wherein *e*- are carried along Fe-S centers in the protein complexes and by mobile carriers (e.g., ubiquinone and cytochrome *c*) from one ETS complex to another while the protons are expelled into the IMS. It is only at the last step where O_2 attracts 2 *e*- and with $2H^+$ unites to make H_2O. During this fast process, *e*- are protected from H_2O by hydrophobic amino acids and tight coupling of the mobile carriers. However, an occasional *e*- escapes and is taken up prematurely by O_2.

This singlet *e*- leak can occur via the fully reduced flavin of the flavin mononucleotide (FMN) moiety at complex I or from the semi-ubiquinone radical ($CoQH^{\bullet-}$) in the ubiquinone binding site of complex I, and from the Q_o site of complex III [1–4], as depicted in the simple cartoon of Fig. 6.1 [5], and in a more detailed manner on structural models of complexes I-IV of Fig. 6.2 [6]. The result can be partially reduced O_2 at complex IV to form the unstable superoxide anion radical $O_2^{\bullet-}$. If the next *e*- is not rapidly transferred at complex IV, this $O_2^{\bullet-}$ rapidly reacts with nearby stable molecules to form other downstream free radicals, or with other free radicals like $NO^{\bullet}$, to form strong, non-free radical reactants, like peroxynitrite ($ONOO^-$) [3, 7, 8], or is scavenged by the superoxide dismutases SOD1 and SOD2 to form hydrogen peroxide, H_2O_2 [7], as shown in Fig. 6.3 [1].

All reactive O_2 species (ROS) in lifeforms begin with $O_2^{\bullet-}$, principally within the mitochondrial ETS as discussed also in Sects. 1.3, 2.3, 5.2, 6.1, 7.2, 7.3, 7.4 and 8.4. There are other sources of ROS [7], e.g., xanthine oxidases (XO), lipo-oxygenases (LOX), and polymorphic cytochrome P450 enzymes (CYPs), as shown in Fig. 5.2; these sources of $O_2^{\bullet-}$ are not discussed further. The amount and rate of

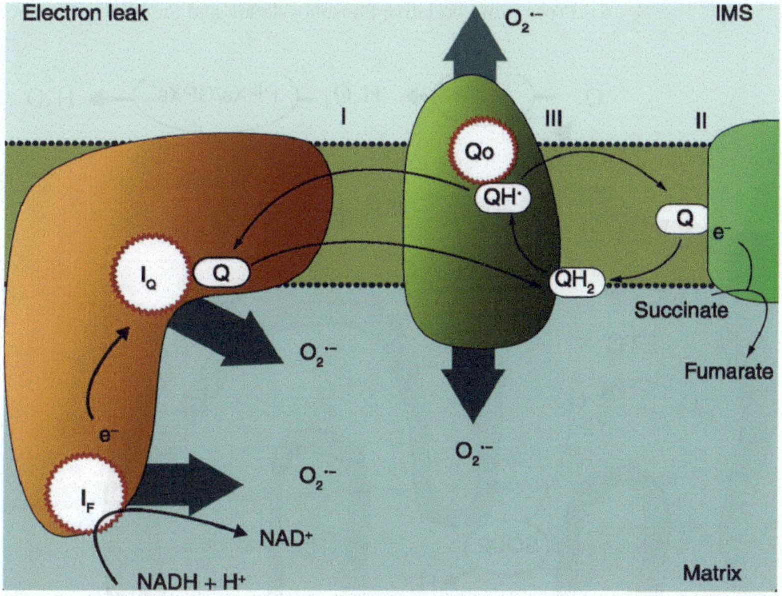

Fig. 6.1 Sites of electron leak (loss) from ETS complexes during *e*- transport: Electrons carried by NADH are transferred to the flavin mononucleotide (I_F) site in complex I, where they normally pass down a chain of Fe-S centers to the ubiquinone binding site (I_Q). At both the I_F and I_Q sites, these *e*- react with O_2, forming superoxide ($O_2^{\bullet-}$) within the matrix. In complex III, QH_2 binds to the Q_O site, where its *e*- can bypass their normal transfer in the Q-cycle and react directly with O_2 to form $O_2^{\bullet-}$ that is released to both sides of the mitochondrial inner membrane. (Cartoon (Fig. 3) and its modified legend resourced with permission from copyright owner of Jastroch et al. [5] (essay))

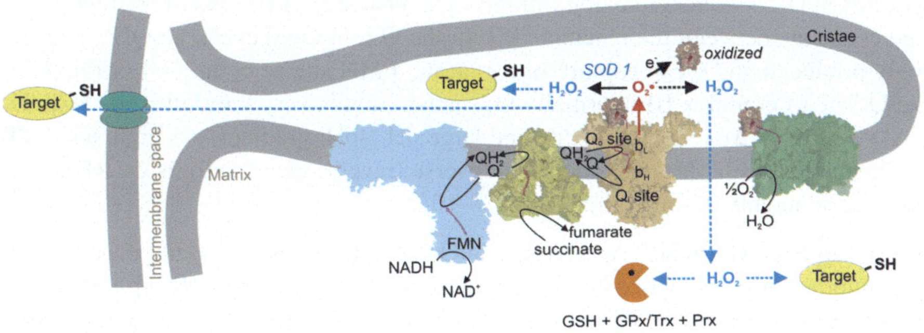

Fig. 6.2 Topology of ROS release from the Q_o site and implications for redox signaling: In the model, the primarily formed superoxide ($O_2^{\bullet-}$) is almost completely released from the Q_o site into the intermembrane space and the cristae lumen that encompass the main portion of the IMM in intact mitochondria. The membrane-impermeable $O_2^{\bullet-}$ can either react with oxidized cytochrome *c* or be readily dismutated to H_2O_2. H_2O_2 can diffuse across membranes thereby reaching the matrix where the bulk is probably consumed by the antioxidative system (especially by the thioredoxin (Trx)/peroxiredoxin (Prx) system and glutathione (GSH)/glutathione in the IMS and in the matrix by complex III derived ROS (Bleier et al. unpublished results). (Cartoon (Fig. 4) and its modified legend resourced with permission from copyright owner of Bleier and Drose [6] (review))

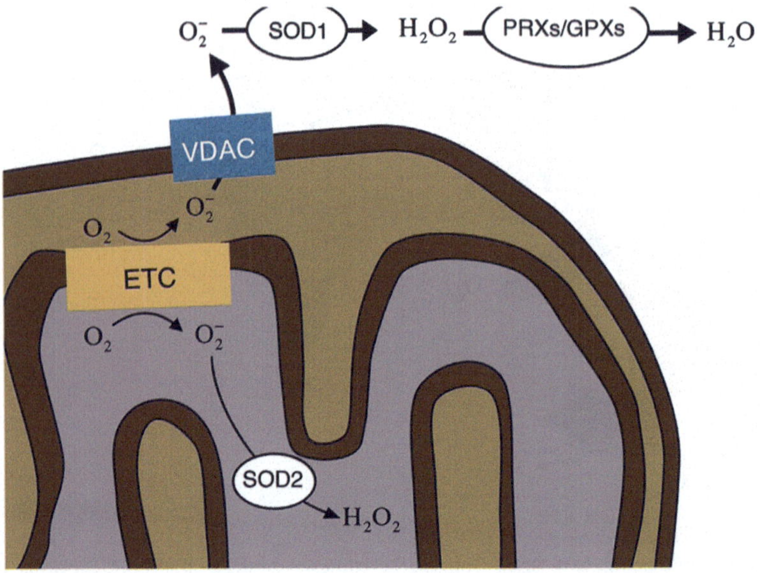

Fig. 6.3 Mitochondrial ROS generation: Superoxide ($O_2^{\bullet-}$) Is produced as an intermediate in the electron transport system (ETS). In mitochondria matrix excess $O_2^{\bullet-}$ is reduced to H_2O_2 by SOD_2. $O_2^{\bullet-}$ in the inter-membrane space (IMS) can also be exported to the cytoplasm through VDAC. It is then transformed into H_2O_2 by SOD_1 and finally reduced to H_2O through catalase, peroxiredoxins, and glutathione peroxidases. (Cartoon (Fig. 1b) resourced with license (https://creativecommons.org/licenses/BY-NC-ND/4.0/) from Dunn et al. [1] (review))

$O_2^{\bullet-}$ generated within the ETS depend on several factors: concentration and proximity of local O_2, *e*- donors like ubiquinone, cytochrome *c*, FMN, and the second order rate constants between the reactants [3]. In the Q (quinone) cycle, $CoQH^{\bullet-}$ is fleetingly produced so O_2 can capture one *e*- at the FMN site of complex I or one *e*- at the Q_o site of complex III. Normally $O_2^{\bullet-}$ generation is very low, but if the NADH/NAD^+ ratio and Δp_m (charged $\Delta\Psi_m$ and large pH_m–pH_c gradient) are high, and ATP production is low, *e*- transfer can go in reverse (reverse electron transfer, RET) and produce abundant $O_2^{\bullet-}$ at complex I [3, 7–10].

Assessing Free Radicals An increase in $O_2^{\bullet-}$ can be assessed by dihydroethidium blue fluorescence in isolated hearts during ischemia reperfusion (IR) injury [11–14] as shown in Fig. 6.4 [13]. This data figure demonstrates that two, five minute pulses of global ischemia (ischemic preconditioning, IPC), applied before 30 min ischemia, greatly reduces the levels of reactive ethidium fluoroescence during and after ischemia. That the ethidium signal represents $O_2^{\bullet-}$ is validated by the reduced ethidium levels if the mitochondrial $O_2^{\bullet-}$ scavenger Mn^{3+} tetrakis (4-benzoic acid) porphyrin (Mn-TBAP) is given before IPC. Computer modeling also suggests that complex II and III are normally the sites of small, physiological *e*- leaks, whereas complex I is the site of a major *e*- leak when forward electron transfer (FET) is blocked and reverse electron transfer (RET) prevails [15].

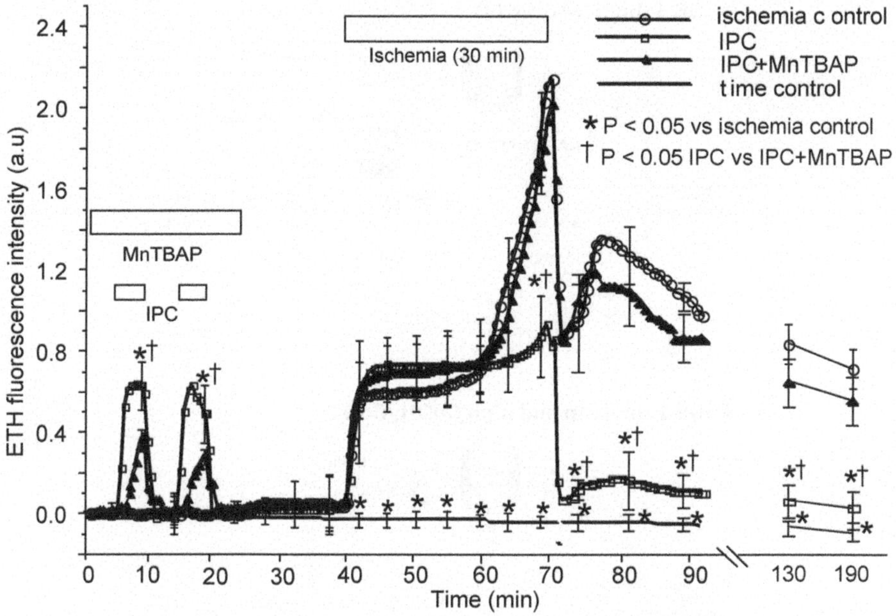

Fig. 6.4 Assessment of a marker of superoxide in guinea pig isolated hearts during ischemia-reperfusion injury: Fluorescence intensity (FI) in arbitrary units (a.u.) was normalized to zero at baseline. The signals are of the oxidation product of dihydroethidium (DHE), i.e., ethidium (Eth), an indicator of intracellular reactive O_2 species (ROS), primarily superoxide. Eth was measured spectrofluorometrically (excitation wavelength 540 nm and emission wavelength 590 nm) in intact hearts via a fiber-optic probe placed against the left ventricular free wall. Manganese (III) tetrakis (4-benzoic acid) porphyrin chloride (MnTBAP) is a mitochondrial ROS scavenger. MnTBAB alone group data are not shown because they were not different from those of the ischemia-control group. Sampling rate was five recordings/min, but to enhance clarity, this figure shows only 1.6 recordings/min. Intracellular ROS generation occurred during ischemic preconditioning (IPC) pulses but was attenuated in the presence of MnTBAP. During later ischemia and throughout reperfusion, ROS generation was markedly less in the IPC group than in the other ischemia groups. (Summarized data plot (Fig. 4) and its modified legend resourced with permission from copyright owner of Kevin et al. [13])

Certain conditions are known to lead to enhanced release of $O_2^{\bullet-}$ into the matrix at complex III with subsequent formation and release of H_2O_2. Again, these conditions include low ATP synthesis with high $NADH/NAD^+$, high Δp_m, as in complex I, and with high $CoQH_2/CoQ$ as shown by three operating modes of mitochondria in Fig. 6.5 [3]. In other words, at high OxPhos rates, $O_2^{\bullet-}$ generation is minimal. For example, if Δp_m is reduced due to increased respiration and ADP is being phosphorylated, as in the beating heart, $O_2^{\bullet-}$ generation is largely kept in check.

$O_2^{\bullet-}$ to H_2O_2 Although $O_2^{\bullet-}$ can reduce Fe^{3+} to Fe^{2+} and react with other ligands to form other radicals, its usual fate is rapid dismutation ($K_d = 1.6 \cdot 10^{-9}$ moles/s) to H_2O_2 by Mn-superoxide dismutase (Mn-SOD) in the mitochondrial matrix [7, 16]. H_2O_2 is not actually a ROS, but it is a quite stable oxidant, and because it is not

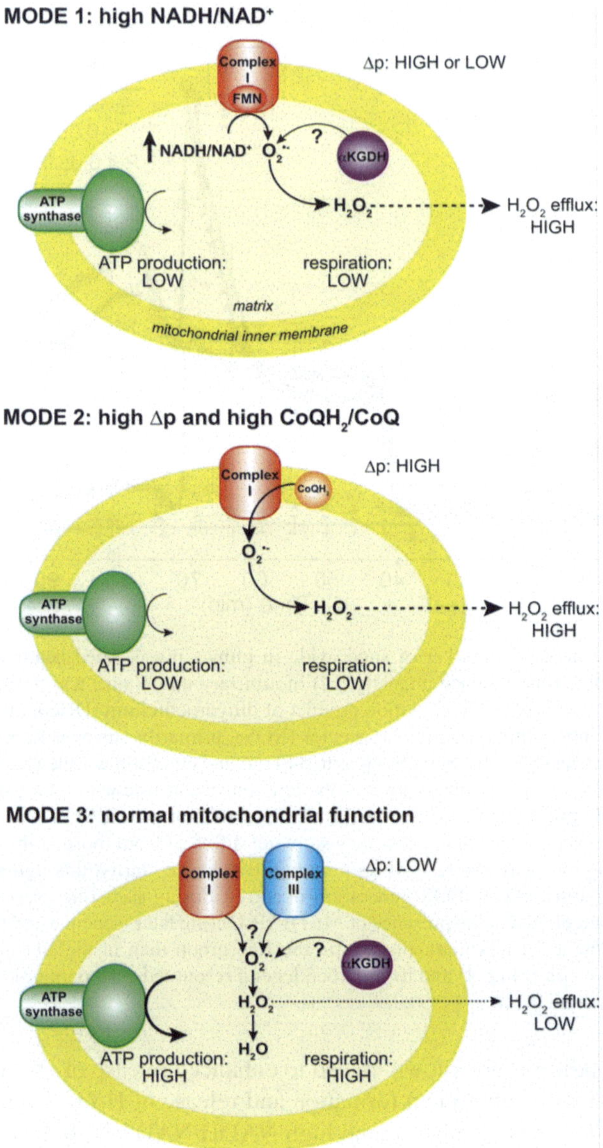

Fig. 6.5 Modes of mitochondrial operation that lead to $O_2^{\bullet-}$ production: In mode 1, the NADH pool is reduced, for example by damage to the respiratory chain, loss of cytochrome *c* during apoptosis, or low ATP demand. This leads to a rate of $O_2^{\bullet-}$ formation at the FMN of complex I that is determined by the extent of FMN reduction which is in turn set by the NADH/NAD+ ratio. Other sites such as αKGDH may also contribute. In mode 2, there is no ATP production and there is a high Δp_m and a reduced CoQ pool which leads to reverse electron transfer (RET) through complex I, producing large amounts of $O_2^{\bullet-}$. In mode 3, mitochondria are actively making ATP and consequently have a lower Δp_m than in mode 2 and a more oxidized NADH pool than in mode 1. Under these conditions, the flux of $O_2^{\bullet-}$ within mitochondria is far lower than in modes 1 and 2, and the $O_2^{\bullet-}$ sources are unclear. (Cartoons (Fig. 3) and its modified legend resourced with license (https://creativecommons.org/licenses/BY-NC/4.0/) from Murphy [3] (review))

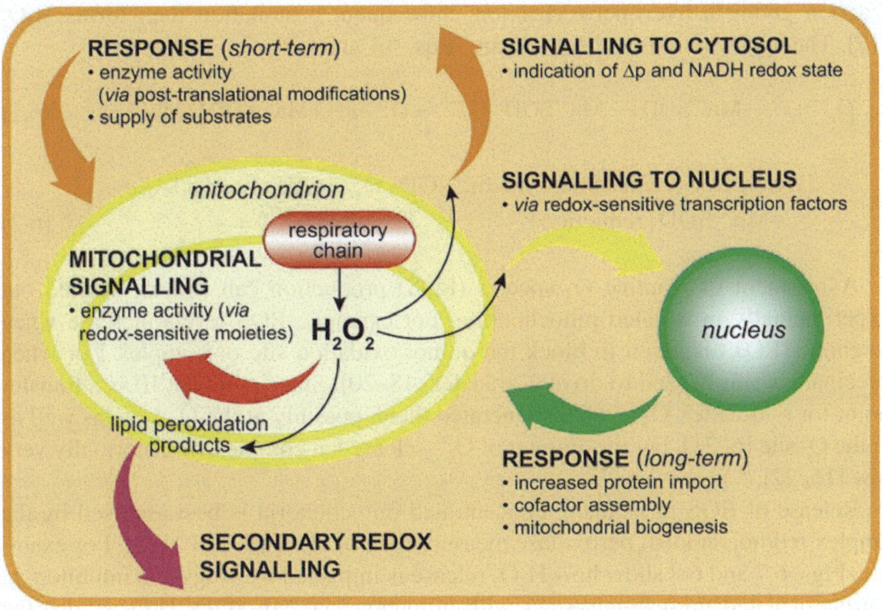

Fig. 6.6 Possible mechanisms of mitochondrial redox signaling: The production of H_2O_2 from mitochondria is a redox signal. H_2O_2 generated by mitochondria can reversibly alter the activity of proteins with critical protein thiols by modifying them to intra- or inter-protein disulfides, or to mixed disulfides with GSH. These modifications can occur on mitochondrial, cytosolic, or nuclear enzymes, and on carriers or transcription factors, which transiently alters their activities. The change in activity can be reversed by reducing the modified protein thiol by endogenous thiol reductants such as GSH or thioredoxin. As the extent of H_2O_2 production from mitochondria depends on factors such as Δp_m, or the redox state of the NADH pool, it can act as a retrograde signal to the rest of the cell, and so report on mitochondrial status. This signal can then lead to the short-term modification of pathways supplying substrates to the mitochondria. Alternatively, longer-term modifications can occur through modifying redox-sensitive transcription factors that adjust the production of mitochondrial components. In addition, external signals may modify $O_2^{\bullet-}$ production by the respiratory chain by post-translational modifications. Alteration of the activity of mitochondrial peroxidases could also modulate H_2O_2 efflux from mitochondria to the rest of the cell. It is also possible that secondary redox signals, such as lipid peroxidation products derived from H_2O_2, could act as secondary redox signals. (Cartoon with schemes (Fig. 5) and its modified legend resourced with license (https://creativecommons.org/licenses/BY-NC/4.0/) from Murphy [3] (review))

charged like $O_2^{\bullet-}$, it passes easily through the IMM into other cell compartments or out of the cell. This makes H_2O_2 perfect as a localized, cellular messenger. At low concentrations, H_2O_2 is a signaling molecule within the mitochondrial matrix as well as in the cytosol and nucleus as presented in Fig. 6.6 [3], where H_2O_2 acts on redox-sensitive moieties and secondarily via lipid peroxidation products. With the resting state of the enzyme as the Mn^{3+} form, *e*- transfer to the metal from an $O_2^{\bullet-}$ ion in the initial half-reaction yields O_2 and enzyme-bound Mn^{2+}. This, in turn, is oxidized back to Mn^{3+} by a 2nd molecule of $O_2^{\bullet-}$ in a proton-coupled *e*- transfer reaction giving a hydroperoxyl anion. Subsequent protonation then forms H_2O_2 [17]. The half-reactions of MnSOD are (Eqs. 6.1 and 6.2):

$$O_2^{\bullet-} + e- + Mn^{3+}SOD \rightarrow Mn^{3+}SOD : O_2^{\bullet-} \rightarrow O_2 + e- + Mn^{2+}SOD(\text{reduction}) \quad (6.1)$$

$$O_2^{\bullet-} + e- + Mn^{2+}SOD \rightarrow Mn^{+2}SOD : O_2^{\bullet-} + 2H^+ \rightarrow H_2O_2 + e- + Mn^{3+}SOD(\text{reduction}) \quad (6.2)$$

Assessment of reactive O_2 species (ROS) production can also be carried out experimentally in isolated mitochondria. For example, ROS levels increase when rotenone (ROT) is given to block the quinol oxidation site of complex I or when succinate is singly used to drive *e*- transfer [18–20]. Since complex III is a transfer point for *e*- transfer, $O_2^{\bullet-}$ can be generated there, possibly at the Q_{bc1} site as well as at the Q_o site [6, 21], but the amount of $O_2^{\bullet-}$ released at complex III is normally very low [15, 22].

Release of ROS in isolated, concentrated mitochondria is best assessed by the amplex red/horseradish peroxidase measurement of H_2O_2 [7, 9, 10, 18]. For example, Figs. 6.7 and 6.8 show how H_2O_2 release is impacted by drugs and inhibitors in cardiac isolated mitochondria. As with mitochondrial Mn-SOD, H_2O_2 is also the product of cytosolically released $O_2^{\bullet-}$ by the enzyme Cu-SOD present in the cytosol (Sect. 2.3). By blocking the individual mitochondrial respiratory complexes, or the specific *e*- transfer sites (Fig. 5.20), useful information is inferred on the direction of *e*- transfer and the mechanisms of *e*- transfer by assessment of ROS [6, 7, 9, 15, 21, 23]. The dual roles of ROS in providing cardiac protection, cardioprotection, and in inducing cardiac ischemic damage, are further discussed in Sects. 7.2 and 8.4.

6.2 Proton (H^+) Pumping and Uptake to Drive Mitochondrial Oxidative Phosphorylation

Respiration (substrate oxidation with O_2 reduced to H_2O) and phosphorylation (ADP phosphorylation to ATP) (Sect. 5.2) are two distinct processes that are linked, but not inextricably. They form the basis of the chemiosmosis hypothesis proposed by Mitchell in 1961 [24]. Recall that chemiosmosis (Sects. 1.3, 2.2, 3.3, 4.1 and 5.2) is a process by which the movement of an ion against its concentration gradient across a membrane impermeable to that ion creates both a charge and a chemical gradient that act together as a battery to store energy, which can then be used if the ion is selectively allowed to recross the membrane. In practice, Mitchell theorized that the pumping of protons from the energy stored in NADH and $FADH_2$, and the reentry of protons into the matrix, drives an engine (ATP synthase) to make high energy ATP from ADP. Figure 6.9 [25] provides a simple diagram of the *e*- transported with the proton pumping and proton re-uptake circuit linked to ATP generation (OxPhos), and the linked transporters for P_i and nucleotides (ACC, i.e., ANT).

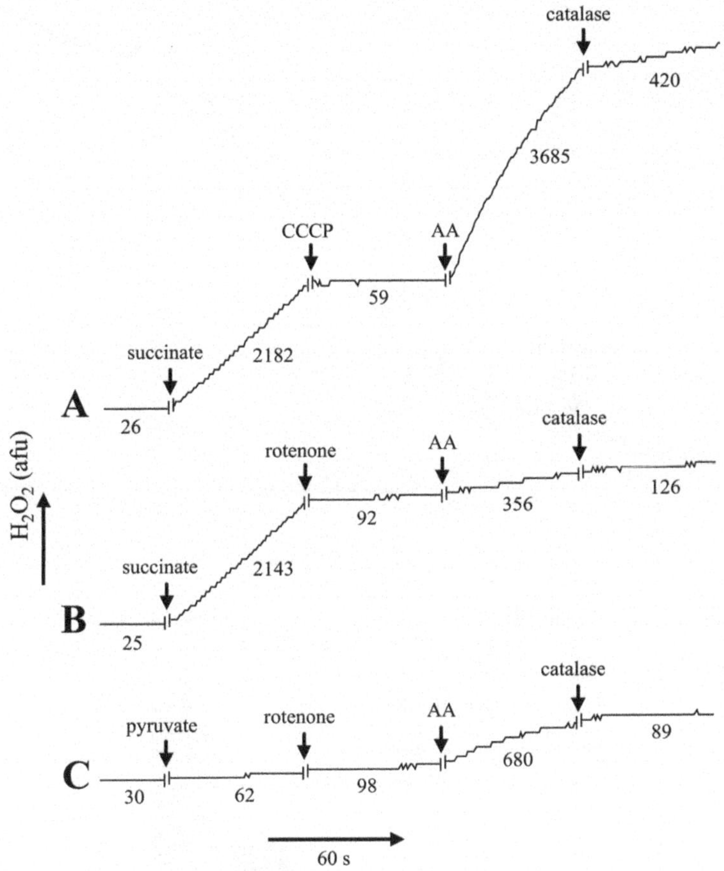

Fig. 6.7 H_2O_2 release rate for guinea pig cardiac isolated mitochondria: Representative traces of H_2O_2 release rates during succinate (10 mM) supported respiration. H_2O_2 generation was abrogated by addition of 4 μM CCCP (panel **a**), a mitochondrial uncoupler, or 4 μM rotenone (panel **b**), a complex I blocker. Panel **c** shows the H_2O_2 generation rate during pyruvate (complex I substrate, 10 mM) supported mitochondrial respiration. Catalase (300 U/ml) was added to confirm H_2O_2 production. AA = antimycin A (complex III blocker, 5 μM). afu = arbitrary fluorescence units. The numbers below lines represent H_2O_2 release rate in afu per minute. (Data plot (Fig. 4) and its modified legend resourced with license (https://creativecommons.org/licenses/BY-NC/4.0/) from Heinen et al. [9])

Bioelectricity The proton circuit is analogous to an electrical circuit [26]. The proton circuit requires an aqueous environment so that protons, in the form of H_3O^+ (Sect. 1.3), can move around in H_2O (Sects. 1.2, 1.3 and 5.1) from the anode (e.g., substrate delivered at cell membrane) to the cathode (e.g., production of H_2O at complex IV). As in a battery, ion flux in a liquid is much slower than the electromagnetic wave (*e*- flux) in a copper wire (Sect. 4.3). When adding an oxidizable substrate to isolated, buffered mitochondria that are depolarized, it takes several seconds to recharge the $\Delta\Psi_m$ [27].

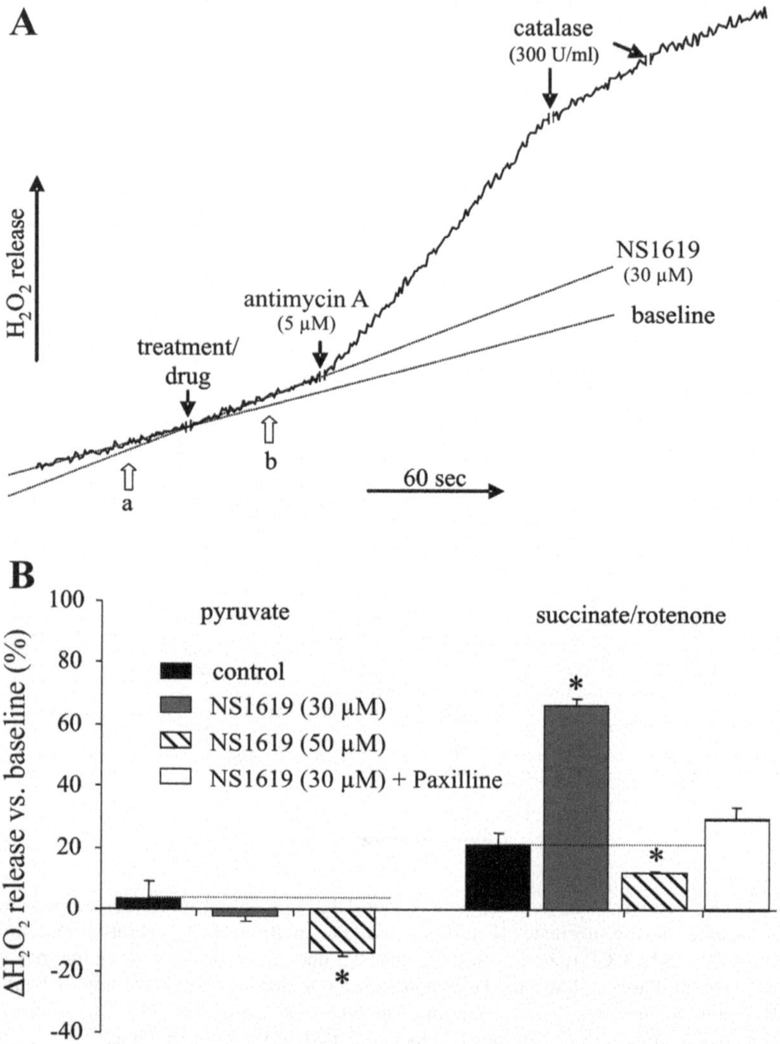

Fig. 6.8 Mitochondrial H_2O_2 release rate. Panel 1: Representative trace for H_2O_2 release rate with succinate as substrate. Panel 1a: Control; panel 1B: 30 μM NS1619 (BK_{Ca} channel opener); panel 1c: 5 μM paxilline (BK_{Ca} channel blocker) + 30 μM NS1619. Panel 2: Summarized data for H_2O_2 release rate as a % of time controls. A 10% change represents a change in H_2O_2 release rate of approximately 1.5 pmol • min^{-1} H_2O_2 • mg^{-1} protein. All treatment effects are compared to the baseline of the same experiment. ($^*P < 0.05$ vs. control; $^\#P < 0.05$ vs. NS1619; $n = 6$ for each group). (Data plot (Fig. 5) and its modified legend resourced with license (https://creativecommons.org/licenses/BY-NC/4.0/) from Heinen et al. [10])

Mitochondrial bioelectricity, e.g., power, P = IV, can be construed as a proton motive force of about −120 to −200 mv ($pmf = \Delta p_m = \Delta\Psi_m + (-59)\,(pH_m - pH_c)$) (at 25 °C) that provides the energy (battery, in V) utilized when protons are carried across the IMM (current, in I) into the matrix at the F_o site of ATP synthase and ATP

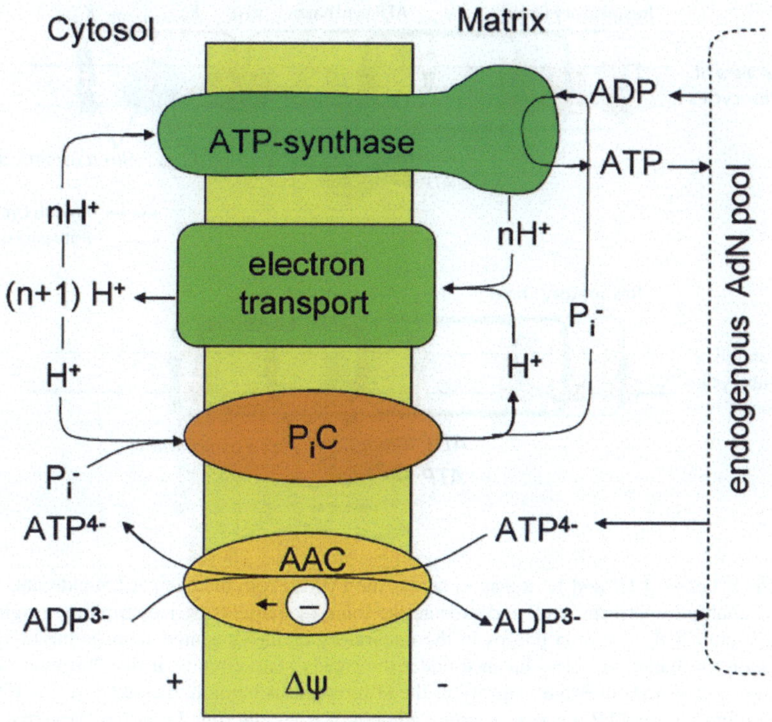

Fig. 6.9 H^+ transport balance for synthesis of cytosolic ATP: Out of ($n + 1$) H^+ pumped by *e*- transport, nH^+ are used for synthesis of endogenous ATP and 1 H^+ for the $\Delta\Psi_m$ driven export of ATP. This H^+ is returned by neutralizing the P_i^- import and thus compensates the export of one negative charge in the ADP^{3-}–ATP^{4-} exchange. (Diagram (Fig. 4, modified from text Ref. [81]) and its modified legend resourced with license (https://creativecommons.org/licenses/BY-NC-ND/4.0/) from Klingenberg [25] (review))

is generated (power P, in watts) at the F_1 site of ATP synthase [28]. It should be noted again that some ancient and modern anaerobic bacteria derive energy only by fermentation (glycolysis). Or they might use alternatives to O_2 as the final *e*- acceptor, e.g., less electronegative nitrogen or sulfur (Fig. 1.10), and in alkaline environments, they might use an alternative ion motive force, e.g., K^+ instead of H^+.

Hence, there is recent compelling evidence that the mammalian complex V can utilize K^+ uptake (Fig. 6.10b) as well as H^+ uptake (Fig. 6.10a) to drive ATP synthesis [29]. These diagrams describe the generally regarded view of the proton and K^+ circuits (a) where protons enter the matrix via the F_o site but also via the exchanger mKHE, secondary to K^+ entry via mK^+ channels. The exchangers mNHE, mCHE, and P_iC can also promote inward proton leaks.

Another view (b) is based on evidence that the F_o site of ATP synthase can also utilize K^+ for phosphorylation, thus also functioning as a K^+ channel, but initiating a proton leak via KHE. However, the F_o site of ATP synthase is 10^6 more selective for H^+ than for K^+, and $m[K^+]$ is 10^6 greater than the $m[H^+]$. It is estimated that the

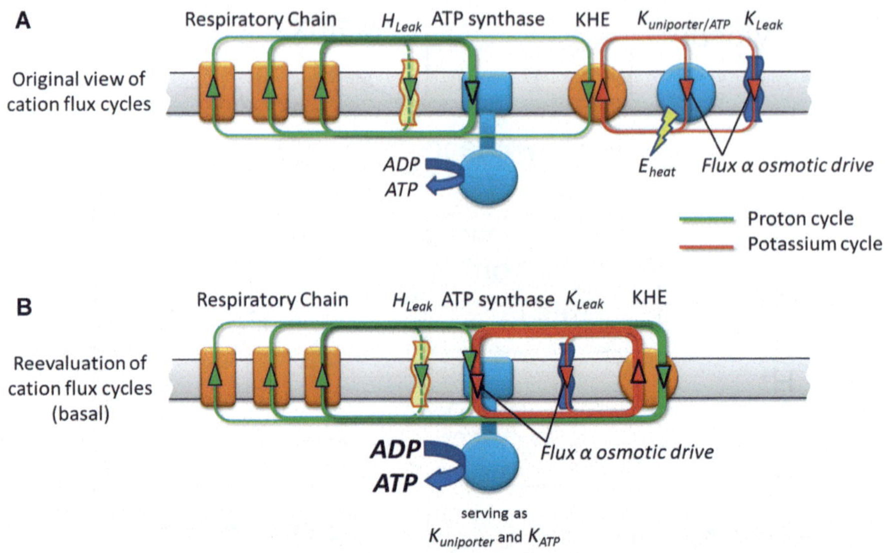

Fig. 6.10 Scheme of H^+ and K^+ transport across the IMM: From the energetic standpoint, all the energy available to perform work and execute the ionic movements derives from the original H^+ gradient established by proton pumps in the respiratory chain. A central point is the obligatory preservation of charge and mass balance under the steady state circuits. In the "original view of cation flux cycles" (**a**), a certain majority of the H^+ gradient is being harnessed by F_oF_1-ATP synthase directly to make ATP, whereas a certain amount of K^+ is entering the matrix through an ordinary K^+ channel mechanism (e.g., a "mK_{ATP}-uniporter" channel), driven by $\Delta\Psi_m$, and extruded via KHE utilizing the energy remaining in the fraction of the H^+ gradient not harnessed by F_oF_1. The equivalent energy of this fraction being used to extrude K^+, and a large fraction of that non-ATP-producing energy would essentially be dissipated as heat in the constant cycle of K^+ recirculation (red circuit in **a**). In our new proposed mechanism (**b**), the same amount of energy available in the original H^+ gradient but largely lost as heat is entirely available to produce ATP, simply by having the mK_{ATP}-uniporter mechanism reside inside, and as natural part of, F_oF_1 with the traffic of H^+ or K^+ contributing its energy to producing ATP. The remainder of the H^+ gradient energy is now utilized to remove all the K^+ that entered via F_oF_1 (n.b., this exchange process of extruding K^+, restoring $\Delta\mu_K$, is the way that the H^+ gradient energy is still the original, entire driving force for ATP production). However, the gain is that more ATP is produced for the same input energy by not wasting some of that energy on maintaining what was originally thought to be a separate K^+ cycle that does not/cannot generate any ATP. Engineered this way, it is a better, tightly coupled system of energy supply-demand matching through the K^+ cycle utilizing F_oF_1 because the matrix influx of K^+ is truly directly proportional to ATP synthesis. (Cartoon (Fig. 6) and its legend resourced with permission of owner (Copyright 2024, American Physiological Society) of Juhaszova et al. [29])

energy of the K^+ IMM gradient could contribute to ATP synthesis by 3.5 K^+ for every H^+ taken up, but at the expense of a 2.6 fold increase in respiration [29], likely triggered by enhanced proton leak. The consequence of proton leak via H^+ ion exchangers is also discussed in Sect. 7.2. It should be noted that the movement of ions across ion exchangers is not dependent on the actual concentrations of each ion but rather on the ratio of the individual ion concentrations across the IMM [30].

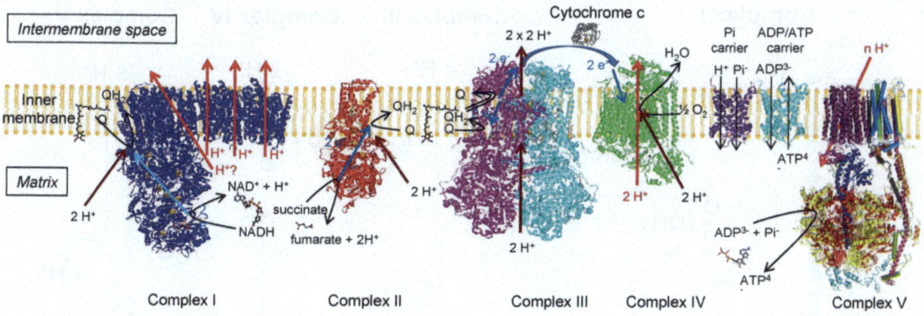

Fig. 6.11 Mitochondrial OxPhos: Pumped protons are in red. Chemical protons are in brown. Electrons transfers are in blue. The scheme is designed with transfer of two e-, even for complex IV (half reaction). Color codes are: complex I (blue); complex II (red); complex III (dimer in magenta and cyan); complex IV (green); cytochrome c (cyt *c*) (grey); ADP/ATP carrier (cyan); P_i carrier (purple); and ATP synthase/ase (multiple colors). [Figure can be viewed in the online issue, available at wileyonlinelibrary.com]. (Cartoon (Fig. 1) and its modified legend resourced with permission from copyright owner of Mazat et al. [34] (review))

Stoichiometry of Bioenergetics As introduced earlier, in nascent lifeforms, only glycolysis was available to transduce C-based nutrients into energy, usually in the form of ATP. Later with bacterial proteins (protomitochondria) having been incorporated into cells as highly condensed organelles with tightly packed cristae [31], the 2-C acetyl-CoA (Figs. 3.8 and 5.6) now entered the mitochondrial matrix to condense with 4-C oxaloacetate to form 6-C citrate (Figs. 5.7 and 5.8) at the beginning of the TCA cycle (Sect. 5.1, Eq. 5.4). One molecule of glucose generates 4 molecules of ATP but only 2 net ATP molecules are available through glycolysis and the TCA cycle. However, from oxidation of one molecule of glucose completely to H_2O, 24 *e*- (Eq. 5.6) are harnessed for transfer down the ETS during respiration, i.e., 12 *e*- pairs, as each O_2 molecule accepts 2 *e*-, and 10 H^+ are available for proton pumping by complexes I, III, and IV for each O_2 consumed (H^+/O ratio of 10:1) [32, 33]. Figure 6.11 effectively summarizes proton and *e*- transfer across the mitochondrial respirasome using structural models of each complex [34].

During OxPhos (Figs. 4.2, 5.13 and 6.11), the influx of approximately 4 protons is required for each mole of ATP produced, and if 10 protons are pumped out by complexes I, III and IV per mole of O_2 consumed, the P/O ratio would be maximal at 2.5:1, but this ratio actually varies from about 2.3:1 to 2.8:1 depending on the substrate. Moreover, these ratios are attenuated from their maximums by proton [34] and electron leaks [3]. In addition, the number of protons required to produce one ATP (H^+/ATP) at complex V is estimated from the theoretical ratio of 4:1 to be closer to 8:3 when tightly coupled as depicted in Fig. 6.12 [36]. However, during proton leak this can go up to 13:3, indicating inefficacy of OxPhos due to uncoupling with release of some of the energy as heat [33].

ATP Synthase Complex V (F_oF_1-ATP synthase/ase) phosphorylates ADP at the F_1 site and H^+ influx rotates the proton-driven motor at the F_o site (Figs. 3.23, 6.11 and 6.12), but the reaction at complex V is reversible under critical conditions. F_oF_1-ATP

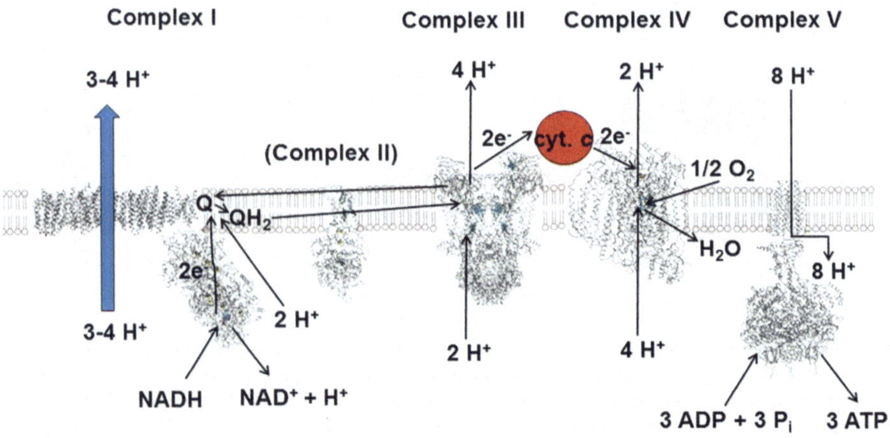

Fig. 6.12 Respiratory chain and ATP synthase: The respiratory chain is depicted in a phospholipid membrane, as it is found in most mitochondria and in many bacteria. ATP synthase is shown to translocate eight protons per three synthesized ATP molecules, as observed in animal mitochondria. (Diagram (Fig. 1) and its legend resourced with permission of owner (Copyright 2024, American Chemical Society) of Wikstrom et al. [35])

synthase/ase directionality is governed by both $\Delta\Psi_m$ and ΔpH_m, and the "reversal potential," $E_{REV-ATPase}$, which in turn is dependent on the concentration of the reactants ATP/ADP, and H^+ [37–40]. $E_{REV-ATPase}$ can occur when $\Delta\Psi_m$ falls between −120 and −100 mV, depending on the matrix $[ATP]_m/[ADP]_m$ ratio, $[P_i]_m$, ΔpH_m, and the coupling ratio H^+/ATP. Thus, if $\Delta\Psi_m$ is below a threshold, the direction is right to left as ADP is made from ATP with outward proton pumping at complex V; if $\Delta\Psi_m$ is maximal, the direction is left to right as ATP is made from ADP with inward flux of protons. Hydrolysis of 1 mole of ATP to ADP yields energy ($\Delta G^{o\prime}$) between −31 and −57 KJ/mol (8–14 kcal/mol) of energy [41] depending on conditions such as temperature and the presence of Mg^{2+}. The left, right reversible reaction is (Eq. 6.3):

$$\text{ADP} + \text{P}_i + 4\text{H}^+\left(\text{IMS}\right)\left(+\text{free energy}\right) \rightleftharpoons \text{ATP} + \text{H}_2\text{O} + 4\text{H}^+\left(\text{matrix}\right) \quad (6.3)$$

Causes of Proton Leak Completely coupled OxPhos would occur only if each *e*- transferred in the ETS were used for proton pumping ($H^+/e-$ = 1:1) and all the protons were used to phosphorylate ADP (H^+/ATP), nominally 8:3, i.e., 2.67 as discussed above [36, 42, 43], but this never occurs in normal in situ or even in isolated mitochondria [34]. Uncoupling can be described as *e*- leak (Figs. 5.17 and 6.2) and/or proton leak (Figs. 5.14 and 6.13) [5, 43, 44], but proton "leak" is not necessarily a deleterious situation; it is a controlled and powerful stimulus to enhance respiration when the Δp_m falls. Although coupling efficiency is greatest when mitochondria have suitable oxidizable substrates (suitable NADH/NAD^+ ratio) and are

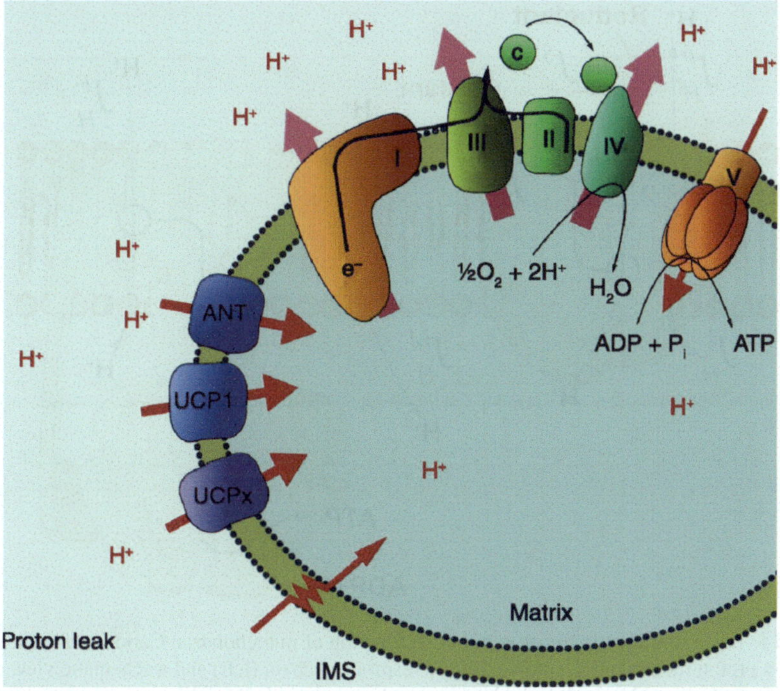

Fig. 6.13 Proton pumping and leak across the mitochondrial inner membrane: Protons are pumped out of the matrix into the intermembrane space (IMS) by complexes I, III, and IV of the ETS. This establishes a proton motive force (Δp_m) across the inner membrane. Proton re-entry through the ATP synthase (complex V) couples the release of Δp_m to ATP synthesis. All other means of proton re-entry constitute proton leak, as Δp_m derived from substrate oxidation is depleted without catalyzing ATP synthesis. Mechanisms of proton leak include direct movement of protons across the phospholipid membrane (the "water wires" model), diffusion through or around integral membrane proteins, or inducible transport through the adenine nucleotide translocase (ANT) or uncoupling proteins (UCP1, UCPx). (Cartoon (Fig. 1) and its legend resourced with permission from copyright owner of Jastroch et al. [5] (essay))

called upon to generate ATP for cellular needs (Fig. 5.14), this activity tends to reduce $\Delta\Psi_m$ as protons enter the matrix via complex V [36].

When mitochondria are charged but OxPhos is low, proton leak (H^+ influx outside of ATP synthase) is known to occur, but the mechanism has been unclear. While inward proton leak reduces the *pmf* (a.k.a., Δp_m), this stimulates respiration as a control mechanism to restore $\Delta\Psi_m$, possibly at the expense of utilizing the ΔpH_m gradient as discussed in more detail in Sect. 7.2. Proton leak can be associated with increased *e*- leak [3, 36]. Figure 6.14 describes a model for *e*- transfer that is linked or not linked (*e*- slip, J_e^S) to H^+ pumping across the IMS, and intrinsic H^+ leak (J_H^L) across the IMS that is not mediated via the F_o site [36]. At least at low levels, excess *e*- leak can trigger cardioprotective mechanisms starting with $O_2^{\bullet -}$, as addressed in Sect. 6.1, and in more detail in Sect. 8.4.

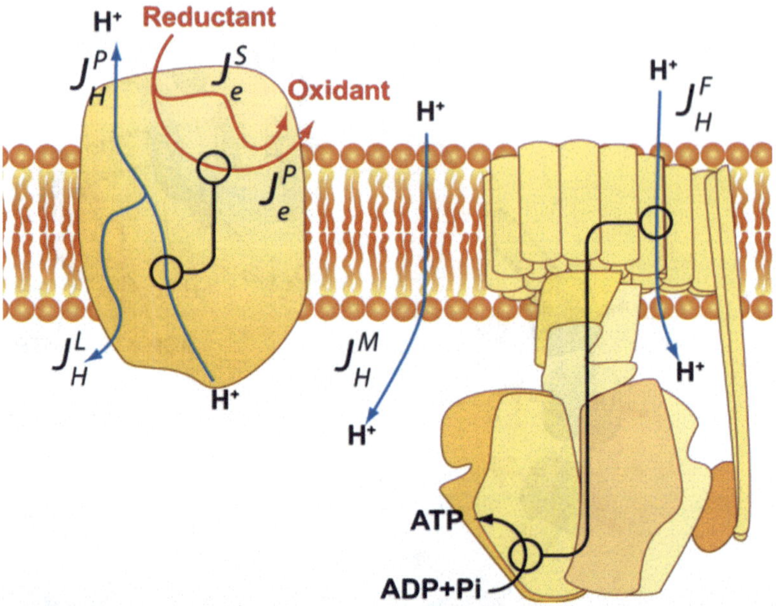

Fig. 6.14 Leaks and slips in the proton motive function of mitochondria: Cartoon shows a prototype of a proton motive redox complex of the respiratory chain (left) and a schematic view of the F_oF_1 ATP synthase (right), both embedded in the phospholipid IMM. Red arrows depict *e*- transfer from reductant to oxidant, either linked (J_e^P) to H^+ pumping across the IMM (J_H^P) or decoupled from it ("electron slip", J_e^S). J_H^L depicts an intrinsic H^+ leak in the pump mechanism ("proton slip"). J_H^M stands for loss of proton motive force (*pmf*) by a H^+ leak across the IMM. H^+ influx across the F_o domain of the ATP synthase (J_H^F) is assumed to be fully coupled to synthesis of ATP without significant slips (see text). The stoichiometric aspects (H^+/*e*- and H^+/ATP ratios) are omitted for clarity (see text). (Cartoon (Fig. 2) and its modified legend resourced with license (https://creativecommons.org/licenses/BY-NC-SA/4.0/) from Wikstrom and Springett [36])

The passive permeability of the lipid IMM for protons normally remains extremely low; most of the natural proton "leak" is attributed to activity of proton cation exchangers (Fig. 5.15) in the IMM that maintain ionic homeostasis (see also Sects. 2.1, 2.3, 4.1, 5.2, 7.1, 8.1 and 8.3) [27, 40, 45–50], and by mitochondrial anion carrier proteins, e.g., adenine nucleotide translocase (ANT) (Figs. 4.4 and 5.15) located in the IMM [5]. Proton leak is also related to uncoupling proteins (UCPs) in the IMM, and to a smaller extent, by the fatty acid composition of the IMM [42]. Proton leak vs. no leak can be estimated in isolated cells or mitochondria by plotting O_2 consumption (respiration rate) as a function of $\Delta\Psi_m$ as explained in Fig. 6.15 [43], where "X" treatment, which increases O_2 consumption at the highest common $\Delta\Psi_m$, indicates proton leak compensated by increased respiration (A), and

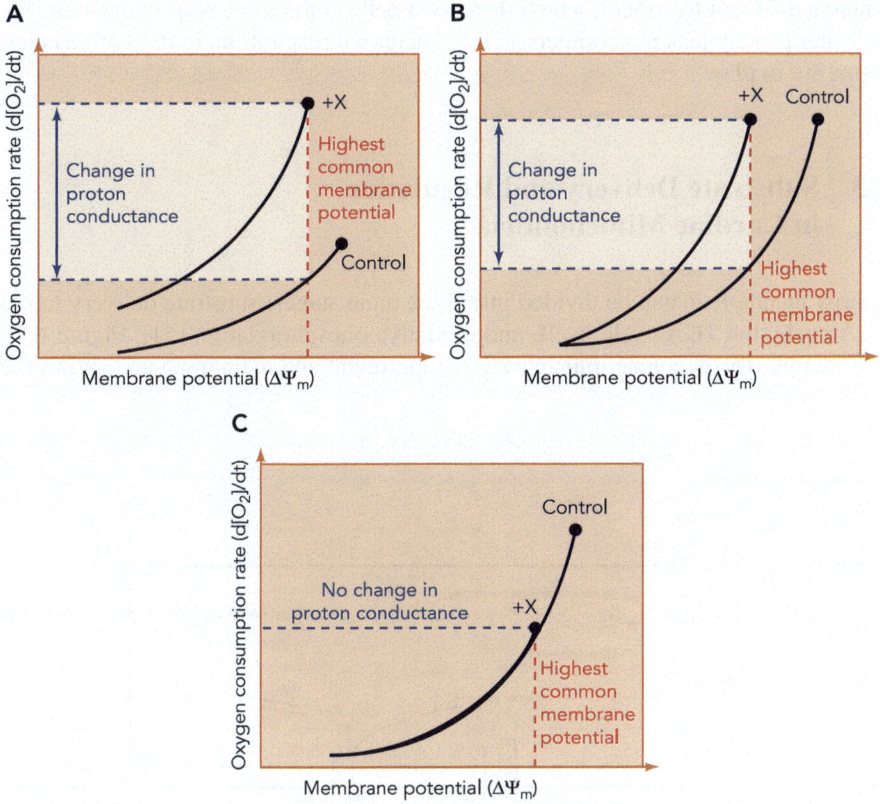

Fig. 6.15 Measurement of mitochondrial proton (H^+) leak kinetics: Simultaneous titrations of respiration and $\Delta\Psi_m$ with respiratory inhibitors in the absence of ATP synthesis reveals non-ohmic H^+ flux across the IMM. A proper quantification of H^+ flux involves measuring the respiration used to drive H^+ leak at a defined $\Delta\Psi_m$. (**a**) Treatment with "X" causes an increase in H^+ flux relative to the control, as shown by the increase in respiration rate at the highest common $\Delta\Psi_m$. The uninhibited state 4 respiration rates are also different (black dots). This is indicative of systems with high oxidative capacity, such as brown adipose tissue mitochondria, capable of increasing substrate oxidation to defend the $\Delta\Psi_m$ quite effectively. This, however, is not universally true. (**b**) A system treated with "X" and the control have similar maximal state 4 respiration rates but dramatically different H^+ fluxes, indicated by the disparity between respiration rates at the highest common $\Delta\Psi_m$. Such a system is indicative of low oxidative capacity that is incapable of responding to the change in H^+ flux. A simple comparison of state 4 respiration misses the underlying change in H^+ conductance; instead, there is a change in $\Delta\Psi_m$. Alternatively, "X" may both increase H^+ conductance as in **a** and by a separate mechanism decrease substrate oxidation as in **c**, giving this composite result. (**c**) A system treated with X and the control have different state 4 respiration rates but similar H^+ leak kinetics. Addition of "X" in this case does not change H^+ flux but rather depresses substrate oxidation, a property not explicitly revealed by an isolated state 4 measurement. This result of a change in H^+ leak rate following depression of substrate oxidation could be misinterpreted as a change in H^+ flux if only state 4 respiration rate was measured. (Cartoon plots (Fig. 2) and its modified, abridged legend resourced with permission from copyright owner of Divakaruni and Brand [43] (review))

where a different treatment, which decreases $\Delta\Psi_m$ at the same respiratory rate (B), indicates protein leak not compensated by increased respiration; in (C) both mechanisms are in play.

6.3 Substrate Delivery and Regulation in Cardiac Mitochondria

Energy metabolism can be divided into three main stages: substrate delivery to the TCA cycle, the TCA cycle itself, and oxidative phosphorylation [51]. Figure 6.16 [51] summarizes key reactions in each compartment and membrane and shows the

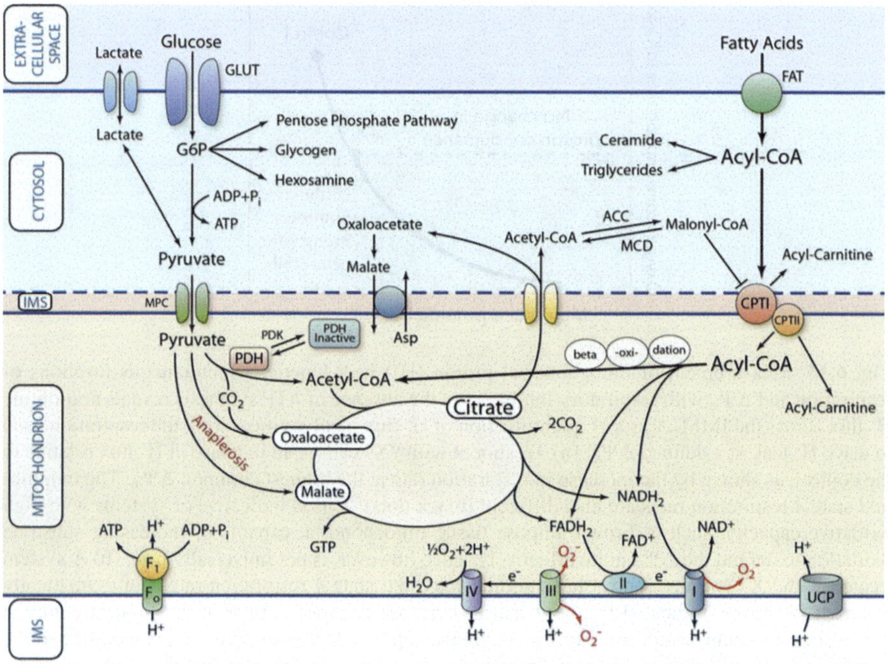

Fig. 6.16 Schematic representation of classic pathways of cardiac metabolism: Substrates are transported across the extracellular membrane into the cytosol and are metabolized in various ways. For oxidation, the respective metabolic intermediates (e.g., pyruvate or acyl-CoA) are transported across the IMM by specific transport systems. Once inside the mitochondrion, substrates are oxidized or carboxylated (anaplerosis) and fed into the TCA cycle for the generation of reducing equivalents (NADH and FADH$_2$) and GTP. The reducing equivalents are used by the ETS to generate a proton gradient, which in turn is used for production of ATP. *IMS* mitochondrial intermembrane space, *GLUT* glucose transporter, *FAT* fatty acid transporter, *MPC* mitochondrial pyruvate transporter. (Illustration Credit: Ben Smith. Diagram (Fig. 1) and modified legend resourced with permission from copyright owner of Doenst et al. [51] (review))

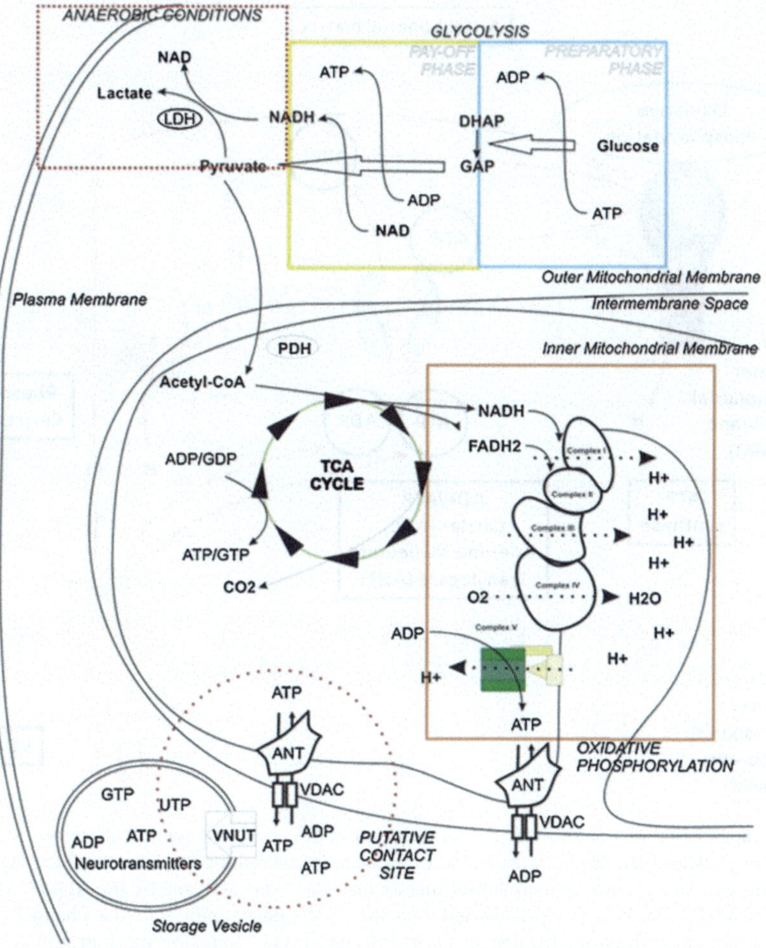

Fig. 6.17 ATP management within the cell: Schematic representation of mechanisms of ATP synthesis and storage inside the cell. Glycolysis is represented in the yellow and blue boxes, the TCA cycle by the green circle, and oxidative phosphorylation in the orange box. Reduction of pyruvate to lactate is represented inside the red dotted rectangle. Hypothetical contacts between ATP storage vesicles and mitochondria, with preferential ATP transfer, are shown within the red dotted circle. (Diagram (Fig. 1) and its legend resourced with permission from copyright owner of Bonora et al. [53])

key roles of acetyl-CoA (Sects. 1.4, 3.1, 3.3, 5.1) and acyl-CoA for FA oxidation. Each of these processes has multiple control points, and each exerts control over the other stages [52]. Sections 5.1, 5.2 and 5.3 and 6.2 describe the processes of glycolysis, TCA cycling, and OxPhos, with *e-* transfer and proton flux across the IMM to provide NADH and $FADH_2$ as reactants, and synthesis and storage of ATP [53], mostly as phosphocreatine (Fig. 1.14), as the energy rich molecule needed for the transduction of energy to work so that myofilament contraction/relaxation, and

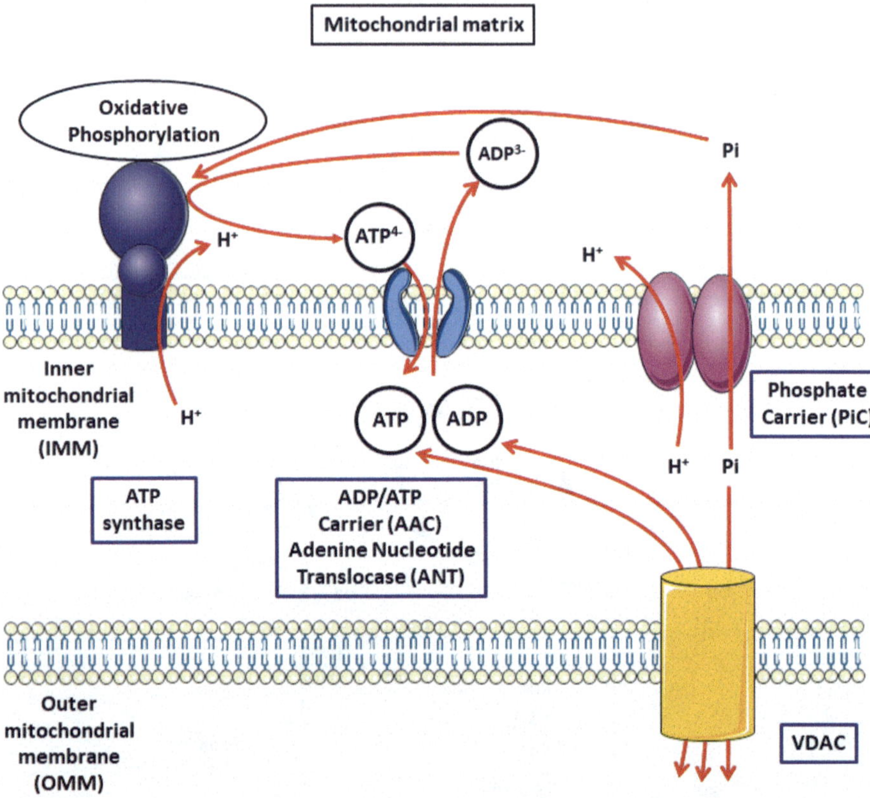

Fig. 6.18 Scheme of functions of carriers regulating the adenosine nucleotide pools: ADP/ATP transport is characterized as an exchange of exogenous and endogenous nucleotides. ADP/ATP exchange across the inner mitochondrial membrane (IMM) is catalyzed by the ADP/ATP carrier (AAC or ANT). The P_i carrier protein and the voltage-dependent anion selective channel (VDAC) are transmembrane proteins located in the IMM and OMM catalyzing the transport of P_i. The process of OxPhos permits generation of ATP from ADP, and P_i drives ATP synthesis by ATP synthase through H^+ reuptake. (Diagram (Fig. 2) and modified legend resourced with permission from copyright owner (MDPI) of Rochette et al. [54] (review))

molecular repair and replacement can occur (Sects. 7.4, 8.1, 9.2 and 9.4). Figure 6.17 [53] depicts synthesis, transport, and vesicular storage of ATP in the cytosol and mitochondria of a nerve cell.

Regulating Oxidation The need for mitochondrial phosphorylation of ADP is met by the abundant presence of ANT in the IMM, and by utilization of the phosphocreatine shuttle and other carriers across the IMM that provide for the varying demands for ATP by cells doing varying amounts of work. The cartoon of Fig. 6.18 [54] presents a simplified view of ADP, ATP, P_i, and H^+ transported via carriers, channels, and ATP synthase/ase, but the mechanisms responsible for control of respiration with ATP demand are more complicated and debated [52, 55–57] than simply discussed here. In the healthy heart, fatty acid β-oxidation (Figs. 5.9 and 6.16) provides over 90% of the acetyl-CoA for entry into the TCA cycle [58]. β-oxidation is largely regulated by the NADH/NAD^+ redox couple, and the acetyl-CoA/CoA ratio, in which low ratios stimulate β-oxidation. Inhibition of mitochondrial fatty acid acyl-CoA transport by malonyl-CoA also regulates β-oxidation [59, 60].

Although glucose is not the preferred substrate for oxidation in the heart (it is the free fatty acids, FFA), glycolysis may provide a small amount of ATP near sites of IMM ion channels and sarcoplasmic reticulum (SR) Ca^{2+} channel pumps [52]. Glycolysis can be limited by a decrease in the availability of cytosolic NAD^+ availability at glyceraldehyde 3-phosphate dehydrogenase (Fig. 5.1, cGPDH) [57]. Hexokinase controls metabolism of glucose via product inhibition by glucose 6-phosphate, and phosphorylated glycogen phosphorylase catalyzes the conversion of glycogen to glucose 1-phosphate faster than its dephosphorylated form.

The pyruvate dehydrogenase complex (PDHC) is allosterically inhibited by increased levels of ATP, and by acetyl-CoA and fatty acids, both substrates for the TCA cycle [61]. PDHC is activated by NAD^+, ADP, and pyruvate, whereas PD kinases (PDKs) are conversely activated by NADH, ATP, and high acetyl-CoA levels [61]. In addition to allosteric regulation, the PDHC is also regulated by covalent modification due to differential phosphorylation of three serine residues in PDH; these effects can be reversed by phosphatases [62].

Regulating TCA Cycle Function The TCA cycle is the final step for oxidation of all cellular nutrients; these include FAs, glucose, and amino acids, all of which enter the TCA cycle only as acetyl-CoA, which condenses with oxaloacetate (OAA) to form citrate (Fig. 5.7). As discussed in Sects. 5.1, 5.2 and 5.3, the TCA cycle strips *e*- from acetyl-CoA and transfers them to the *e*- carriers, NAD^+ and FAD, which then funnel *e*- onto the ETS. The TCA cycle also provides chemical intermediates for biosynthetic reactions. Pyruvate and TCA cycle reactants, succinyl-CoA, OAA, citrate, and α-ketoglutarate (αKG), provide carbon structures for the biosynthesis of lipids, glucose, nucleic acids, amino acids, and cofactors.

Although the TCA cycle pathway (Fig. 5.7) is often simplified as a mechanism to break down FAs and glucose, many of the reactions in the TCA cycle can go in either direction (amphibolic pathways) depending on substrate levels at each step. Examples in Fig. 5.9 [61] of cataplerosis are (1) pyruvate to gluconeogenesis and lactate; (2) succinyl-CoA to porphyrins (Fe-containing carbon rings as found in hemoglobin); (3) OAA to gluconeogenesis, and with glutamate, to aspartate; (4) citrate to FA synthesis; and (5) αKG to glutamate and glutamine (via transamination).

From an evolutionary viewpoint, particularly in archaean anaerobes, a reverse (r) rTCA cycle could have operated to cleave citrate in two steps to form acetyl-CoA and to regenerate OAA by citryl-CoA synthase (CS) and citryl-CoA lyase (CCL) in the presence of high CO_2 levels [61]. The emergence of atmospheric O_2 likely facilitated forward oxidation of the TCA reactants with forward transfer of *e*- onto NAD^+ and FAD for eventual acceptance by O_2 with proton pumping by ETS complexes.

Flux through the TCA cycle is tightly regulated by positive and negative allosteric enzymes (Figs. 5.7 and 5.9) [61]. Rate controlling enzymes of the TCA are citrate synthase (CS), isocitrate dehydrogenase (IDH), and oxoglutarate dehydrogenase (OGDH); each is allosterically inhibited by high NADH/NAD^+ ratios. Without this control, unneeded intermediates oxidizing to excess NADH and $FADH_2$ can lead to excess *e*- delivered to the ETS and result in excess production of $O_2^{\bullet -}$ and other damaging reactants. High ATP/ADP levels also allosterically inhibit IDH, and succinyl-CoA allosterically inhibits OGDH and CS, whereas OAA allosterically inhibits succinate dehydrogenase (SDH) [61]. A mathematical model on the catalytic mechanism of NAD-linked IDH suggests that under physiological conditions, the ADP/ATP ratio plays a more significant role than m[Ca^{2+}] in regulating IDH activity [63].

6.4 Modulation of Respiration by the Mitochondrial Proton (H^+) Motive Force

Mitochondrial respiration is governed primarily by the proton motive force Δp_m; when high, respiration deceases and when low, respiration increases. Thus, control of respiration is both $\Delta\Psi_m$ and ΔpH_m -dependent as $\Delta p_m = \Delta\Psi_m + (-59)\Delta pH_m$. In practice, in the presence of a substrate, and increased ATP consumption by a myocyte, ADP enters mitochondria via ANT with outward exchange of ATP. ADP stimulates production of ATP with entry of protons at complex V (ATP synthase) as also described in Sect. 6.2 and 7.4. In isolated mitochondria, from a control state (substrate only, state 2) to an activated state (added ADP, state 3), followed by complete ADP conversion to ATP (state 4), respiration is much faster in state 3 due to a fall in Δp_m. State 3/state 4 respiration is called the respiratory control rate or index (RCI). RCI is a common measure to gauge the bioenergetic health of mitochondria; a high index indicates good coupling of OxPhos.

Measuring O_2 Consumption Mitochondrial respiration is commonly measured as O_2 consumed/time using an O_2 electrode in a sealed container of concentrated mitochondria and buffer. For example, Fig. 6.19 [64] shows results from experiments designed to assess states 2, 3, and 4 respiration under two substrate conditions and to compare effects of the general anesthetic isoflurane at complex I with complex I inhibitor rotenone (ROT) to assess if isoflurane might act on complex I [64]. The RCI is substrate dependent and is generally faster with a complex I substrate (e.g., pyruvate), and slower with a complex II substrate (e.g., succinate). In cardiomyocytes the RCI is dependent on cellular ATP consumption, mitochondrial ATP/ADP ratios, creatine kinase and phosphatase activity, the trans-IMM pH gradient, the delivery of different substrates, reaction intermediates and products (e.g., NADH, $FADH_2$), temperature, proton and *e*- leaks, creatine availability, and other factors [27, 65–69].

Regulating Respiration In addition to the direct coupling of Δp_m to respiration, another influential mechanism is allosteric regulation, which allows for fine-tuning of enzymatic activity based on the concentrations of reaction substrates, intermediates, and products as mentioned in the TCA cycle (Sects. 5.1, 6.3 and 7.4). In the

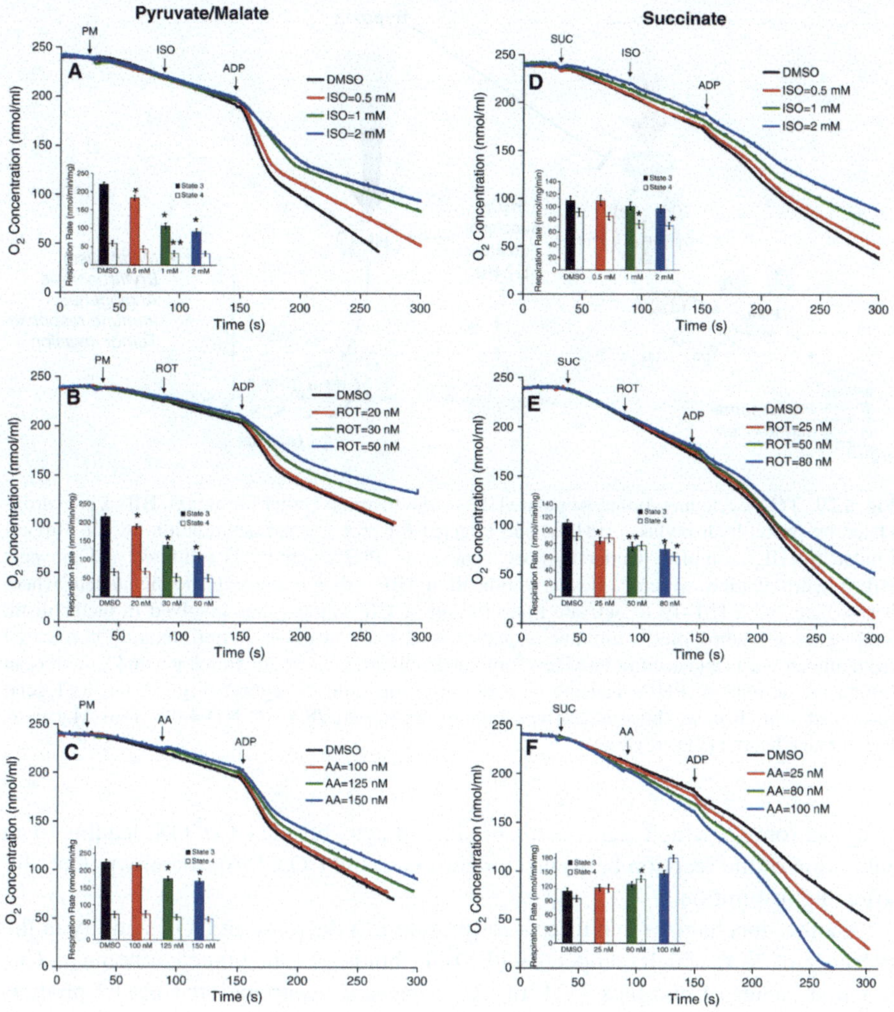

Fig. 6.19 Mitochondrial O_2 concentration during states 2, 3, and 4 respiration: Isolated mitochondria were treated with different substrates and either the volatile anesthetic isoflurane (ISO) or a blocker of ETS complexes. ISO, rotenone (ROT), or antimycin A (AA) was added after pyruvate malate (PM) (A–C) or succinate (SUC) (D–F) and before adding ADP. Adding ADP induced a transient depolarization. The insets summarize effects of added ISO, ROT and AA on state 3 and state 4 respiration rates (nmol • min^{-1} • mg^{-1} protein). Note similarities and differences of responses to each drug. Data represent mean ± SD of 3 replicates. Data represent mean ± SD of three replicates. (Data plots (Fig. 5) and its legend resourced with permission from copyright owner of Agarwal et al. [64])

ETS, cytochrome *c* is the one *e*- carrier within the IMM that shuttles *e*- from bc_1 complex III to complex IV (cytochrome *c* oxidase, COX) (Figs. 5.17, 6.1, 6.11 and 6.12) (Sects. 5.2 and 5.3). A regulatory mechanism found in COX and cytochrome *c* is the binding of adenine nucleotides. At physiological concentrations, ATP binds

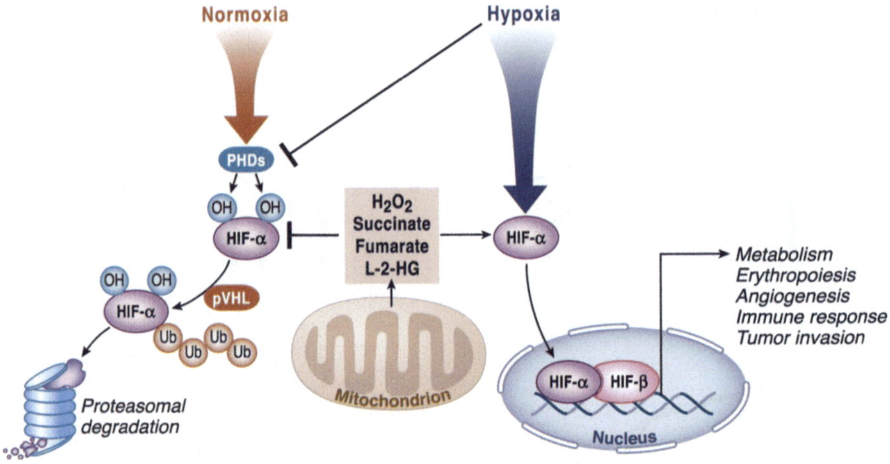

Fig. 6.20 TCA cycle metabolites regulate HIF-α stabilization: Under normoxia, HIF-α is hydroxylated by prolyl-hydroxylases (PHDs) and targeted for proteasomal degradation by von Hippel-Lindau (pVHL) complex. Under hypoxic conditions, PHDs activity is inhibited and prevents HIF-α hydroxylation, which causes its stabilization. HIF-α then translocates to the nucleus where it associates with HIF-1β to activate transcription of HIF target genes involved in metabolism, angiogenesis, erythropoiesis, immune responses, and tumor invasion. In normoxia, ROS released from mitochondria and accumulated levels of the metabolites succinate, fumarate, and L-2-HG can inhibit the activity of PHDs causing a pseudo-hypoxia state. (Diagram (Fig. 5) and its legend resourced with license (https://creativecommons.org/licenses/BY-NC-ND/4.0/) from Martinez-Reyes and Chandel [71] (review))

to cytochrome *c*, which changes the binding of cytochrome *c* to COX, leading to an inhibition of the reaction between cytochrome *c* and COX [56], thereby potentially affecting respiration.

Another mechanism that affects respiration is a decrease in COX activity in the presence of NO•, which competes with O_2 for binding at the binuclear heme a_3–Cu_B reaction center of complex IV [56]. A biophysical computational model predicts that P_i also plays a role in regulation of overall OxPhos [70]. On the other hand, TCA cycle intermediates may act as cell signaling molecules to control chromatin modifications, DNA methylation, the hypoxic response, and immunity [71]. For example, hypoxia-inducible factor 1α (HIF1α) is associated with upregulation of glycolytic genes, e.g., phosphoglycerate kinase (PGK) and lactate dehydrogenase (LDH), which, metabolically adapt glycolysis during O_2 deprivation with anaerobic ATP synthesis. Figure 6.20 illustrates how during normoxia, HIF1α is kept under control and eventually deactivated [71], but during hypoxia, HIF1α is induced to translocate to the nucleus where it associates with HIF1β to activate the transcription of HIF target genes involved in control of metabolic processes, including modulation of function by the respiratory complexes [72].

References

1. Dunn JD, Alvarez LA, Zhang X, Soldati T. Reactive oxygen species and mitochondria: A nexus of cellular homeostasis. Redox Biol. 2015;6:472–85. Epub 20150910. https://doi.org/10.1016/j.redox.2015.09.005. PubMed PMID: 26432659; PMCID: PMC4596921.
2. Kussmaul L, Hirst J. The mechanism of superoxide production by NADH:ubiquinone oxidoreductase (complex I) from bovine heart mitochondria. Proc Natl Acad Sci USA. 2006;103(20):7607–12. Epub 20060508. https://doi.org/10.1073/pnas.0510977103. PubMed PMID: 16682634; PMCID: PMC1472492.
3. Murphy MP. How mitochondria produce reactive oxygen species. Biochem J. 2009;417(1):1–13. https://doi.org/10.1042/BJ20081386. PubMed PMID: 19061483; PMCID: PMC2605959.
4. Fridovich I. Oxygen: how do we stand it? Med Princ Pract. 2013;22(2):131–7. Epub 20120628. https://doi.org/10.1159/000339212. PubMed PMID: 22759590; PMCID: PMC5685332.
5. Jastroch M, Divakaruni AS, Mookerjee S, Treberg JR, Brand MD. Mitochondrial proton and electron leaks. Essays Biochem. 2010;47:53–67. https://doi.org/10.1042/bse0470053. PubMed PMID: 20533900; PMCID: PMC3122475.
6. Bleier L, Drose S. Superoxide generation by complex III: from mechanistic rationales to functional consequences. Biochim Biophys Acta. 2013;1827(11–12):1320–31. Epub 20121223. https://doi.org/10.1016/j.bbabio.2012.12.002. PubMed PMID: 23269318.
7. Stowe DF, Camara AK. Mitochondrial reactive oxygen species production in excitable cells: modulators of mitochondrial and cell function. Antioxid Redox Signal. 2009;11(6):1373–414. https://doi.org/10.1089/ars.2008.2331. PubMed PMID: 19187004; PMCID: PMC2842133.
8. Lambert AJ, Brand MD. Superoxide production by NADH:ubiquinone oxidoreductase (complex I) depends on the pH gradient across the mitochondrial inner membrane. Biochem J. 2004;382(Pt 2):511–7. https://doi.org/10.1042/BJ20040485. PubMed PMID: 15175007; PMCID: PMC1133807.
9. Heinen A, Aldakkak M, Stowe DF, Rhodes SS, Riess ML, Varadarajan SG, Camara AK. Reverse electron flow-induced ROS production is attenuated by activation of mitochondrial Ca^{2+}-sensitive K^{+} channels. Am J Physiol Heart Circ Physiol. 2007;293(3):H1400–7. Epub 20070518. https://doi.org/10.1152/ajpheart.00198.2007. PubMed PMID: 17513497.
10. Heinen A, Camara AK, Aldakkak M, Rhodes SS, Riess ML, Stowe DF. Mitochondrial Ca^{2+}-induced K^{+} influx increases respiration and enhances ROS production while maintaining membrane potential. Am J Physiol Cell Physiol. 2007;292(1):C148–56. Epub 20060726. https://doi.org/10.1152/ajpcell.00215.2006. PubMed PMID: 16870831.
11. Camara AK, Aldakkak M, Heisner JS, Rhodes SS, Riess ML, An J, Heinen A, Stowe DF. ROS scavenging before 27 °C ischemia protects hearts and reduces mitochondrial ROS, Ca^{2+} overload, and changes in redox state. Am J Physiol Cell Physiol. 2007;292(6):C2021–31. Epub 20070207. https://doi.org/10.1152/ajpcell.00231.2006. PubMed PMID: 17287367.
12. Camara AK, Riess ML, Kevin LG, Novalija E, Stowe DF. Hypothermia augments reactive oxygen species detected in the guinea pig isolated perfused heart. Am J Physiol Heart Circ Physiol. 2004;286(4):H1289–99. Epub 20031126. https://doi.org/10.1152/ajpheart.00811.2003. PubMed PMID: 14644763.
13. Kevin LG, Camara AK, Riess ML, Novalija E, Stowe DF. Ischemic preconditioning alters real-time measure of O_2 radicals in intact hearts with ischemia and reperfusion. Am J Physiol Heart Circ Physiol. 2003;284(2):H566–74. Epub 20021031. https://doi.org/10.1152/ajpheart.00711.2002. PubMed PMID: 12414448.
14. Kevin LG, Novalija E, Riess ML, Camara AKS, Rhodes SS, Stowe DF. Sevoflurane exposure generates superoxide but leads to decreased superoxide during ischemia and reperfusion in isolated hearts. Anesth Analg. 2003;96(4):949–55. https://doi.org/10.1213/01.ANE.0000052515.25465.35. PubMed PMID: 12651639.
15. Duong QV, Levitsky Y, Dessinger MJ, Strubbe-Rivera JO, Bazil JN. Identifying site-specific superoxide and hydrogen peroxide production rates from the mitochondrial electron transport

system using a computational strategy. Function (Oxf). 2021;2(6):zqab050. Epub 20210920. https://doi.org/10.1093/function/zqab050. PubMed PMID: 35330793; PMCID: PMC8788716.
16. Buettner GR. Superoxide dismutase in redox biology: the roles of superoxide and hydrogen peroxide. Anti Cancer Agents Med Chem. 2011;11(4):341–6. https://doi.org/10.2174/187152011795677544. PubMed PMID: 21453242; PMCID: PMC3131414.
17. Borgstahl GE, Parge HE, Hickey MJ, Beyer WF Jr, Hallewell RA, Tainer JA. The structure of human mitochondrial manganese superoxide dismutase reveals a novel tetrameric interface of two 4-helix bundles. Cell. 1992;71(1):107–18. https://doi.org/10.1016/0092-8674(92)90270-m. PubMed PMID: 1394426.
18. Aldakkak M, Stowe DF, Dash RK, Camara AK. Mitochondrial handling of excess Ca^{2+} is substrate-dependent with implications for reactive oxygen species generation. Free Radic Biol Med. 2013;56:193–203. Epub 20120923. https://doi.org/10.1016/j.freeradbiomed.2012.09.020. PubMed PMID: 23010495; PMCID: PMC3542420.
19. Gerdes HJ, Yang M, Heisner JS, Camara AKS, Stowe DF. Modulation of peroxynitrite produced via mitochondrial nitric oxide synthesis during Ca^{2+} and succinate-induced oxidative stress in cardiac isolated mitochondria. Biochim Biophys Acta Bioenerg. 2020;1861(12):148290. Epub 20200820. https://doi.org/10.1016/j.bbabio.2020.148290. PubMed PMID: 32828729; PMCID: PMC7560995.
20. Lindsay DP, Camara AK, Stowe DF, Lubbe R, Aldakkak M. Differential effects of buffer pH on Ca^{2+} induced ROS emission with inhibited mitochondrial complexes I and III. Front Physiol. 2015;6:58. Epub 20150310. https://doi.org/10.3389/fphys.2015.00058. PubMed PMID: 25805998; PMCID: PMC4354303.
21. Brand MD. The sites and topology of mitochondrial superoxide production. Exp Gerontol. 2010;45(7–8):466–72. Epub 20100111. https://doi.org/10.1016/j.exger.2010.01.003. PubMed PMID: 20064600; PMCID: PMC2879443.
22. Chen Q, Lesnefsky EJ. Time to target mitochondrial reactive oxygen species generation from complex I. Function (Oxf). 2022;3(2):zqac010. Epub 20220323. https://doi.org/10.1093/function/zqac010. PubMed PMID: 35359911; PMCID: PMC8951057.
23. Chen Q, Vazquez EJ, Moghaddas S, Hoppel CL, Lesnefsky EJ. Production of reactive oxygen species by mitochondria: central role of complex III. J Biol Chem. 2003;278(38):36027–31. Epub 20030702. https://doi.org/10.1074/jbc.M304854200. PubMed PMID: 12840017.
24. Mitchell P. Coupling of phosphorylation to electron and hydrogen transfer by a chemi-osmotic type of mechanism. Nature. 1961;191:144–8. https://doi.org/10.1038/191144a0. PubMed PMID: 13771349.
25. Klingenberg M. The ADP and ATP transport in mitochondria and its carrier. Biochim Biophys Acta. 2008;1778(10):1978–2021. Epub 20080502. https://doi.org/10.1016/j.bbamem.2008.04.011. PubMed PMID: 18510943.
26. Nicholls DG. Mitochondrial membrane potential and aging. Aging Cell. 2004;3(1):35–40. https://doi.org/10.1111/j.1474-9728.2003.00079.x. PubMed PMID: 14965354.
27. Malas KM, Lambert DS, Heisner JS, Camara AKS, Stowe DF. Time and charge/pH-dependent activation of K^+ channel-mediated K^+ influx and K^+/H^+ exchange in guinea pig heart isolated mitochondria; role in bioenergetic stability. Biochim Biophys Acta Bioenerg. 2022;1863(8):148908. Epub 20220809. https://doi.org/10.1016/j.bbabio.2022.148908. PubMed PMID: 35961396.
28. Neupane P, Bhuju S, Thapa N, Bhattarai HK. ATP synthase: structure, function and Inhibition. Biomol Concepts. 2019;10(1):1–10. Epub 20190307. https://doi.org/10.1515/bmc-2019-0001. PubMed PMID: 30888962.
29. Juhaszova M, Kobrinsky E, Zorov DB, Nuss HB, Yaniv Y, Fishbein KW, de Cabo R, Montoliu L, Gabelli SB, Aon MA, Cortassa S, Sollott SJ. ATP synthase K^+- and H^+-fluxes drive ATP synthesis and enable mitochondrial K^+-"uniporter" function: I. Characterization of ion fluxes. Function (Oxf). 2022;3(2):zqab065. Epub 20211213. https://doi.org/10.1093/function/zqab065. PubMed PMID: 35229078; PMCID: PMC8867323.

30. Dzbek J, Korzeniewski B. Control over the contribution of the mitochondrial membrane potential ($\Delta\psi_m$) and proton gradient (ΔpH) to the protonmotive force (Δp). In silico studies. J Biol Chem. 2008;283(48):33232–9. Epub 20080811. https://doi.org/10.1074/jbc.M802404200. PubMed PMID: 18694940; PMCID: PMC2662253.
31. Hackenbrock CR. Energy-linked condensed-orthodox ultrastructural transformations in mitochondria. Chemotherapy. 1981;27(Suppl 2):21–6. https://doi.org/10.1159/000238042. PubMed PMID: 7285635.
32. Grochala W. First there was hydrogen. Nat Chem. 2015;7(3):264. https://doi.org/10.1038/nchem.2186. PubMed PMID: 25698337.
33. Mookerjee SA, Gerencser AA, Nicholls DG, Brand MD. Quantifying intracellular rates of glycolytic and oxidative ATP production and consumption using extracellular flux measurements. J Biol Chem. 2017;292(17):7189–207. Epub 20170307. https://doi.org/10.1074/jbc.M116.774471. PubMed PMID: 28270511; PMCID: PMC5409486.
34. Mazat JP, Ransac S, Heiske M, Devin A, Rigoulet M. Mitochondrial energetic metabolism-some general principles. IUBMB Life. 2013;65(3):171–9. https://doi.org/10.1002/iub.1138. PubMed PMID: 23441039.
35. Wikstrom M, Sharma V, Kaila VR, Hosler JP, Hummer G. New perspectives on proton pumping in cellular respiration. Chem Rev. 2015;115(5):2196–221. Epub 20150219. https://doi.org/10.1021/cr500448t. PubMed PMID: 25694135.
36. Wikstrom M, Springett R. Thermodynamic efficiency, reversibility, and degree of coupling in energy conservation by the mitochondrial respiratory chain. Commun Biol. 2020;3(1):451. Epub 20200818. https://doi.org/10.1038/s42003-020-01192-w. PubMed PMID: 32811895; PMCID: PMC7434914.
37. Chinopoulos C. Mitochondrial consumption of cytosolic ATP: not so fast. FEBS Lett. 2011;585(9):1255–9. Epub 20110407. https://doi.org/10.1016/j.febslet.2011.04.004. PubMed PMID: 21486564.
38. Chinopoulos C, Adam-Vizi V. Mitochondria as ATP consumers in cellular pathology. Biochim Biophys Acta. 2010;1802(1):221–7. Epub 20090826. https://doi.org/10.1016/j.bbadis.2009.08.008. PubMed PMID: 19715757.
39. Chinopoulos C, Gerencser AA, Mandi M, Mathe K, Torocsik B, Doczi J, Turiak L, Kiss G, Konrad C, Vajda S, Vereczki V, Oh RJ, Adam-Vizi V. Forward operation of adenine nucleotide translocase during F_0F_1-ATPase reversal: critical role of matrix substrate-level phosphorylation. FASEB J. 2010;24(7):2405–16. Epub 20100305. https://doi.org/10.1096/fj.09-149898. PubMed PMID: 20207940; PMCID: PMC2887268.
40. Haumann J, Camara AKS, Gadicherla AK, Navarro CD, Boelens AD, Blomeyer CA, Dash RK, Boswell MR, Kwok WM, Stowe DF. Slow Ca^{2+} efflux by Ca^{2+}/H^+ exchange in cardiac mitochondria Is modulated by Ca^{2+} re-uptake via MCU, extra-mitochondrial pH, and H^+ pumping by F_OF_1-ATPase. Front Physiol. 2018;9:1914. Epub 20190204. https://doi.org/10.3389/fphys.2018.01914. PubMed PMID: 30804812; PMCID: PMC6378946.
41. O'Brien WC. ATP-Adenosine Triphosphate. LibreTexts Biology. 2023. Available from: https://bio.libretexts.org/Bookshelves/Microbiology/Microbiology_(Boundless)/02%3A_Chemistry/2.05%3A_Organic_Compounds/2.5.06%3A_ATP-_Adenosine_Triphosphate.
42. Brand MD, Chien LF, Ainscow EK, Rolfe DF, Porter RK. The causes and functions of mitochondrial proton leak. Biochim Biophys Acta. 1994;1187(2):132–9. https://doi.org/10.1016/0005-2728(94)90099-x. PubMed PMID: 8075107.
43. Divakaruni AS, Brand MD. The regulation and physiology of mitochondrial proton leak. Physiology (Bethesda). 2011;26(3):192–205. https://doi.org/10.1152/physiol.00046.2010. PubMed PMID: 21670165.
44. Rolfe DF, Brand MD. The physiological significance of mitochondrial proton leak in animal cells and tissues. Biosci Rep. 1997;17(1):9–16. https://doi.org/10.1023/a:1027327015957. PubMed PMID: 9171916.
45. Hill BG, Benavides GA, Lancaster JR Jr, Ballinger S, Dell'Italia L, Jianhua Z, Darley-Usmar VM. Integration of cellular bioenergetics with mitochondrial quality control and autophagy.

Biol Chem. 2012;393(12):1485–512. https://doi.org/10.1515/hsz-2012-0198. PubMed PMID: 23092819; PMCID: PMC3594552.

46. Natarajan GK, Glait L, Mishra J, Stowe DF, Camara AKS, Kwok WM. Total matrix Ca^{2+} modulates Ca^{2+} efflux via the Ca^{2+}/H^{+} exchanger in cardiac mitochondria. Front Physiol. 2020;11:510600. Epub 20200916. https://doi.org/10.3389/fphys.2020.510600. PubMed PMID: 33041851; PMCID: PMC7526510.
47. Luongo TS, Lambert JP, Gross P, Nwokedi M, Lombardi AA, Shanmughapriya S, Carpenter AC, Kolmetzky D, Gao E, van Berlo JH, Tsai EJ, Molkentin JD, Chen X, Madesh M, Houser SR, Elrod JW. The mitochondrial Na^{+}/Ca^{2+} exchanger is essential for Ca^{2+} homeostasis and viability. Nature. 2017;545(7652):93–7. Epub 20170426. https://doi.org/10.1038/nature22082. PubMed PMID: 28445457; PMCID: PMC5731245.
48. Hassan MT, Lytton J. Potassium-dependent sodium-calcium exchanger (NCKX) isoforms and neuronal function. Cell Calcium. 2020;86:102135. Epub 20191209. https://doi.org/10.1016/j.ceca.2019.102135. PubMed PMID: 31846835.
49. Blomeyer CA, Bazil JN, Stowe DF, Pradhan RK, Dash RK, Camara AK. Dynamic buffering of mitochondrial Ca^{2+} during Ca^{2+} uptake and Na^{+}-induced Ca^{2+} release. J Bioenerg Biomembr. 2013;45(3):189–202. Epub 20121207. https://doi.org/10.1007/s10863-012-9483-7. PubMed PMID: 23225099; PMCID: PMC4065551.
50. Aldakkak M, Stowe DF, Heisner JS, Spence M, Camara AK. Enhanced Na^{+}/H^{+} exchange during ischemia and reperfusion impairs mitochondrial bioenergetics and myocardial function. J Cardiovasc Pharmacol. 2008;52(3):236–44. https://doi.org/10.1097/FJC.0b013e3181831337. PubMed PMID: 18806604; PMCID: PMC2659015.
51. Doenst T, Nguyen TD, Abel ED. Cardiac metabolism in heart failure: implications beyond ATP production. Circ Res. 2013;113(6):709–24. https://doi.org/10.1161/CIRCRESAHA.113.300376. PubMed PMID: 23989714; PMCID: PMC3896379.
52. Jafri MS, Dudycha SJ, O'Rourke B. Cardiac energy metabolism: models of cellular respiration. Annu Rev Biomed Eng. 2001;3:57–81. https://doi.org/10.1146/annurev.bioeng.3.1.57. PubMed PMID: 11447057.
53. Bonora M, Patergnani S, Rimessi A, De Marchi E, Suski JM, Bononi A, Giorgi C, Marchi S, Missiroli S, Poletti F, Wieckowski MR, Pinton P. ATP synthesis and storage. Purinergic Signal. 2012;8(3):343–57. Epub 20120412. https://doi.org/10.1007/s11302-012-9305-8. PubMed PMID: 22528680; PMCID: PMC3360099.
54. Rochette L, Meloux A, Zeller M, Malka G, Cottin Y, Vergely C. Mitochondrial SLC25 carriers: novel targets for cancer therapy. Molecules. 2020;25(10). Epub 20200522. https://doi.org/10.3390/molecules25102417. PubMed PMID: 32455902; PMCID: PMC7288124.
55. Kay L, Nicolay K, Wieringa B, Saks V, Wallimann T. Direct evidence for the control of mitochondrial respiration by mitochondrial creatine kinase in oxidative muscle cells in situ. J Biol Chem. 2000;275(10):6937–44. https://doi.org/10.1074/jbc.275.10.6937. PubMed PMID: 10702255.
56. Huttemann M, Helling S, Sanderson TH, Sinkler C, Samavati L, Mahapatra G, Varughese A, Lu G, Liu J, Ramzan R, Vogt S, Grossman LI, Doan JW, Marcus K, Lee I. Regulation of mitochondrial respiration and apoptosis through cell signaling: cytochrome c oxidase and cytochrome c in ischemia/reperfusion injury and inflammation. Biochim Biophys Acta. 2012;1817(4):598–609. Epub 20110713. https://doi.org/10.1016/j.bbabio.2011.07.001. PubMed PMID: 21771582; PMCID: PMC3229836.
57. Kobayashi K, Neely JR. Control of maximum rates of glycolysis in rat cardiac muscle. Circ Res. 1979;44(2):166–75. https://doi.org/10.1161/01.res.44.2.166. PubMed PMID: 216503.
58. Iso T, Kurabayashi M. Fatty acid uptake by the heart during fasting. In: Preedy VP, editor. Handbook of famine, starvation, and nutrient deprivation. Cham: Springer; 2019. p. 1179–97. https://doi.org/10.1007/978-3-319-55387-0_13.
59. Dong S, Qian L, Cheng Z, Chen C, Wang K, Hu S, Zhang X, Wu T. Lactate and myocardiac energy metabolism. Front Physiol. 2021;12:715081. Epub 20210817. https://doi.org/10.3389/fphys.2021.715081. PubMed PMID: 34483967; PMCID: PMC8415870.

60. Eaton S, Bartlett K, Pourfarzam M. Mammalian mitochondrial beta-oxidation. Biochem J. 1996;320(Pt 2):345–57. https://doi.org/10.1042/bj3200345. PubMed PMID: 8973539; PMCID: PMC1217938.
61. Arnold PK, Finley LWS. Regulation and function of the mammalian tricarboxylic acid cycle. J Biol Chem. 2023;299(2):102838. Epub 20221226. https://doi.org/10.1016/j.jbc.2022.102838. PubMed PMID: 36581208; PMCID: PMC9871338.
62. Chen L, Zhou M, Li H, Liu D, Liao P, Zong Y, Zhang C, Zou W, Gao J. Mitochondrial heterogeneity in diseases. Signal Transduct Target Ther. 2023;8(1):311. Epub 20230823. https://doi.org/10.1038/s41392-023-01546-w. PubMed PMID: 37607925; PMCID: PMC10444818.
63. Qi F, Chen X, Beard D. Detailed kinetics and regulation of mammalian NAD-linked isocitrate dehydrogenase. Biochim Biophys Acta. 2008;1784(11):1641–51. Epub 20080711. https://doi.org/10.1016/j.bbapap.2008.07.001. PubMed PMID: 18672100; PMCID: PMC2584801.
64. Agarwal B, Dash RK, Stowe DF, Bosnjak ZJ, Camara AK. Isoflurane modulates cardiac mitochondrial bioenergetics by selectively attenuating respiratory complexes. Biochim Biophys Acta. 2014;1837(3):354–65. Epub 20131217. https://doi.org/10.1016/j.bbabio.2013.11.006. PubMed PMID: 24355434; PMCID: PMC4084852.
65. Tomar N, Zhang X, Kandel SM, Sadri S, Yang C, Liang M, Audi SH, Cowley AW, Jr., Dash RK. Substrate-dependent differential regulation of mitochondrial bioenergetics in the heart and kidney cortex and outer medulla. Biochim Biophys Acta Bioenerg. 2022;1863(2):148518. Epub 20211203. https://doi.org/10.1016/j.bbabio.2021.148518. PubMed PMID: 34864090; PMCID: PMC8957717.
66. Riess ML, Camara AK, Heinen A, Eells JT, Henry MM, Stowe DF. K_{ATP} channel openers have opposite effects on mitochondrial respiration under different energetic conditions. J Cardiovasc Pharmacol. 2008;51(5):483–91. https://doi.org/10.1097/FJC.0b013e31816bf4a4. PubMed PMID: 18437094; PMCID: PMC3010203.
67. Jiang C, Moorthy BT, Patel DM, Kumar A, Morgan WM, Alfonso B, Huang J, Lampidis TJ, Isom DG, Barrientos A, Fontanesi F, Zhang F. Regulation of mitochondrial respiratory chain complex levels, organization, and function by arginyltransferase 1. Front Cell Dev Biol. 2020;8:603688. Epub 20201221. https://doi.org/10.3389/fcell.2020.603688. PubMed PMID: 33409279; PMCID: PMC7779560.
68. Chen Q, Camara AK, Stowe DF, Hoppel CL, Lesnefsky EJ. Modulation of electron transport protects cardiac mitochondria and decreases myocardial injury during ischemia and reperfusion. Am J Physiol Cell Physiol. 2007;292(1):C137–47. Epub 20060913. https://doi.org/10.1152/ajpcell.00270.2006. PubMed PMID: 16971498.
69. Riess ML, Camara AK, Chen Q, Novalija E, Rhodes SS, Stowe DF. Altered NADH and improved function by anesthetic and ischemic preconditioning in guinea pig intact hearts. Am J Physiol Heart Circ Physiol. 2002;283(1):H53–60. https://doi.org/10.1152/ajpheart.01057.2001. PubMed PMID: 12063274.
70. Beard DA. A biophysical model of the mitochondrial respiratory system and oxidative phosphorylation. PLoS Comput Biol. 2005;1(4):e36. Epub 20050909. https://doi.org/10.1371/journal.pcbi.0010036. PubMed PMID: 16163394; PMCID: PMC1201326.
71. Martinez-Reyes I, Chandel NS. Mitochondrial TCA cycle metabolites control physiology and disease. Nat Commun. 2020;11(1):102. Epub 20200103. https://doi.org/10.1038/s41467-019-13668-3. PubMed PMID: 31900386; PMCID: PMC6941980.
72. Huang X, Zhao L, Peng R. Hypoxia-inducible factor 1 and mitochondria: An intimate connection. Biomolecules. 2022;13(1). Epub 20221227. https://doi.org/10.3390/biom13010050. PubMed PMID: 36671435; PMCID: PMC9855368.

Chapter 7
Ionic and Molecular Regulation of Mitochondrial Bioenergetics

Abstract The present-day cardiac myocyte mitochondrion is much more complex than the basic machinery at the beginnings of chemiosmosis and oxidative phosphorylation. Indeed, it acts in constant and extensive symbiosis with its host cell sarcoplasm and with several organelles where information is transmitted back and forth to benefit the survival of the cell, its tissues, and the species. This includes not only mitochondrial biogenesis, fission and fusion, but also elimination of dysfunctional mitochondria and cells by apoptotic and necrotic pathways. Acute mitochondrial control is modulated by many factors including fast Ca^{2+} and K^+ influx through mitochondrial channels, with slow Ca^{2+} and K^+ efflux via mitochondrial exchangers, both of which modulate mitochondrial respiration, which in turn can modify the response to oxidant or ischemic stress. Mitochondrial Ca^{2+} levels can also modulate enzyme activity in the tricarboxylic acid cycle, but too much Ca^{2+} loading can stress the respiration machinery enough to fail with loss of the proton motive force. This can occur when the mitochondrial Ca^{2+} buffering capacity is exceeded. How this happens is inferred by experimental studies. Mild and temporary mitochondrial stress promotes the generation of hydrogen peroxide, which is linked to protection in isolated and in situ mitochondria against a subsequent, greater stress. Studies indicate that the opening of mitochondrial Ca^{2+}-sensitive K^+ channels during Ca^{2+} loading or mitochondrial stress induces the generation of superoxide radicals and hydrogen peroxide and causes mitochondrial swelling that is relieved by removal of the excess K^+ in exchange for H^+ influx. Evidence suggests this causes a proton leak (mitochondrial uncoupling) which stimulates respiration to enhance forward electron transfer to boost mitochondrial bioenergetic activity. In addition to mitochondrial protection against acute stress, mitochondria adapt to chronic changes (quality control) by induction of biogenesis, fission and fusion, and mitophagy. This results in a heterogeneity between normal and stressed or damaged mitochondria with differential expression of mitochondrial enzymes involved in bioenergetics, quality control, and regulation. The signals emitted by mitochondria induce nuclear and sarcoplasmic factors within the mitochondrial-associated membranes to target pathways that protect, propagate, and repair mitochondria, but also to trigger their demise and elimination.

D. F. Stowe, *Evolution of Bioenergetics from Elements to Life*,
https://doi.org/10.1007/978-3-031-76668-8_7

Keywords Antioxidants · Aquaporins · ATP synthase · Ca^{2+}-ATPase · Cardiac ischemia/reperfusion · Cardiac stress · Cardioprotection · Cell apoptosis · Cell necrosis · Cell Na^+-K^+-ATPase · Concentration gradient · Eukaryotes · Facilitated diffusion · Forward electron transfer · Hydrogen peroxide · Outer mitochondrial membrane · Oxidative stress · Matrix swelling · Membrane permeability transition pore · Mitochondrial apoptosis channel · Mitochondrial-associated membranes · Mitochondrial Ca^{2+}-sensitive K^+ channels · Mitochondrial biogenesis · Mitochondrial Ca^{2+} · Mitochondrial Ca^{2+} buffering capacity · Mitochondrial Ca^{2+} uniporter · Mitochondrial dynamics · Mitochondrial energetic adaption · Mitochondrial fission/fusion · Mitochondrial heterogeneity · Mitochondrial Ca^{2+}-H^+ exchanger · Mitochondrial K^+-H^+ exchanger · Mitochondrial Na^+-Ca^{2+} exchanger · Mitochondrial Na^+-H^+ exchanger · Mitochondrial quality control · Mitohormesis · Mitophagy · Myocyte contraction/relaxation · Passive diffusion · Prokaryotes · Protein nitration · Reactive oxygen species · Sarcoplasmic reticulum · Selectivity filter · Superoxide radical · Transmembrane transport · Voltage dependent anion channel · Voltage gradient

7.1 Diffusion and Transport of Ions and Molecules

Uncharged gases like N_2, O_2, CO_2, and $NO^\bullet$, but not $O_2^{\bullet-}$, dissolve and easily transit (diffuse) through plasma and organelle lipid bilayer membranes. Lipid, non-polar molecules, also transit membranes relatively unhindered. H_2O, a polar molecule, passes through pores called aquaporins as shown in this simple scheme of Fig. 7.1 [1]. The transmembrane transport of larger polar molecules, e.g., sugars, nucleotides, and amino acids, and all charged ions, is selectively carried out by transmembrane proteins. The protein structure dictates the selectivity for which ions or polar molecules can diffuse. Diffusion, as guided by the protein's structure, can be passive down a concentration and/or voltage gradient, i.e., mCa^{2+} uptake via the IMM mitochondrial Ca^{2+} uniporter (MCU) channels or mK^+ uptake by K_{Ca} channels, or can be facilitated, meaning that energy is required to pump the ion or molecule against a concentration or voltage gradient, i.e., Na^+-K^+-ATPase (Fig. 1.12) or Ca^{2+}-ATPase (sarcoendoplasmic reticular Ca^{2+} ATPase (SERCA) (see Sects. 8.1, 8.2, 9.2 for Figs. 8.1, 8.4 and 9.4).

Mitochondrial Channels and Exchangers Channel proteins, initially linear and unfolded, are incorporated into a membrane where they often form several transmembrane helices, or loops, with polar amino acids that form a hydrophilic (H_2O filled) channel to allow entry of hydrated ions or polar molecules when the channel is activated. Examples are the BK_{Ca} and SK_{Ca} channels, which have a "selectivity filter" (SF), or pore (P), that is opened when the channels becomes activated as depicted in Fig. 7.2 [2, 3]. The SF is a specific sequence of conserved amino acids in the narrowing of the hydrophilic channel. Negatively ("−") charged amino acids attract the positively ("+") K^+ ion and the side groups (carbonyls) of the amino acids

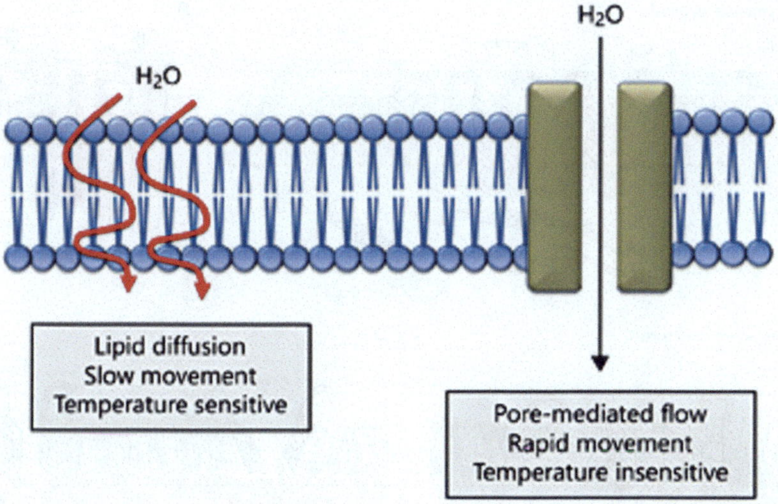

Fig. 7.1 Pathways by which (polar) water can cross a membrane lipid bilayer: Water can diffuse slowly across the bilayer between the lipid molecules (shown in blue). This slow process is highly temperature dependent, as in most diffusional processes. In contrast, water traverses the membrane much more rapidly through aqueous channels (aquaporins—shown in brown), and this process is much less temperature-sensitive. Aquaporins selectively allow rapid movement of water molecules, in single file, by osmosis into and out of cells, while ions are abrogated from aquaporins. Osmosis is essential for homeostasis of most vital physiological processes. (Cartoon (Fig. 1) and its abridged, modified legend resourced with license (https://creativecommons.org/licenses/BY-NC-ND/4.0/) from Brown [1] (review))

strip the H_2O molecules away from the K^+ so K^+ ions are allowed to pass through the narrow SF and across the membrane [3]. The peptide sequence of a given channel is generally highly conserved across species and throughout evolution. However, due to the structural plasticity of various tertiary conformations, the protein may have various ligands that activate or inactivate the channel opening, so conservation of sequences in the SF is the key molecular determinate of the ion channel gating behavior [4].

In difference to activated channels where ions are rapidly moved across membranes, ions are moved much more slowly, across membranes, often in the opposite direction, via carriers or exchangers (Figs. 2.2 and 4.4), for example, NCE and NHE. During a cycle of conformational changes, ions or molecules bind to the protein and then dissociate on the opposite side of the membrane. The conformational cycle can be coupled to energy sources like pre-existing ion gradients or ATP hydrolysis, thus allowing substrates or ions to be moved "uphill" against their concentration gradients [5].

All transporters must effectively have two "gates" that control access from either side of the membrane to the substrate-binding sites as well as a conformational cycle that prevents both these gates from being open at the same time. To prevent an exchanger from working like a channel, typically one side of the gate will not open

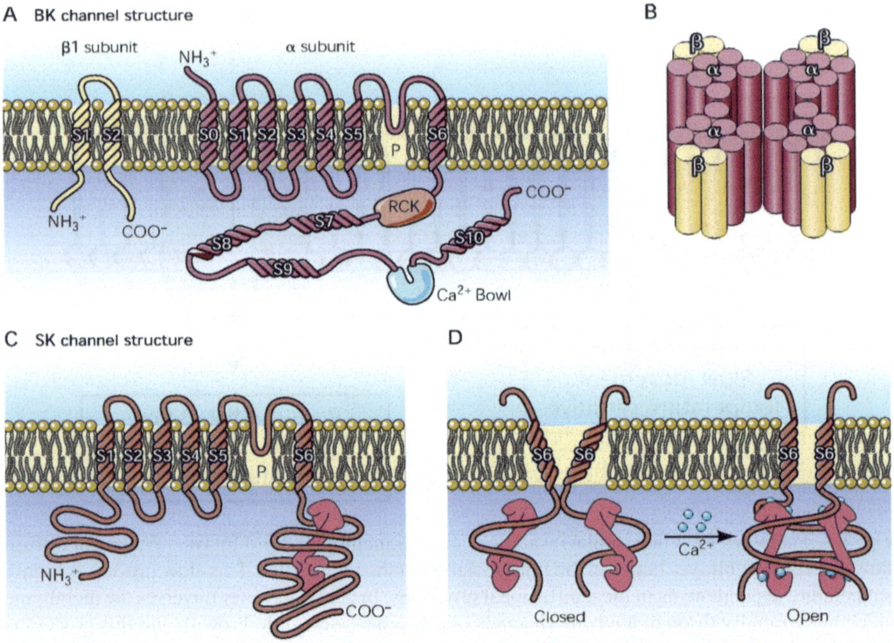

Fig. 7.2 Schematic diagrams of BK_{Ca} and SK_{Ca} channels: (**a**) Structure of the α and β1-subunits of the BK_{Ca} channel. The β1-subunit consists of 2 transmembrane domains and the α-subunit of 11 (S0–S10) hydrophobic domains, with S0–S6 located in the cytoplasmic membrane and the pore region (P) between S5 and S6. The RCK domain is the cytosolic high affinity Ca^{2+} sensing site. (**b**) Association of four α and four β1-subunits forms the native BK_{Ca} channel. (**c**) Structure of the SK_{Ca} channel consists of six transmembrane domains, a pore region (P) between S5 and S6, and calmodulin (pink) interacting with the channel's intracellular COOH terminus. (**d**) Theoretical gating mechanism of the SK_{Ca} channel by Ca^{2+} and calmodulin. (Cartoon structures (Fig. 2) and its modified legend resourced with permission from copyright owner of Ledoux et al. [2] (review))

until the other side gate is closed. Cation-proton exchangers originated in prokaryotes such as blue-green algae and bacteria along with the development of ion channels; mitochondrial exchangers were derived from archaeans or bacteria incorporated into primitive cells during symbiogenesis [6]. The conformational distinction of channels and carriers/exchangers may not be as different as earlier supported [5, 7].

7.2 Modulation of Bioenergetics, Homeostasis, and Apoptosis by Mitochondrial Ca^{2+} and K^+

Messenger Ca^{2+} Ca^{2+} became the best ion for communicating signals in early lifeforms in part because of its unique chemical properties to easily bind and unbind to complex molecules including on "Ca^{2+} sensors" in some channel proteins (see EF hand proteins below) [8]. The sensitivity of Ca^{2+} required the lowering of its free concentration in cells and its organelles to levels in the nM range (Sect. 1.2) (as with nM [H^+]) that would avoid its precipitation with phosphates in solution [8]. Figure 7.3 [8] lists a multitude of functions mediated by nM [Ca^{2+}]. The role of Ca^{2+} during evolution was important since soluble phosphates like ATP in mM concentrations became the currency of energetics (see Sects. 1.4, 2.3, 3.1 and 3.3). When $\Delta\Psi_m$ is sufficiently negative (charged) and cytosolic Ca^{2+} is elevated, Ca^{2+} enters the IMS through voltage dependent anion channels (VDAC), and Ca^{2+}, as a positive charge, enters the mitochondria matrix via MCU and mRyR (ryanodine) channels from SERCA. Figure 7.4 [9] depicts Ca^{2+} ions (red dots) diffusing across membranes to tightly interconnect Ca^{2+} signaling across the sarcoplasmic reticulum (SR), mitochondrial associated membranes (MAM), VDAC, (RyR channels, MCU, cytosol, and Ca^{2+} ion exchangers (NCE, CHE). For example, a large increase in matrix [Ca^{2+}] can directly or indirectly regulate respiration by action on linked TCA intermediates (Fig. 7.4) [10–17].

Specifically, it was long ago reported that an increase in cytosolic c[Ca^{2+}] enhances the activities of pyruvate dehydrogenase (PDH), NAD^+-isocitrate dehydrogenase (IDH), and oxoglutarate dehydrogenase (OGDH) in intact mitochondria at $k_{0.5}$ values of about 1 μM [18]. Moreover, the direct binding of mCa^{2+} to IDH and OGDH lowers the K_m's for isocitrate and αKG, and increased m[Ca^{2+}] indirectly activates PDHC by activating PDH-phosphatase subunit 1 (PDP1), which activates the PDHC by dephosphorylation [12, 13].

It should be noted that these cell Ca^{2+} concentrations are five to ten times higher than normal and would lead to higher m[Ca^{2+}] than is normal in healthy mitochondria, so this mechanism may not be physiological. However, Ca^{2+} may play an important role on these TCA enzymes in stressed mitochondria during Ca^{2+} overloading. Interestingly, in rat heart mitochondria, αKG dehydrogenase is suggested to be a more physiologically relevant target of Ca^{2+} action than PDH [19]. In addition, respiration may be regulated via the cytoskeleton with uptake of cell Ca^{2+} by sarcolemmal L-type Ca^{2+} channels in close approximation with mCa^{2+} uptake via MCU channels [11]. Although increases in m[Ca^{2+}] can enhance the activity of TCA intermediates and stimulate respiration by decreasing $\Delta\Psi_m$, which stimulates ATP synthesis, if the Ca^{2+} buffering capacity is exceeded, the excess in Ca^{2+} induces mitochondrial and cell death by apoptosis and or necrosis [15]. The buffering of Ca^{2+} in the matrix, as well as the SR, protects the organelles from Ca^{2+} overload so that the amount of free, cationic Ca^{2+} available for signaling remains in the nM range.

Modulators of mCa^{2+} Activity Figure 7.5 demonstrates that when $CaCl_2$ is added to the buffer solution of mitochondria isolated from cardiomyocytes, the rise in nM m[Ca^{2+}] is enhanced over twofold when ADP is added [20]. It was posited that this occurs because the increase in matrix ADP displaces ATP, which in the matrix must have a greater Ca^{2+} buffering capacity than ADP. Once synthesized, ATP is rapidly removed from the matrix via ANT and nM m[Ca^{2+}] falls; this increase in nM m[Ca^{2+}] by ADP can be blocked by an ANT inhibitor (carboxyatractyloside, CATR) and by

A selection of important functions modulated by the calcium signal in eukaryotic cells
Metabolism: generation of fuels
Glycogenolysis (phosphorylase b kinase)
α-Glycerophosphate dehydrogenase
Pyruvate dehydrogenase phosphate phosphatase
NAD-dependent isocitric dehydrogenase
α-Ketoglutarate dehydrogenase
NADH dehydrogenase (plant mitochondria)
β-Hydroxybutyrate dehydrogenase
Lipases and phospholipases
Membrane-linked functions
Excitation-contraction coupling
Excitation-secretion coupling (*e.g.* neurotransmitter release)
Action potentials
Tight junctions
Cell contact
Calcium-transporting ATPases
Channels in the plasma membrane and organelles
Plasma membrane-vesicles fusion
Hormonal regulation
Formation/degradation of cyclic AMP and GMP
Release of several hormones
Contractile and motile systems
Muscle myofibrils
Cilia and flagella
Microtubules and microfilaments
Cytoplasmic streaming
Pseudopod formation
Miscellaneous functions
Fertilization
Bone metabolism
Proteases
Protein kinases
Protein phosphatase (calcineurin)
Production of messengers (*e.g.* NO)
Gene expression
Neurogenesis and memory
Vision
Apoptosis

Fig. 7.3 A selection of important functions modulated by Ca^{2+} signaling in eukaryotic cells. (Table resourced with its legend with license (https://creativecommons.org/licenses/BY/4.0/) from Carafoli and Krebs [8] (review))

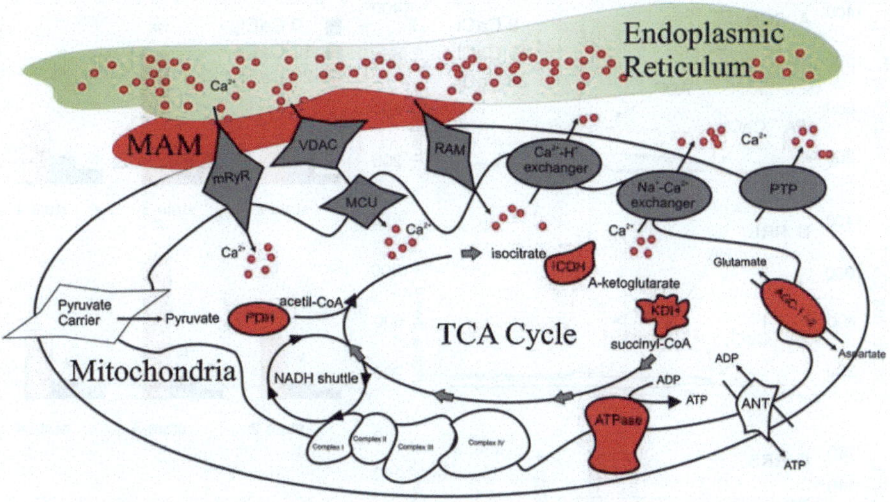

Fig. 7.4 Mitochondrial Ca^{2+} dynamics and stimulation of the TCA cycle and OxPhos during mitochondrial ATP production: The direct transfer of Ca^{2+} (red dots) from the ER to mitochondria supports OxPhos during mitochondrial ATP production. There is metabolic crosstalk between these organelles that characterizes the importance of ER/SR–mitochondria contacts. Schematic mCa^{2+} uptake and efflux mechanisms are displayed in grey. All enzymes and protein complexes stimulated or influenced by Ca^{2+} during mitochondrial ATP production are displayed in red circular forms. Abbreviations: *MCU* mitochondrial calcium uniporter, *mRyR* mitochondrial ryanodine receptor, *RAM* rapid mode, *VDAC* voltage-dependent anionic channel, *PTP* permeability transition pore, *ANT* adenine nucleotide transporter, *PDH* pyruvate dehydrogenase, *ICDH* isocitrate dehydrogenase, *αKDH* α-ketogluterate dehydrogenase, *AGC1/2* aralar and citrin aspartate/glutamate carriers, *MAM* mitochondrial–endoplasmic reticulum associated membranes. (Cartoon (Fig. 3) and its modified legend resourced with permission from copyright owner of Bonora et al. [9])

an ATP synthase inhibitor (oligomycin, OMN) [20]. Ruled out effects of ADP on $m[Ca^{2+}]$ are an enhanced uptake of Ca^{2+} by the MCU, an increase in mNCE activity, and an increase in P_i levels [20]. Ca^{2+} buffering is discussed in more detail in Sect. 8.3.

Endogenous $NO^{\bullet}$, as a natural inhibitor of cytochrome *c* oxidase (complex IV), can also modulate respiration stimulated by increased $m[Ca^{2+}]$ [17]. As a counter cation, Mg^{2+} can affect responses to Ca^{2+}. In isolated cardiac mitochondria the presence of extra-mitochondrial $e[MgCl_2]$ reduces the stepwise uptake of mCa^{2+} when $CaCl_2$ is added [20]. Adding $MgCl_2$ attenuates the slow but not the initial fast MCU-mediated mCa^{2+} uptake and $m[Ca^{2+}]$ [21]. However, adding $MgCl_2$ does not increase $m[Mg^{2+}]$; instead, it alters the bioenergetics (NADH, respiration) by its direct effect to decrease mCa^{2+} uptake. Moreover, at a given $[Ca^{2+}]_m$, state 3 respiration is incrementally attenuated, and state 4 respiration is enhanced, at higher $e[Mg^{2+}]$ [21]. This suggests that a higher $e[Mg^{2+}]$, without a change in $m[Mg^{2+}]$, can modulate bioenergetics independently of MCU-mediated Ca^{2+} transport [21].

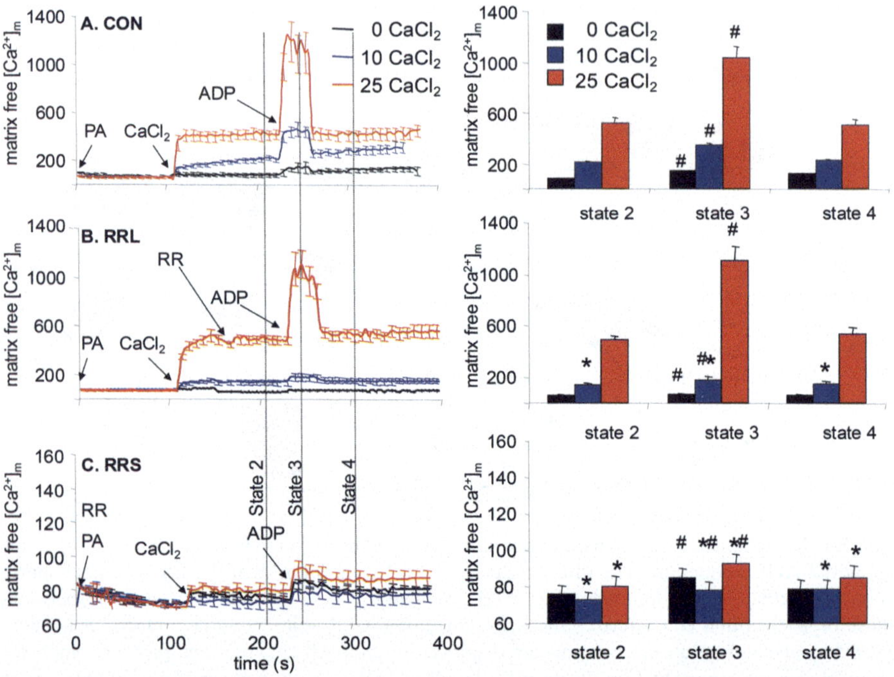

Fig. 7.5 Effects of ADP and ruthenium red (RR) on $[Ca^{2+}]_m$ over time in guinea pig isolated cardiac mitochondria: $[Ca^{2+}]_m$ was measured by indo-1 AM fluorescence. Dynamic, time-dependent changes (left column) and mean changes (right column) taken at the vertical dotted lines in the left column are displayed. (**a**) Adding 10 and 25 μM $CaCl_2$ in energized mitochondria (after adding pyruvic acid, PA) caused abrupt, graded increases in nM $[Ca^{2+}]_m$ to steady-state values. Adding ADP caused an abrupt but transient rise in nM $[Ca^{2+}]_m$ above the values established with added $CaCl_2$. (**b**) Adding RR to block the mitochondrial Ca^{2+} uniporter (MCU) after 25 μM $CaCl_2$ (RRL) did not blunt the rise in $Ca^{2+}]_m$ induced by adding ADP. (**c**) Adding RR before $CaCl_2$ (RRS) blocked Ca^{2+} uptake, but the ADP-induced increase in nM $[Ca^{2+}]_m$ remained evident, although much smaller (note scale) after adding 25 μM $CaCl_2$. Note that the buffer contained 40 μM EGTA. In the right column, for $P < 0.05$, # indicates state 3 versus states 2 and 4 respiration; * indicates RRL and RRS versus the CON group at the same respiration state and $CaCl_2$. (Data plots (Fig. 3) and its modified legend resourced with permission from copyright owner of Haumann et al. [20])

Modulator mK⁺ Several types of cell K^+ channels that carry out various functions in the sarcolemmal membrane (SLM) during phasic contraction/relaxation in cardiac cells are discussed in Sects. 2.1, 4.1 and 8.1. There are also isoforms (variants) of these K^+ channels in the IMM that have different ligands and functions [22]. Like Ca^{2+}, K^+ enters mitochondrial via various IMM K^+ channels down an electrochemical gradient (Figs. 7.2 and 7.6) [22, 23]. The cartoon of Fig. 7.6 displays several of these mK^+ channels and several ligands that activate these channels [22]. Big Ca^{2+}-

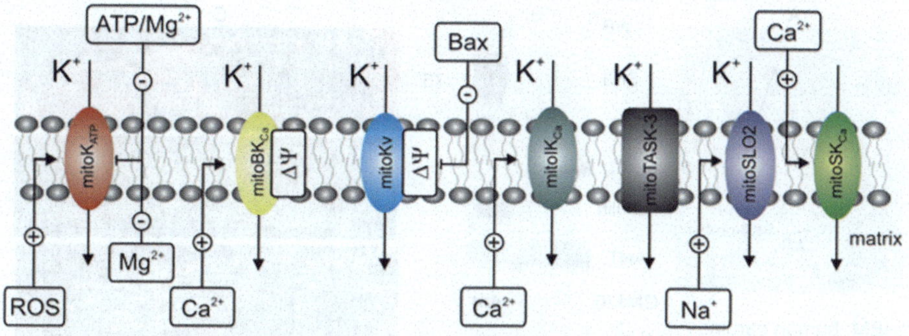

Fig. 7.6 Mitochondrial K^+ channels and some physiological modulators: From left: ATP-regulated channel (mK_{ATP}) expressed in liver, heart, T lymphocytes, skeletal muscle, and brain; large conductance Ca^{2+}-regulated channel (mBK_{Ca}) expressed in glioma cells, cardiac myocytes, skeletal muscle, brain, astrocytoma and endothelial cells; voltage-gated channel (mKv) expressed in T lymphocytes, hippocampal neurons; intermediate-conductance Ca^{2+}-regulated channel (mIK_{Ca}) found in colon tumor cells; two-pore domain channel (mTASK-3) expressed in keratinocytes; large-conductance channel (mSLO2) expressed in cardiac myocytes; small-conductance Ca^{2+}-regulated channel (mSK_{Ca}) expressed in cardiac myocytes and neuronal cells. This indicates that splice variants of many cell membrane K^+ channels are also expressed in their mitochondria. (Cartoon (Fig. 1) and its modified legend resourced with permission from copyright owner of Laskowski et al. [22] (review))

sensitive K^+ channels (BK_{Ca}) [24–26] and small Ca^{2+}-sensitive K^+ channels (SK_{Ca}) [27–29] are found both in the SLM and IMM (Fig. 7.2) [22]. Figure 7.7 demonstrates first evidence of a mitochondrial SK_{Ca} channel splice variant in human and rat cardiomycytes [27].

Function of Ca^{2+} Sensitive mK^+ Channels Mitochondrial BK_{Ca} channels are sensitive to voltage ($\Delta\Psi$) and Ca^{2+} levels, whereas mSK_{Ca} channels are sensitive only to Ca^{2+}. The most important of these "Ca^{2+} sensors" are the so-called EF-hand proteins as depicted in Fig. 7.8 [8] that first appeared in prokaryotes. They sense Ca^{2+} and decode the message, and in the case of K_{Ca} channels admit K^+ into the cell or mitochondrial matrix. These channels are active during cardiac stress, e.g., when cell and mCa^{2+} overloading occurs during ischemia-reperfusion (IR) injury [25, 28–30]. For example, Fig. 7.9 shows how a drug-induced opening of the SK_{Ca} channel with DCEB before IR injury attenuates mCa^{2+} loading after cardiac IR injury [29]. This figure also demonstrates that the drug effect can be reversed by adding the Mn-superoxide dismutase (SOD) mimetic TBAP. This suggests that SK_{Ca} channel opening indirectly leads to some ROS release, possibly by inducing a proton leak as discussed below.

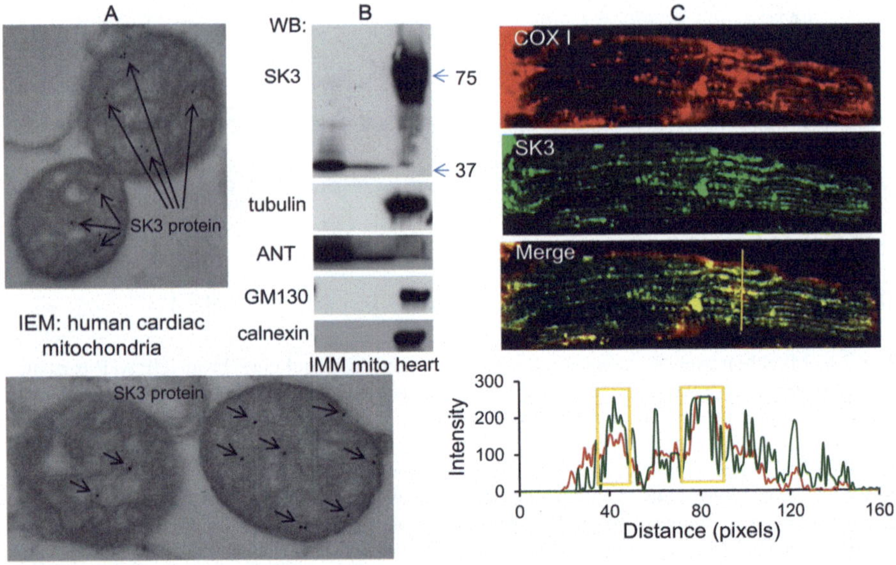

Fig. 7.7 Identity of a SK_{Ca} splice variant in cardiac mitochondria: Location of mSK3 in human heart and rat cardiomyocytes. (**a**) Immuno-electron microscopic (IEM) detection of isolated human ventricular mitochondria immune-stained with specific antibody against SK3. (**b**) Western blot analyses of SK3 expression in human ventricular tissue (heart), mitochondria (mito) and inner mitochondrial membrane (IMM). Tubulin confirmed purity of mitochondrial isolation, adenine nucleotide translocase (ANT) confirmed enrichment of mitochondria and IMM, and GM130 and calnexin confirmed absence of contamination by Golgi or sarcoplasmic reticulum membrane, respectively. (**c**) Confocal fluorescence images of adult rat cardiomyocytes immuno-stained simultaneously with SK3 antibody followed by FITC conjugated secondary antibody (middle panel, Green) and COX 1 (cytochrome *c* oxidase) antibody, followed by Alexa Fluor 546 conjugated secondary antibody (top panel, red). Yellow color in merged image (bottom panel) depicts SK3 localized in mitochondria; the intensity profile panel showed the overlapping degree of SK3 with COX 1. The x-axis in the intensity profile is from top to bottom. The yellow rectangles label the overlap areas of SK3 with COX 1. (Images (Fig. 6) with its modified legend resourced with permission from copyright owner of Yang et al. [27])

Mitochondrial K^+ Flux and ROS Experiments in isolated mitochondria (Fig. 6.8) show that release of H_2O_2 by BK_{Ca} opener NS1619 is dependent on the substrate [24]. In the presence of complex II substrate succinate + complex I inhibitor rotenone, NS1619 markedly increases H_2O_2 release whereas the substrate pyruvate has no effect. On the other hand, if the substrate is succinate but without rotenone (Fig. 7.10) [31], control H_2O_2 release is high due to reverse electron transfer (RET); in this situation NS1619 enhances respiration and greatly attenuates H_2O_2 release. It is suggested this likely occurs because of a compensatory outward flux of K^+ and influx of protons via mitochondrial K^+-H^+ exchange (mKHE), which accelerates forward electron transfer (FET), thus reducing the impact of RET on promoting excess $O_2^{\bullet-}$ generation.

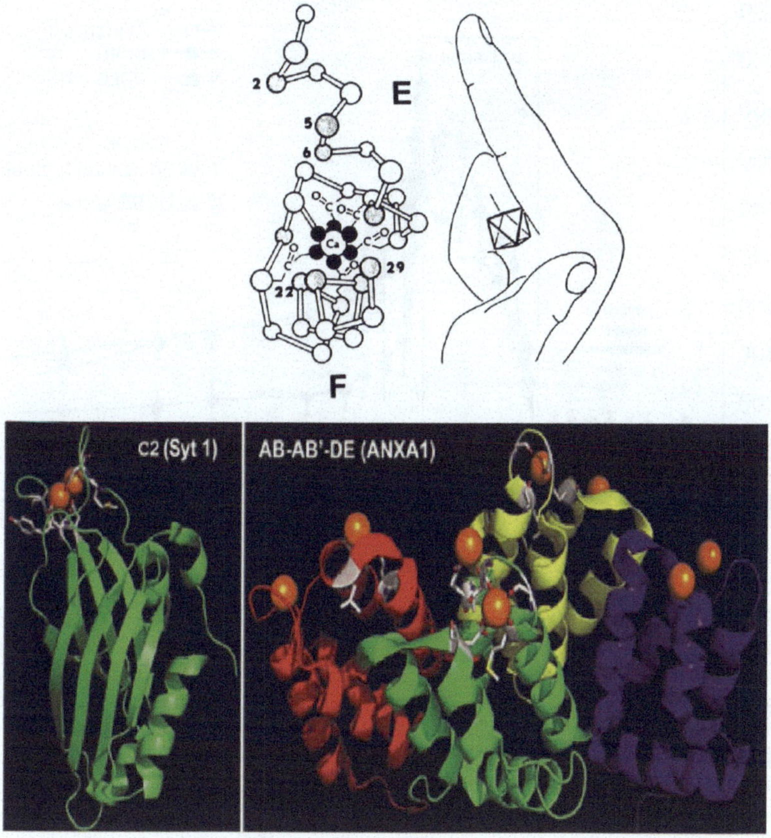

Fig. 7.8 Ca^{2+} binding protein sensors: Top, the EF-hand-binding motif, proposed by Kretsinger et al. (text Ref. [10]) from the crystal structure of parvalbumin, can be represented by the forefinger (helix E) and the thumb (helix F), enclosing the Ca^{2+}-binding loop represented by the bent middle finger. (Adapted from text Ref. [10]). Bottom left, C2b motif of synaptotagmin I (Protein Data Bank (PDB) file 1TJX). Bottom right, full-length annexin A1 (PDB file 1MCX), with repeats 1–4 in red, yellow, purple, and green, respectively. Ca^{2+} depicted as orange spheres, and the residues involved in its coordination are shown as sticks. (Graphic (Fig. 2, adapted from text Ref. [7]) resourced with its legend with license (https://creativecommons.org/licenses/BY/4.0/) from Carafoli and Krebs [8] (review))

Experiments in isolated hearts Fig. 7.11 [27, 28] show improved left ventricular pressure and reduced infarct size on reperfusion after treatment with SK_{Ca} channel opener DCEB before global ischemia. Moreover, ROS scavengers block the protective effects of mBK_{Ca} and mSK_{Ca} channel opening (Fig. 7.11) [25, 27–29] suggesting again that the signaling ROS induced by mK^+ entry plays a protective role in cardiac ischemia reperfusion (IR) injury.

These results in isolated hearts and mitochondria indicate that the protective effects of endogenous or exogenous mBK_{Ca} and mSK_{Ca} channel opening are mediated by an alteration in mitochondrial bioenergetics. These channel activities may

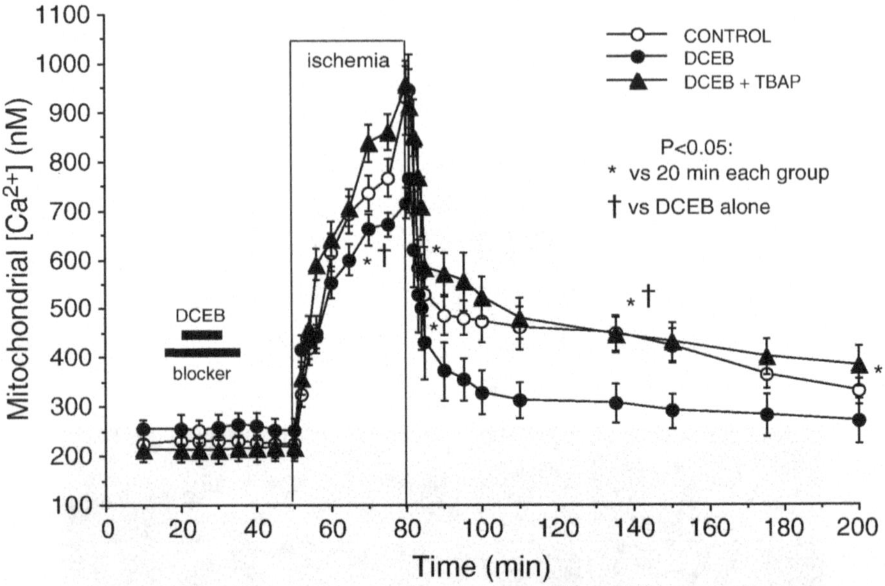

Fig. 7.9 Mitochondrial [Ca^{2+}] measurements in isolated hearts: mCa^{2+} was assessed by indo 1 fluorescence after preconditioning with the SK_{Ca} channel activator DCEB. Note the increases in these signals during ischemia and the slow decline during reperfusion. DCEB attenuated the increase in these signals during ischemia and reperfusion and this was reversed by mitochondrial superoxide dismutase mimetic (Mn)TBAP, (Mn(III)tetrakis(4-benzoic acid)porphyrin chloride). (Data plot (Fig. 4b) and its legend resourced with permission from copyright owner of Stowe et al. [29])

individually or together protect cardiac mitochondria and myocytes. Based on results of several studies, the cartoon of Fig. 7.12) [28] suggests how mBK_{Ca} or mSK_{Ca} channel opening leads to activation of mKHE, i.e., a proton leak, to induce cardioprotection. In the cartoon of Fig. 7.13 [28], possible scenarios are presented on how different mK^+ channels might respond to different sensing Ca^{2+} concentrations and or voltage gradients during oxidative stress-induced injury with the potential involvement of release of a small amount of pre-signaling $O_2^{\bullet-}$ with mKHE activation. Release of a small amount of $O_2^{\bullet-}$ before cardiac ischemia improves cardiac function and reduces $O_2^{\bullet-}$ release (Fig. 6.4) on reperfusion after ischemia [32, 33]. The diagram of Fig. 7.14 [24] suggests a pathway by which opening of mBK_{Ca} channels with NS1619 causes an increase in $O_2^{\bullet-}$ and release of H_2O_2 as a protective signaling molecule.

Importance of mKHE on Bioenergetics An important protective effect of K^+ influx via mK_{Ca} and other mK^+ channels is postulated to be its indirect action to activate the efflux of mK^+ via mKHE, resulting in the of influx of H^+ as a proton leak as portrayed in Figs. 7.12 and 7.13 [28]. It is possible that a proton leak attenuates reverse electron transfer (RET) at complex I [24, 26]. RET at complex I can occur in many pathological situations [34] and RET results in excess release of $O_2^{\bullet-}$

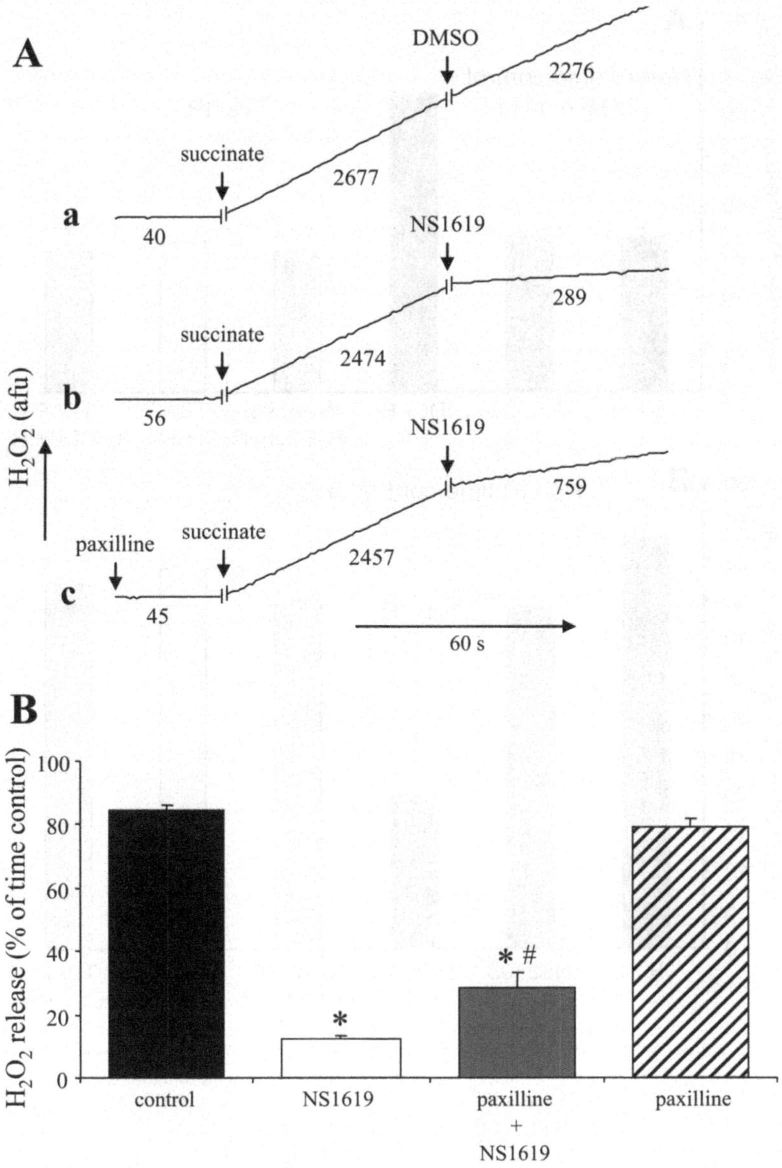

Fig. 7.10 Mitochondrial H_2O_2 release rate during mBK_{Ca} channel opening (NS1619) and blocking (paxilline): (**a**) Representative trace for H_2O_2 release rate with succinate as substrate. **a**a: Control; **a**b: 30 μM NS1619; **a**c: 5 μM paxilline + 30 μM NS1619. Numbers under lines are slopes (H_2O_2 release per time). (**b**) Summarized data for H_2O_2 release rate as a % of time controls. A 10% change represents a change in H_2O_2 release rate of approximately 1.5 pmol • min^{-1} H_2O_2 • mg^{-1} protein. All treatment effects are compared to the baseline of the same experiment. (*$P < 0.05$ vs. control; $^{\#}P < 0.05$ vs. NS1619; n = 6 for each group). (Data plots (Fig. 5) and its modified legend resourced with permission from copyright owner of Heinen et al. [31])

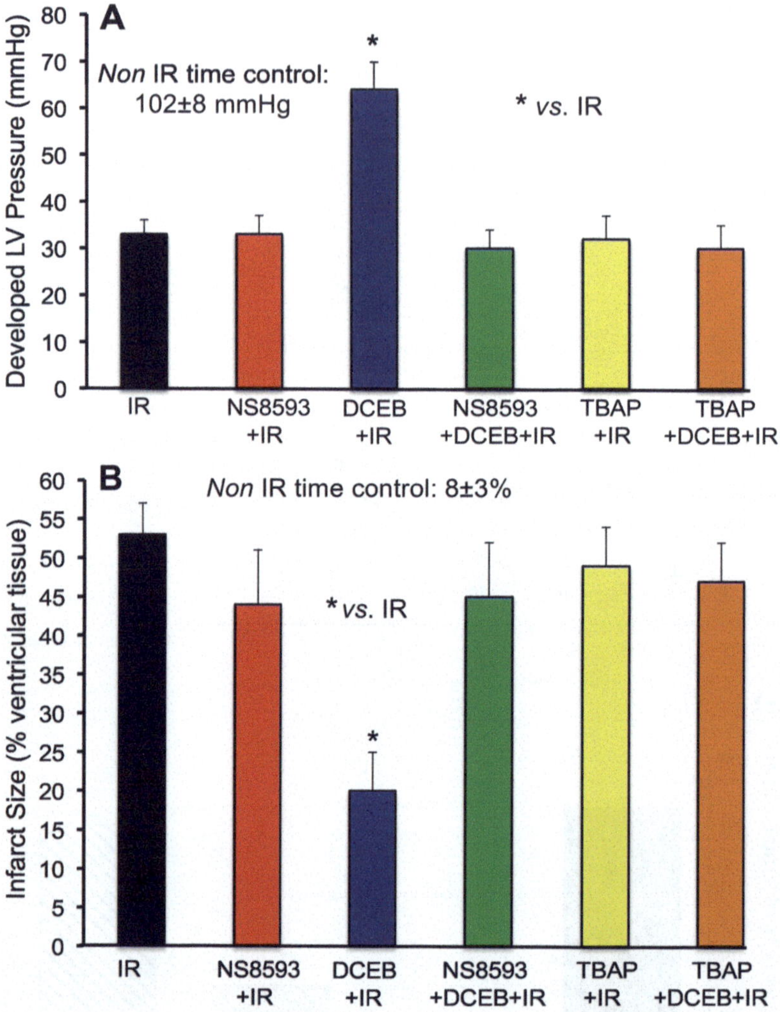

Fig. 7.11 Cardioprotection by SK_{Ca} channel agonist: Graph displays developed (systolic-diastolic) LV pressure (**a**) and infarct size (**b**) after global, no coronary flow ischemia and reperfusion (IR) in isolated guinea pig hearts perfused with vehicle (IR only), and IR + 3 μM DCEB (SK_{Ca} agonist), IR + DCEB + 10 μM NS8593 (SK_{Ca} blocker), or IR + DCEB + 5 μM (Mn)TBAP (dismutator of $O_2^{\bullet-}$). Note that DCEB enhanced developed LVP (**a**) and reduced infarct size (**b**) vs. IR alone. NS8593 and TBAP had no effect when given alone but blocked protective effects of DCEB. *$P < 0.05$ vs. IR. (Data (Fig. 1) and its modified legend resourced with permission from copyright owner of Yang et al. [27])

(Figs. 7.10 and 7.14) [31, 35, 36], and downstream free radicals and reactants that cause mitochondrial and cell damage [36]. Indeed, a major role of opening any mK^+ channel may be to induce a proton leak with release of a small amount of $O_2^{\bullet-}$.

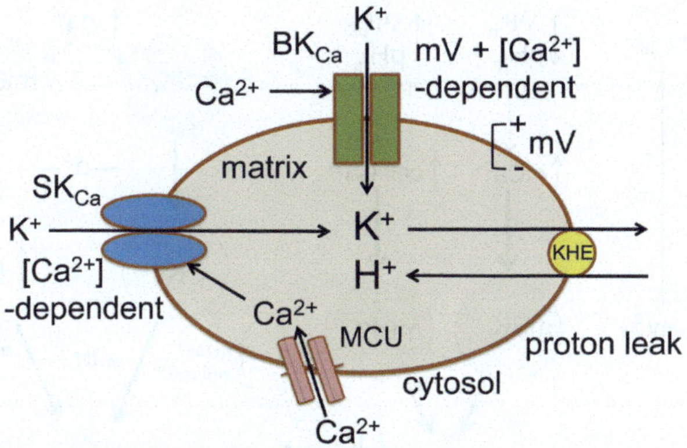

Fig. 7.12 Proposed interaction of mBK_{Ca} and mSK_{Ca} channels: Scheme shows regulation of mitochondrial mK^+ influx stimulated by increases in $\Delta\Psi_m$ or cytosolic $[Ca^{2+}]$, and or an increase in $m[Ca^{2+}]$ with reciprocal mitochondrial H^+ influx (leak) coupled with K^+ efflux via mK^+/H^+ exchange (KHE). (Scheme (Fig. 6) and legend resourced with permission from copyright owner of Stowe et al. [28])

The opening of mK^+ channels may have another important effect that is mediated by mKHE. A recent study [37] demonstrates that activation of mKHE by mK^+ influx via BK_{Ca} or other K^+ channels has an additional effect secondary to stimulation of respiration. This is an effect to rapidly rebolster a decline in $\Delta\Psi_m$ at the expense of the ΔpH_m component of Δp_m [37]. Recall that the ΔpH_m gradient involves only the movement of H^+ across the IMM, whereas the regulation and stabilization of $\Delta\Psi_m$ is essential for transport of all ions and solutes across the IMM. Because rapid Ca^{2+}-induced mK^+ influx, via mBK_{Ca}, mSK_{Ca}, or other mK^+ channels, induces slower mK^+ efflux by mKHE, the resulting gradual H^+ leak stimulates respiration to restore and maintain $\Delta\Psi_m$ by utilizing the energy stored in the ΔpH_m gradient [24, 37].

Details of this study [37] are provided here. Endogenous activation of mKHE is first demonstrated indirectly by energizing non-respiring, cardiac isolated mitochondria in an acidic buffer (pH_e 6.9) with pyruvic acid (PA) [37]. Figure 7.15a, b [37] furnishes data to demonstrate that providing NADH equivalents from PA promptly regenerates $\Delta\Psi_m$ but not ΔpH_m, while also inducing fast K^+ influx (assessed by increased mitochondrial volume) via mK^+ channels (Fig. 7.15a, b) [37]. Rapid regeneration of $\Delta\Psi_m$ indicates rapid proton pumping by the respiratory complexes and changes in pH, and Ca^{2+}, and or ATP, may trigger mK^+ influx. However, the fast recharging of $\Delta\Psi_m$, occurs over a few seconds, while the continued stimulation of respiration in the absence of ADP, the gradual increase in pH_m from 7.05 to 7.55 pH units, and the gradual 20% increase in mK^+ efflux, occur over a period of about 5–7 min. This indicates that the ΔpH_m (Fig. 7.16a) is only slowly restored due to the slow inward H^+ leak (Fig. 7.16c) [37]. In effect, this mechanism stimulates respiration (Figs. 7.16b) to rapidly energize and maintain $\Delta\Psi_m$ (Fig. 7.16a) while much more slowly increasing the ΔpH_m, thus only slowly establishing the overall Δp_m (Fig. 7.15c) [37].

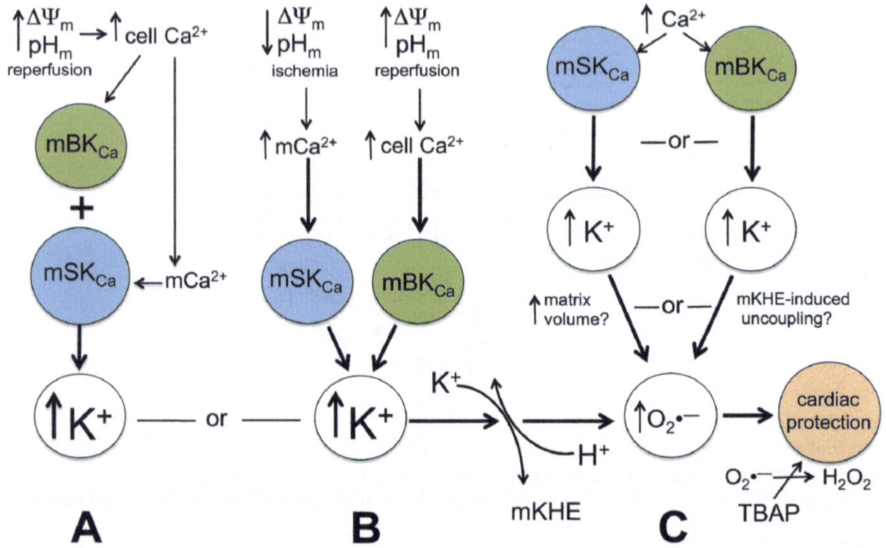

Fig. 7.13 Possible scenarios for the interactions of mSK_{Ca} and mBK_{Ca} channels and their effects in mediating protection against cardiac ischemia-reperfusion (IR) injury: (**a**) mBK_{Ca} channels might be rapidly activated during reperfusion just after ischemia when $\Delta\Psi_m$ rapidly becomes fully charged and there is a surge in cytosolic c[Ca^{2+}], and concomitantly an increase in m[Ca^{2+}]. (**b**). Independently, mSK_{Ca} channels might be activated more slowly during ischemia when m[Ca^{2+}] progressively rises in association to the lower $\Delta\Psi_m$ and pH_m during ischemia In this scenario, BK_{Ca} and SK_{Ca} channels might function rather independently to enhance K^+ influx and trigger K^+-H^+ exchange (KHE), to induce mild uncoupling, stimulate respiration, and induce small amounts of signaling ROS. (**c**) In another alternative pathway, net mK^+ influx due to SK_{Ca} channel opening might be affected by simultaneous BK_{Ca} channel opening to alter either the matrix volume or the degree of mKHE-induced uncoupling. In all possibilities, signaling levels of ROS would need to be due to mK^+ influx to initiate cardiac protection. A = additive or single pathway, B = parallel pathways with convergence, C = parallel pathways with different mechanisms. (Scheme (Fig. 7) and legend resourced with permission from copyright owner of Stowe et al. [28])

Mitochondrial K^+ and Ca^{2+} Extrusion Stimulate Respiration In a similar manner, MCU-mediated mCa^{2+} uptake activates mitochondrial Ca^{2+}-H^+ exchange (mCHE) [38, 39]. Figure 7.17 [38, 39] demonstrates mCHE activity in isolated mitochondria by monitoring how e[Ca^{2+}] and m[H^+] simultaneously increase at different buffer pHs after giving a bolus of $CaCl_2$ followed, or not followed, by blocking the MCU. Figure 7.18 [38] depicts how mCHE is involved in restoring and maintaining $\Delta\Psi_m$ with proton pumping by F_1ATPase when stimulated by partial $\Delta\Psi_m$ depolarization induced by the H^+ ionophore dinitrophenol (DNP). As with mCa^{2+} cycling, mK^+ cycling is another feedback mechanism by which K^+, and Ca^{2+}, increase respiration to maintain $\Delta\Psi_m$. Overall, these studies indicate that both mKHE and mCHE may play important roles in regulating mitochondrial bioenergetics as a consequences of proton leak resulting from channel mediated mCa^{2+} and mK^+ influx.

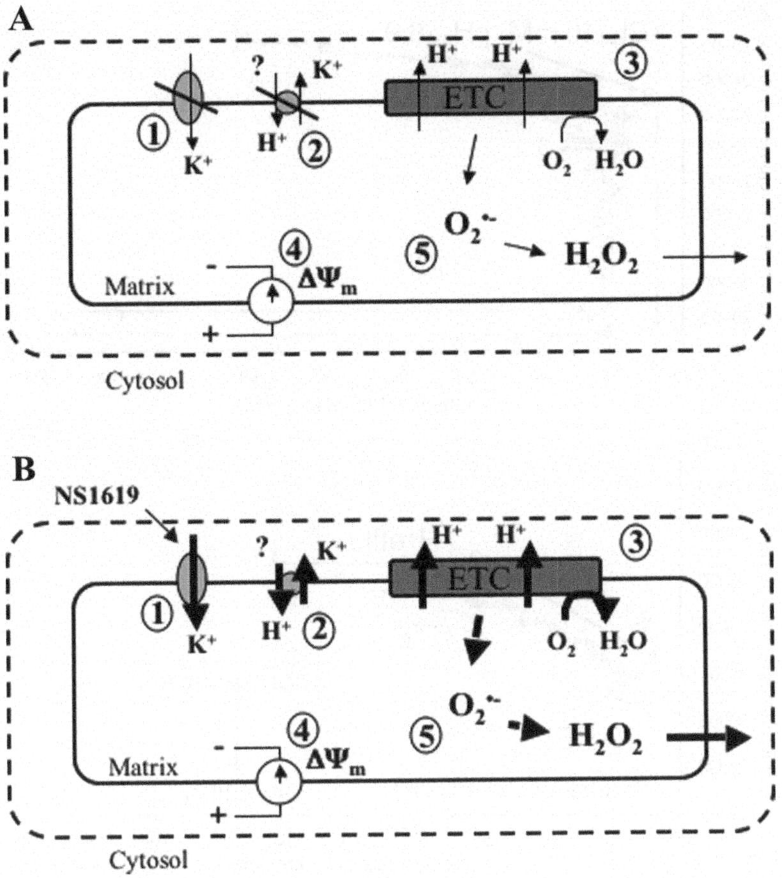

Fig. 7.14 K^+ influx causing proton leak: Proposed effects of submaximal matrix K^+ influx with mBK_{Ca} channel opening (1) on proton leak (2), proton ejection and respiration (3), $\Delta\Psi_m$ (4), and generation of superoxide ($O_2^{\bullet-}$) and H_2O_2 (5). (**a**) mBK_{Ca} channel closed. (**b**) mBK_{Ca} channel open. Net effect of mBK_{Ca} channel opening would be to accelerate *e-* transfer without a change in $\Delta\Psi_m$ due to support by proton leak; maintained $\Delta\Psi_m$ and higher *e-* transfer would lead to more ROS generation. Other promoters of proton leak are not shown. ETC, electron transport chain (system). Bolder arrows depict increased flux or reaction rate. (Scheme (Fig. 7) and its modified legend resourced with permission from copyright owner of Heinen et al. [24])

7.3 Factors Governing Mitochondrial Quality Control

Mitochondria occupy about a third of the volume of cardiomyocytes. This reflects the high energy requirements of the heart. Mitochondria are located mostly in the sub-sarcolemmal (SS), peri-nuclear (PN), and intra-myofibrillar (IM) areas, i.e., within the contraction/relaxation filament regions of cardiomyocytes [40]. These sites are depicted in Fig. 7.19 [41] by cartoon and image. With the high and continuous capacity of cardiac mitochondria to produce energy, the mitochondria must be regularly replaced in these non-dividing cardiomyocytes. Mitochondrial quality

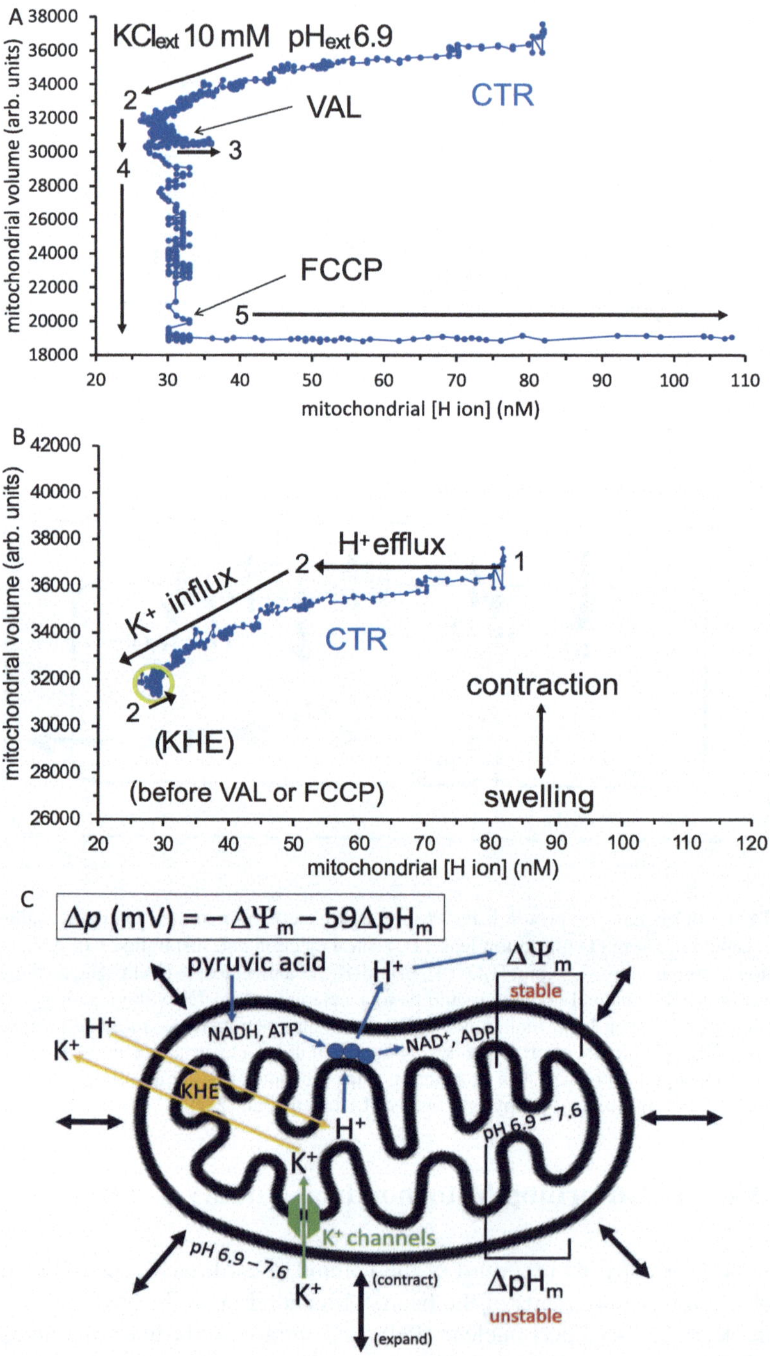

Fig. 7.15 Mitochondrial volume as a function of $[H^+]_m$ at pH_{ext} 6.9, $[K^+]_{ext}$ 10 mM: Reduced light scattering (arbitrary units, y-axis) is used to assess increased mitochondrial volume as a marker of increased matrix K^+ entry. Controls (CTR) show initial effects of charging mitochondria with

control (MQC) (Fig. 7.20) [42] entails roughly three processes: biogenesis, dynamics (fission/fusion) and mitophagy. Figure 7.21 [43] summarizes some of the important factors involved in these three processes.

Mitochondrial Biogenesis This refers to the proliferation of mitochondria when stimulated, e.g., by chronic exercise [44], which activates intracellular pathways, such as PGC-1α (proliferator-activated receptor-γ co-activator 1α), AMPK, and p53, to upregulate mitochondrial function such as mitochondrial biogenesis, ATP production, and mitophagy as depicted in Fig 7.22 [42]. Exercise induces upregulation of PGC-1α [42] as well as the transcription of nuclear genes encoding mitochondrial proteins, such as mitochondrial transcription factor A (TFAM) [45]. After import into mitochondria by protein import machinery (TOM, TIM), TFAM acts on mDNA to upregulate ETS components encoded by mDNA [43].

Mitochondrial Dynamics This refers to the constant fusion/fission/elimination activities that determine mitochondrial movement, size, shape, count, and turnover (Fig. 7.21) [42, 43]. The SR is essential for modulating mitochondrial dynamics. Fission is promoted by Drp1, which when translocated to the outer mitochondrial membrane (OMM), activates proteins (e.g., Mff) that initiate OMM separation. Fusion is coordinated by mitofusins (Mfn 1,2) at the OMM and by Opa1 (and other proteins) at the IMM, allowing for mitofusin oligomerization and fusion of the OMM of adjacent mitochondria [40]. Mitochondrial plasticity is also regulated by cell signaling that converges on networks that operate during conditions of increased (i.e., exercise-induced heart rate increase) and decreased (i.e., rest-induced heart rate decrease) energy requirements.

Some of the initial signals that induce adaptive responses are common to increased or decreased energy demands but differ in their magnitude and temporal pattern, thus producing different mitochondrial phenotypes. For example, in response to exercise (Fig. 7.22) [42], signaling to the peroxisome proliferator-activated receptor (PPAR-γ), PGC-1α, and other regulators, stimulate abundant, high-quality mitochondria with reduced mitochondrial turnover, and a higher

Fig 7.15 (continued) pyruvic acid in step 1 (**a**); additional effects of K^+ ionophore valinomycin (VAL) (step 4) and H^+ ionophore (uncoupler) FCCP (step 5), respectively, provide clarity and validation of endogenous effects on matrix K^+ in flux and matrix $[H^+]$. A closer-up view before adding VAL and FCCP is shown (**b**). Plots are based on time-dependent data of fig. S2 (see text for n's and SEM). Equilibrium coordinates for $[H^+]_m$ and volume are shown by yellow circle (**b**). Arrows show direction of time or time point after first adding pyruvic acid (and later VAL and FCCP). Scheme (**c**) depicts that the role of K^+ influx via K^+ channels is not only to induce rapid swelling, but also to initiate slower K^+-H^+ exchange (KHE) to extrude K^+ in exchange for H^+ while decreasing swelling. The H^+ leak stimulates respiration to pump H^+ and maintain $\Delta\Psi_m$ while the ΔpH_m is only slowly reestablished, so that *pmf* (Δp_m) is initially smaller at external pH 6.9 (vs. 7.6). Electron transport complexes (blue circles) are in the IMM. ATPase pumps H^+. All data generated in guinea pig, cardiac isolated mitochondria. (Data plots (Fig. 4a, b and diagram (Fig. 4c) and modified legends resourced with permission from copyright owner of Malas et al. [37])

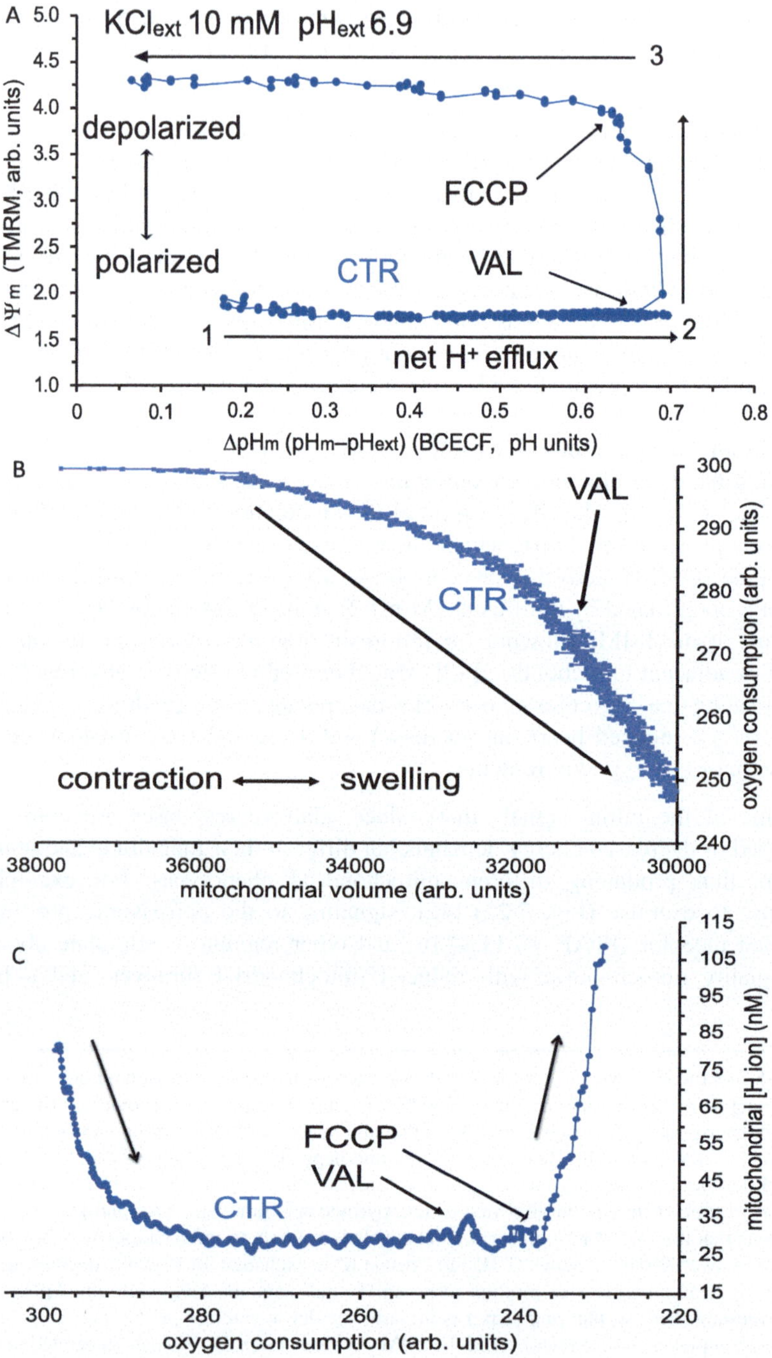

Fig. 7.16 Data supporting role of mK⁺ uptake on maintaining $\Delta\Psi_m$: (**a**) $\Delta\Psi_m$ as a function of ΔpH_m; (**b**) O_2 consumption as a function of volume; and (**c**) $[H^+]_m$ as a function of O_2 consumption at pH_{ext} 6.9, $[K^+]_{ext}$ 10 mM in control (CTR) group. Arrows show direction of time or time point of

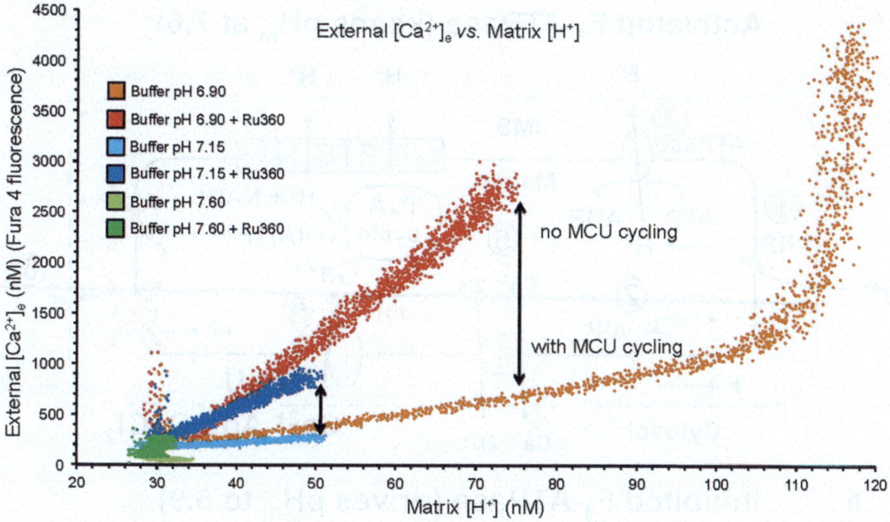

Fig. 7.17 Plots of extra-mitochondrial $[Ca^{2+}]_e$ as a function of $[H^+]_m$ in guinea pig, cardiac isolated mitochondria: These results indicate how increasing extra-mitochondrial $[H^+]$ induces matrix H^+ entry and reciprocal matrix Ca^{2+} extrusion by mCHE. Methods: A bolus of $CaCl_2$ was added at zero time and RU360 90 s later; $[Ca^{2+}]_e$ and $[H^+]_m$ were assessed continuously over 28 min. The data shows Ca^{2+} extrusion from the matrix $_e[Ca^{2+}]$ after the bolus of $CaCl_2$, the effect of blocking MCU by Ru360 to prevent matrix Ca^{2+} reuptake, and these effects of changing Ca^{2+} flux on matrix $_m[H^+]$. Data are derived from the mean data over time (see Fig. 1a, b of text). Buffer and drugs contained no Na^+ to prevent mNCE activity. Arrows denote the difference in net mCa^{2+} efflux at a given $[H^+]_m$ when MCU was or was not blocked. While mCa^{2+} was extruded by mitochondrial Ca^{2+}-H^+ exchange (CHE_m), much of it re-entered via the MCU when it was not blocked. At trans-membrane pH equilibrium (pH 6.9) mCa^{2+} uptake via MCU stopped as the $\Delta\Psi_m$ was nearly depolarized having resulted in a large net extrusion of Ca^{2+}. (Data plots (Fig. 2) and modified legend resourced with license (https://creativecommons.org/licenses/BY/4.0/) from Haumann et al. [38])

mitochondrial content [46]. Injury to cardiomyocytes during acute myocardial IR injury is accompanied by mitochondrial dysfunction. Because mitochondrial fusion and fission proteins modulate mitochondrial biogenesis and mitophagy with IR injury, modulation of these proteins may provide novel strategies to protect the cardiomyocytes by regulating mitochondrial morphology and quality control [47]. Figure 7.23 [47] displays factors involved in cardiac IR injury that regulate mitochondrial fission and fusion that can lead to cardioprotection. With knowledge of the factors involved, mitochondrial fission induced by IR injury could be targeted by

Fig 7.16 (continued) first adding pyruvic acid (VAL and FCCP added later). These results indicate how increasing extra-mitochondrial $[H^+]$ induces matrix H^+ entry with reciprocal matrix H^+ extrusion by mKHE. n's and SEM same as in Supplemental Figs. S1, S3. Data generated in guinea pig cardiac isolated mitochondria. A decrease in arbitrary units of light scattering (B, x-axis) represents increased volume. (Data plots (Fig. 6) and modified legend resourced with permission from copyright owner of Malas et al. [37])

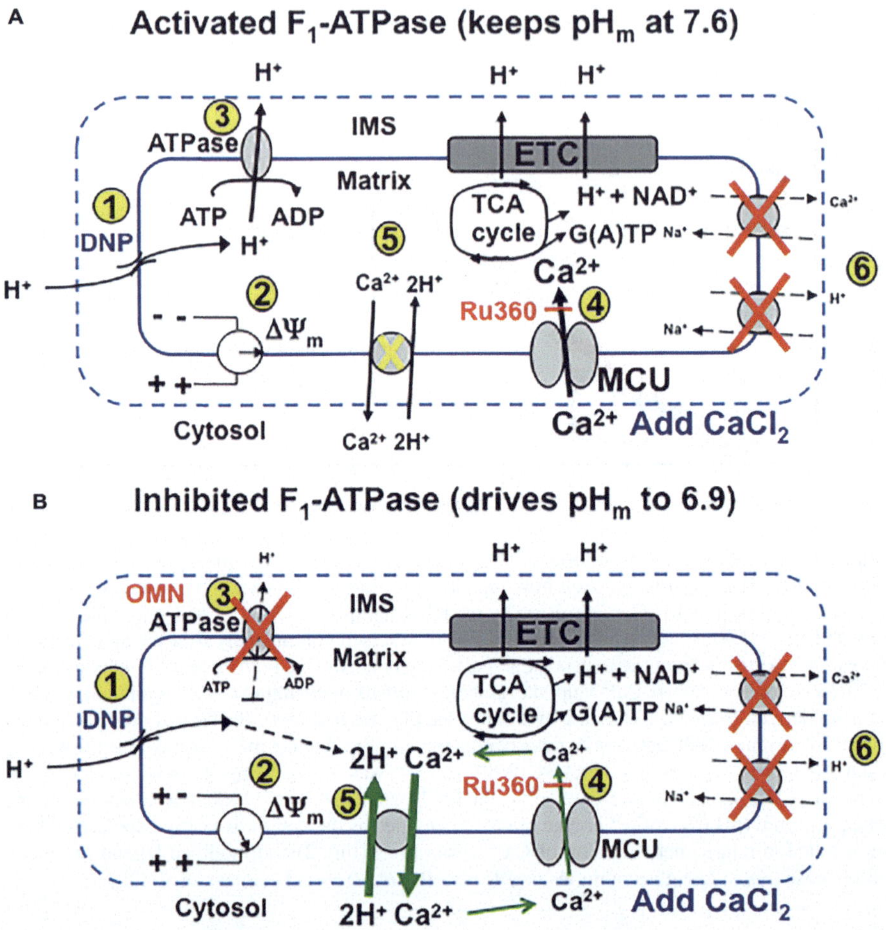

Fig. 7.18 Schema depicting putative role of MCU and CHE_m on mCa^{2+} influx and slow mCa^{2+} efflux, respectively, during stepwise depolarization with dinitrophenol (DNP) with un-inhibited (i.e., minus OMN) (**a**) vs. inhibited (**b**) F_oF_1-ATP synthase(ase) (i.e., plus OMN) after a bolus addition of $CaCl_2$ to the mitochondrial medium (**a**). (1) DNP permits H^+ entry that tends to (2) decrease $\Delta\Psi_m$, which enhances H^+ pumping by respiratory complexes, enhancing (3) F_oF_1-ATP ase, so that pH_m does not decrease appreciably and $\Delta\Psi_m$ is partially supported (2). Adding $CaCl_2$ further depolarizes $\Delta\Psi_m$ by allowing more cationic (Ca^{2+}) charges into the matrix via the MCU (4). Over time, in the range of a 20–60% decline in $\Delta\Psi_m$, pH_m remains unchanged as (5) H^+ is pumped out (3) in exchange for permissive H^+ entry (1) in triggering additional mCa^{2+} uptake by MCU and causing $\Delta\Psi_m$ to decrease further. CHE_m is inhibited by the lack (pH 7.6) of matrix acidity (5), and NCE_m and NHE_m are inactivated by the lack of substrate and buffer Na^+ (6). (**b**) In contrast, when F_oF_1-ATP ase is inhibited (3), matrix acidity gradually increases (pH 6.9), $\Delta\Psi_m$ is less supported (2) and Ca^{2+} slowly exits (CHE_m) in exchange for slow H_+ entry due to step $\Delta\Psi_m$ depolarization by DNP (5). This sequence triggers a net loss of mCa^{2+} even though uptake of Ca^{2+} via the MCU continues, as shown by a greater efflux of mCa^{2+} by CHE_m when additional mCa^{2+} uptake via MCU is blocked by Ru360 (4). Abbreviations: *DNP* dinitrophenol, *ETC* electron transport chain, *IMS* inner membrane space, *MCU* mitochondrial Ca^{2+} uniporter, *OMN* oligomycin, *TCA* tricarboxylic acid. (Diagrams (Fig. 10) and modified legend resourced with license (https://creativecommons.org/licenses/BY/4.0/) from Haumann et al. [38])

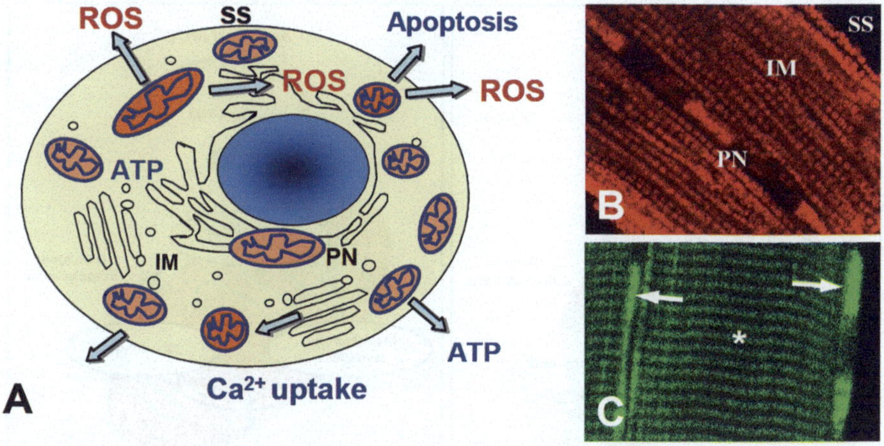

Fig. 7.19 Mitochondrial localization in cardiac myocyte: (**a**) Mitochondria-specific function may be dependent on the particular cell type and on the specific region and surroundings of the cell. (**b**) Subsarcolemmal (SS); intermyofibrillar (IM), and perinuclear (PN) mitochondrial subpopulations in rat cardiomyocyte visualized by TMRM (0.1 μM). (**c**) Subsarcolemmal (arrows) and intermyofibrillar mitochondria (asterisk) in rat skeletal muscle (m. soleus) show very different intensity of flavoprotein autofluorescence, demonstrating different redox state of these subpopulations. (Cartoon and images (Fig. 2) with modified legend resourced with license (https://creativecommons.org/licenses/BY-NC/4.0/) from Kuznetsov and Margreiter [41] (review))

using pharmacological inhibitors of Drp1 and P110, a peptide inhibitor of the interaction between Drp1 and hFis1 (Fig. 7.23) [47]. The other factors that target fusion may also have potential to confer protection via preservation of mitochondrial function.

Mitochondrial Elimination Mitochondria serve not only to activate a highly complex and regulated cell fusion and fission process, but also to activate a mitochondria-induced cell death program [40]. Mitophagy is the elimination of damaged mitochondria. Mitophagy selectively targets dysfunctional mitochondria for removal and is closely linked to mitochondrial fission and fusion (Fig. 7.23) [47, 48]. Figure 7.24 [47] describes mitochondrial fission with degradation. PINK1-Parkin-mediated mitophagy is triggered by mitochondrial depolarization, which results in PINK1 accumulation in the OMM [49]. Mitophagy itself does not necessarily lead to death of the host cell. However, an imbalance in fission, fusion, and mitophagy can result in overall bioenergetic dysfunction in mitochondria of the host cardiomyocyte and lead to cell death by apoptosis (programmed cell death) or necrosis [48, 50].

Mitochondrial Diversity in Cardiomyocytes Cardiac myocyte mitochondria (Fig. 7.19b) [41] are organized into a lattice of parallel rows surrounding the contractile myofilaments and are restrained in their positions forming a regular, crystal-like arrangement that provides for a tight and semi-permanent, bioenergetic

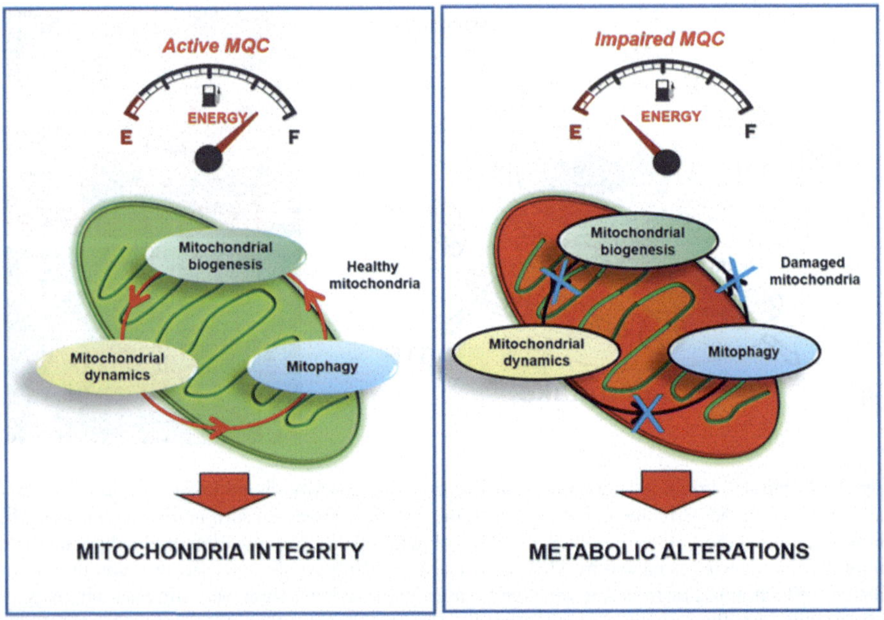

Fig. 7.20 Active mitochondrial quality control (MQC): MQC is essential for maintaining mitochondrial integrity and function as defined in the triad of mitochondrial biogenesis, dynamics, and demise (mitophagy). The impairment of MQC leads to metabolic alterations and irreversible mitochondrial damage. (Scheme (Fig. 1) resourced with license (https://creativecommons.org/licenses/BY/4.0/) from Sorriento et al. [42] (review))

connection between the SR and mitochondria for Ca^{2+} signaling and transfer of ATP from mitochondria to the contraction/relaxation apparatus. In highly metabolic cells, mitochondria cluster (Fig. 7.19a) in the perinuclear (PN) area, subsarcolemmal (SS) area, and intermyofibrillar areas (IM) areas where microdomains have higher local Ca^{2+} levels than in mitochondria-free areas. PN mitochondria interact with the nuclear membrane to localize the transfer of metabolism controlling factors; SS mitochondria localize the delivery of ATP for ATP-driven ion pumps; IM mitochondria localize the rapid exchange of ADP and ATP during each contraction relaxation cycle. In these areas, mitochondria are often in contact with other mitochondria and work temporally together to regulate overall mitochondrial function such as propagation of the Ca^{2+} wave.

In contrast, mitochondria in neurons (as in the heart) are distributed unevenly with different axonal and dendritic densities and morphologies. In neurons there is also a highly dynamic organization of mitochondria that move long distances along cytoskeletal structures, e.g., microfilaments, microtubules, and intermediate filaments and attach to the cytoskeleton using specific motor and connector proteins [41]. Because of the high activity of synapses to transmit information, mitochondria move to, and cluster in nerve cell synapses.

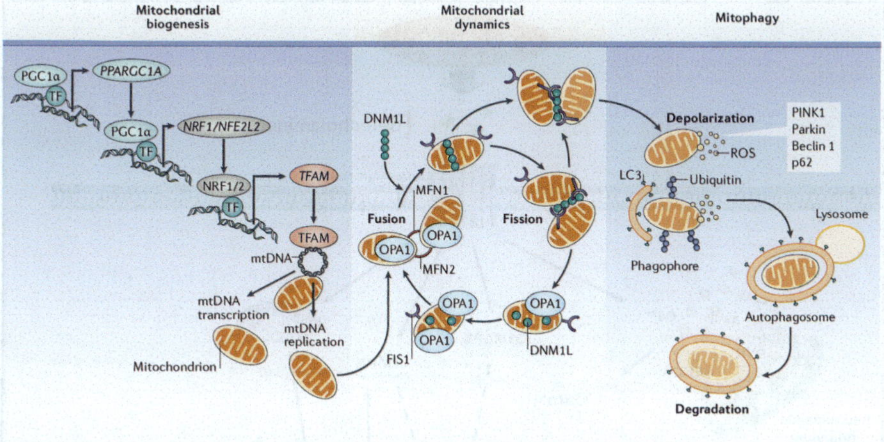

Fig. 7.21 Mitochondrial quality control pathways: Mitochondrial homeostasis is ensured through the coordination of mitochondrial biogenesis, dynamics, and autophagy. After appropriate stimuli (such as an exercise stimulus), the upregulation of proliferator- activated receptor- γ co-activator 1α (PGC1α) and other transcription factors (TFs) activate the transcription of nuclear genes encoding mitochondrial proteins such as mitochondrial transcription factor A (TFAM). TFAM is then imported into mitochondria by the protein import machinery and reaches its final destination on mitochondrial DNA (mtDNA). Here, TFAM upregulates the expression of genes encoding ETS subunits, resulting in increased O_2 consumption, ATP synthesis, and mitochondrial content. Changes in mitochondrial morphology are under the control of fusion (mitofusin 1 (MFN1), MFN2, and mitochondrial dynamin- like 120 kDa protein (OPA1) and fission (dynamin 1-like protein (DNM1L) and mitochondrial fission 1 protein (FIS1)) proteins. These factors regulate mitochondrial turnover by facilitating dilution and clearance of damaged organelles. Mitochondrial components are eventually recycled through a specialized autophagic pathway, known as mitophagy. *LC3* microtubule- associated proteins 1A/1B light chain 3, *NRF1* nuclear respiratory factor 1, *NRF2* nuclear factor erythroid 2 related factor 2, *p62* sequestosome 1, *parkin* E3 ubiquitin-protein ligase parkin, *PINK1* serine/threonine-protein kinase PINK1, *ROS* reactive O_2 species. (Cartoon (Fig. 1) and its legend resourced with permission from copyright owner of Picca et al. [43] (review))

Mitochondrial Heterogeneity During Cardiac Stress Here and in Sects. 6.4, 7.2, 7.4, 8.2 and 8.4, several examples are given of the acute effects of cardiac IR injury on myocyte and mitochondrial function. Some of these are a switch from FA oxidation to glucose oxidation, increased ROS release, mCa^{2+} overloading, and protein nitration associated with dysfunctional complex I activity as depicted in Fig. 7.25 [41]. Several structural changes in mitochondria from normal to dysfunctional are depicted in cartoons of Fig. 7.26 [51].

Chronic heart diseases due to dysrhythmias, hypertension, atherosclerosis, atherothromboses, and valvular stenosis or regurgitation are associated not only with mitophagy, but also with fragmented, dissociated mitochondria with decreased crista surface area, and gene mutations in mitochondrial tRNA and respiratory complexes. These are accompanied with down regulation of mSOD and peroxiredoxins, upregulation of TCA malate dehydrogenase and pyruvate dehydrogenase, and altered expression of proteins associated with TCA metabolism and antioxidant

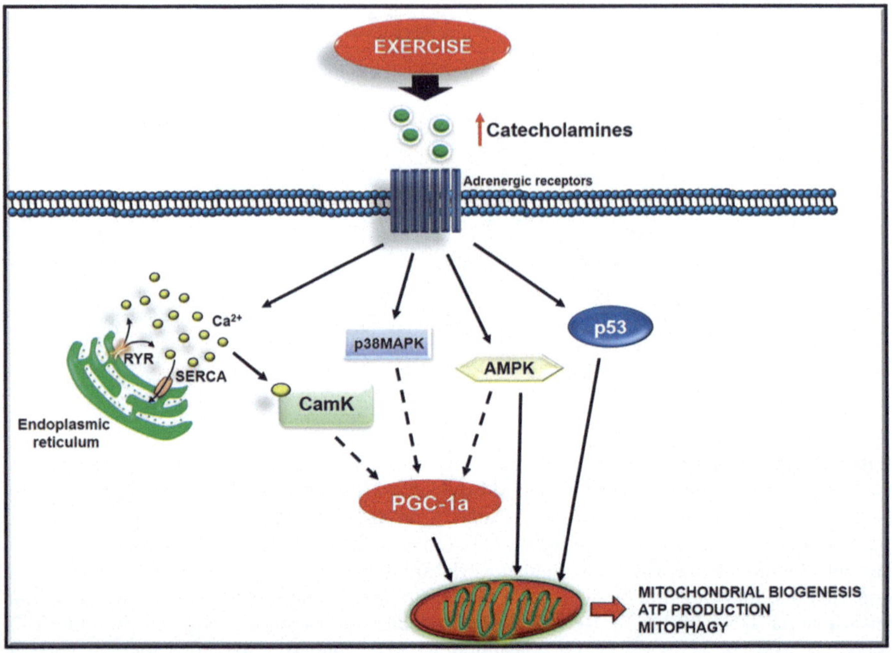

Fig. 7.22 Mitochondrial regulators: Exercise activates several intracellular pathways, such as SERCA, PGC-1α by way of: (**a**) Ca^{2+}-calmodulin -dependent kinases (CAMK), (**b**) p38 mitogen activated protein kinases (p38MAPK), and (**c**) 5′ AMP activated protein kinases (AMPK), and by p53 to regulate mitochondrial function. (Scheme (Fig. 2) resourced with license (https://creativecommons.org/licenses/BY/4.0/) from Sorriento et al. [42] (review))

activity [51]. Figure 7.27 [51] depicts mechanisms that contribute to mitochondrial heterogeneity such as mutation of the mitochondrial genome and import of mitochondrial proteins.

Severe mitochondrial stress secondary to cellular stress can also result in matrix acidity, low $\Delta\Psi_m$, high P_i level, and low ADP level. These and other factors, like mCa^{2+} overload and excess reactive species (ROS), can induce what is called membrane permeability transition pore (MPTP) opening. However, the complete structure of this "pore" is unknown and its components may only aggregate during severe mitochondrial stress. Transient or "flickering" MPTP opening may be an exit pathway for excess Ca^{2+} when $\Delta\Psi_m$ is partially depolarized that does not necessarily lead to cell death.

Mitochondria respond to their environment by sending signals to the rest of the cell that regulate, among others, cell death, inflammation, mitochondrial dynamics, and thiol oxidation as outlined in Fig. 7.28 [52]. A major signaling pathway for apoptosis is mitochondrial release of cytochrome *c* (Fig. 7.28) [52–54]. Other signaling pathway are Smac, Omi/Htr2A, and AIF [55]. Apoptotic stimuli induce relocation of the protein Bax from the cytosol into the OMM. Bax, Bak, and other proteins are reported to oligomerize the voltage dependent anion channel (VDAC)

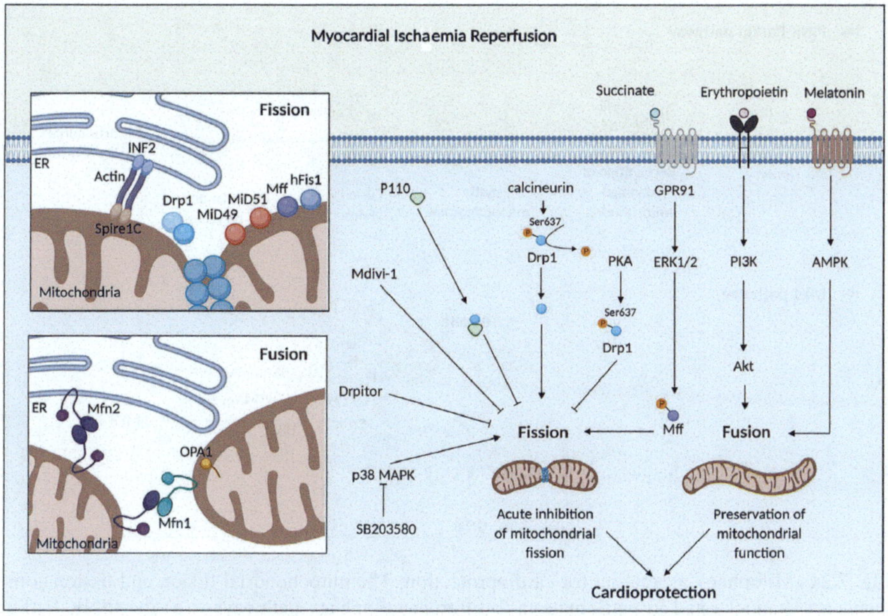

Fig. 7.23 Mitochondrial fission and fusion proteins as targets for cardioprotection: Mitochondrial fission induced by acute myocardial ischemia/reperfusion injury can be targeted by mdivi-1 and Drpitor, pharmacological inhibitors of Drp1 and P110, a peptide inhibitor of the interaction between Drp1 and hFis1 to confer cardioprotection. Other factors (such as SB203580, PKA activators, succinate, erythropoietin and melatonin) have also been shown to confer cardioprotection by targeting the fusion and fission proteins. Inset upper box: This scheme shows the mitochondrial fission machinery, comprising Drp1 and its outer mitochondrial membrane receptors, MiD49/MiD51, Mff and hFis1. Pre-constriction by the endoplasmic reticulum (ER) via INF2, actin and Spire1C initiates the Drp1-driven mitochondrial fission process. Inset lower box: This scheme shows the pleiotropic non-fusion effect of Mfn2 in tethering mitochondria to the ER. (Cartoon (Fig. 1) and legend resourced with license (https://creativecommons.org/licenses/by/4.0/) from Hernandez-Resendiz et al. [47] (review))

and to form a mitochondrial apoptosis channel (MAC) that causes release of cytochrome *c* and other factors. During apoptosis, the release of cytochrome *c* [56] to the cytosol causes pro-caspases to be activated to caspases, which are a family of proteases that rapidly degrade cell proteins and so can terminate myocyte function [56]. Necrotic stimuli lead to increased Ca^{2+} uptake and ROS generation by mitochondria; this can force formation and terminal opening of the MPTP that results in mitochondrial matrix swelling, rupture, and release of Ca^{2+} and other factors. Swelling disrupts the OMM too, while released Ca^{2+} activates proteases, phosphatases, and nucleases that lead to necrotic degradation of the cell [50]. All of these processes result in a given cardiomyocyte, interacting with its neighbors, having just the right amount of healthy, optimally functioning mitochondria despite the demise of irreparably damaged mitochondria.

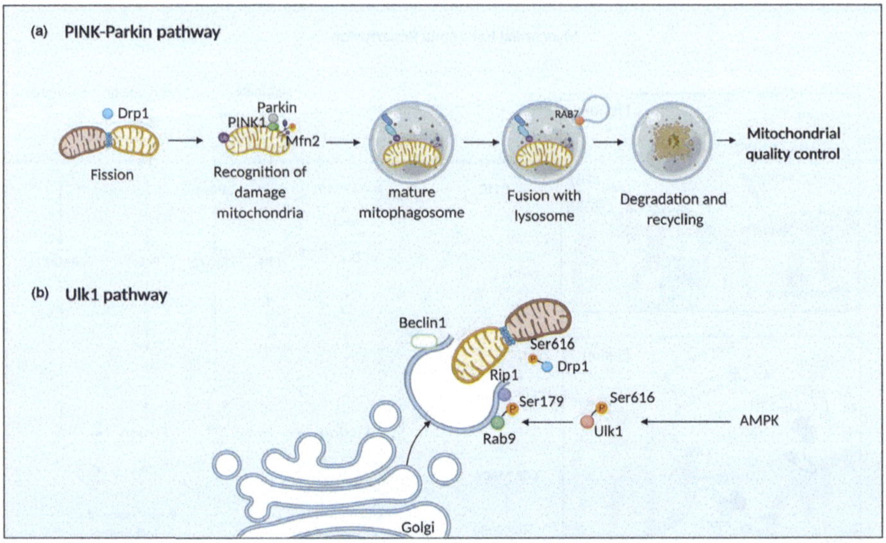

Fig. 7.24 Mitophagy as a target for cardioprotection: The mitochondrial fusion and fission comments are demonstrated to participate in 2 different mitophagy pathways—(**a**) the PINK-Parkin mitophagy pathway and (**b**) the Ulk1 mitophagy pathway. (Modified from text Ref. [121]). (Cartoon (Fig. 2) and legend resourced with license (https://creativecommons.org/licenses/by/4.0/) from Hernandez-Resendiz et al. [47] (review))

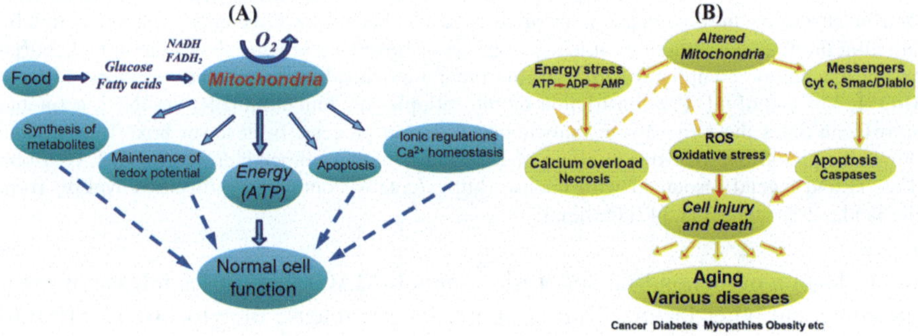

Fig. 7.25 Healthy and damaged mitochondria: (**a**) Healthy mitochondria are not only as the main intracellular source of ATP needed for normal cell function and viability, but are also major controllers in many cellular pathways, where they perform a great variety of other important cell functions. (**b**) Damaged mitochondria are associated with organ injury by several mechanisms, including diminished cellular energy status (energy stress), production of ROS (oxidative stress), disturbance of ionic balance, mCa^{2+} overload, cytochrome *c* release, and induction of apoptosis. (Cartoon (Fig. 1) and its modified legend resourced with license (https://creativecommons.org/licenses/BY-NC/4.0/) from Kuznetsov and Margreiter [41] (review))

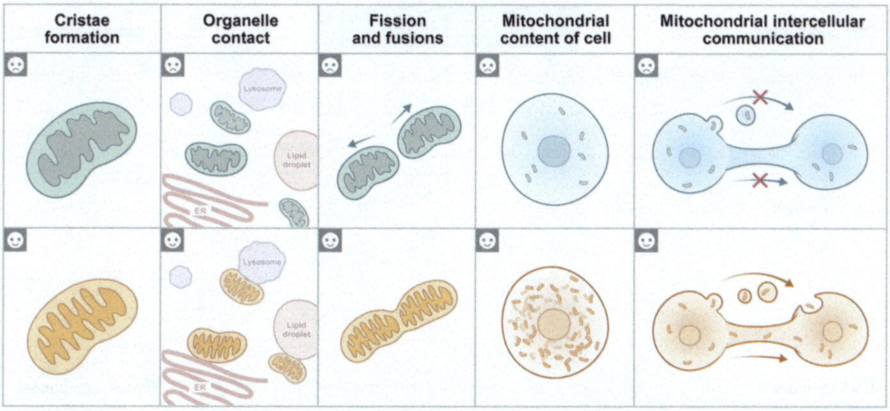

Fig. 7.26 Mitochondrial characteristics under different conditions in upper and lower panels: Low mitochondrial cristae density is associated with a low cellular energy supply (upper), while a high mitochondrial crista density reflects an adaptation to meet cell energy demands (lower). Mitochondrial contact with many additional organelles, such as the ER, lysosomes and lipid droplets, and the number of contacts mitochondria make with a specific organelle can vary dramatically from only a few contacts to hundreds of contacts per cell. A decrease in the number of mitochondrial connections to other organelles is typically a response to an inefficient metabolic pathway (upper), and in contrast, an increase indicates a response to an active cellular energy metabolic pathway (lower). Mitochondrial fission and fusion are the main pathways of mitochondrial morphology regulation, and under stressful environments, mitochondria are active during mitochondrial fission and produce an increased number of punctate mitochondria (upper), while mitochondrial fusion mediates the formation of mitochondrial networks that adapt to the high energy demands of cells (lower). The content of mitochondria in a cell reflects the intensity of cellular metabolism. Low mitochondrial content in cells is usually associated with low metabolic activity (upper), and high mitochondrial content is associated with high cellular metabolic activity (lower). Intercellular mitochondrial communication is extensive under physiological and pathological conditions, and a low frequency of mitochondrial communication is a response to low cellular adaptation to stressful environments (upper), and in contrast, efficient mitochondrial transfer enhances cellular adaptive capacity (lower). (Cartoons (Fig. 1) with modified legend resourced with license (https://creativecommons.org/licenses/BY-NC/4.0/) from Chen et al. [51] (review))

7.4 Regulation of Mitochondrial Oxidative Phosphorylation by Nuclear and Sarcoplasmic Factors

A changing cellular environment and varying energetic needs require a change in mitochondrial responsiveness such as an adjustment in respiratory capacity [57] that can trigger changes in gene, protein translation, protein complex formation, transmembrane transport, enzymatic activities, and metabolite levels. These changes can promote mitochondrial biogenesis and membrane dynamics that in turn support mitochondrial respiration. Regulatory components of mitochondrial energetic adaptation include not only the transcription coactivators PPAR-γ and PGC-1α and associated transcription factors, but also the nutrient sensor, target of rapamycin, (mTOR) and SR stress signaling,TOM70-dependent mitochondrial protein import,

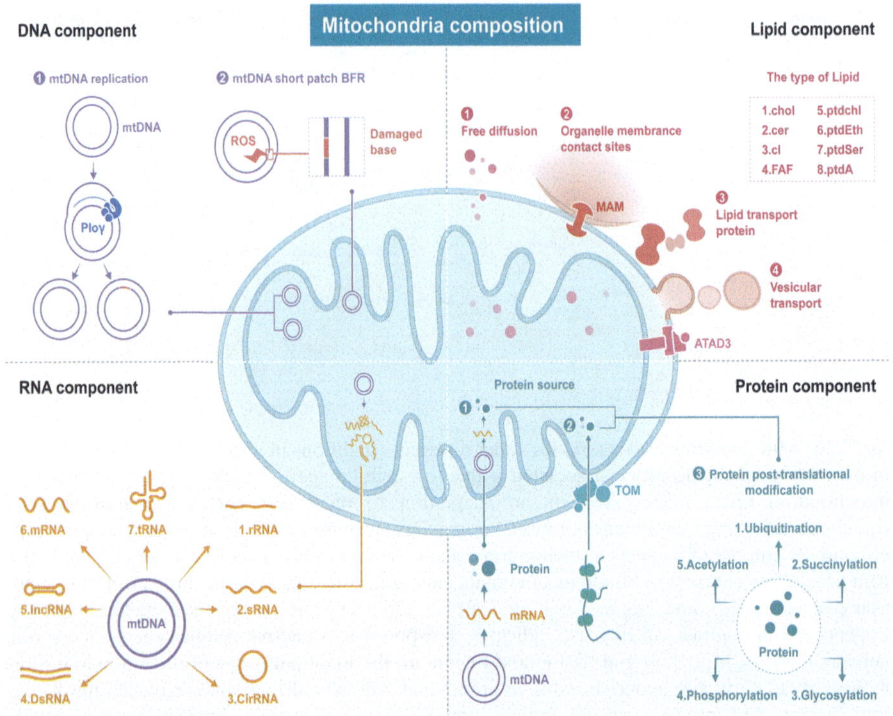

Fig. 7.27 The heterogeneity of mitochondrial components: For the DNA component, the diversity of the mitochondrial genome arises from base-pair mismatches during the replication of the genome and base mutations after mtDNA short-patch base excision repair (BER). For the RNA component, the transcripts from the mitochondrial genome include rRNAs, sRNAs, CircRNAs, dsRNAs, IncRNAs, mRNAs, and tRNAs. For the protein component, the diversity of the mitochondrial proteome arises from the two main protein resources, the nucleic coding protein import pathway and the mtDNA translation pathway. In addition, mitochondrial proteins are altered via a complex posttranslational modification mechanism. The heterogeneity of mitochondrial lipids is a result of four related lipid transport pathways, including free diffusion, organelle membrane contact, lipid transport proteins and vesicular transport. (Scheme (Fig. 4) and its modified legend resourced with license (https://creativecommons.org/licenses/BY-NC/4.0/) from Chen et al. [51] (review))

cristae remodeling factors, including the mitochondrial contact site and cristae organizing system (MICOS) and Opa1, lipid remodeling, and the assembly and metabolite-dependent regulation of respiratory complexes. These are all adaptive molecular and structural mechanisms that regulate respiratory capacity long term [57]. Without these adaptive responses to changing energetic conditions, cell damage with inflammation or senescence would follow, thus compromising nutrient availability, temperature maintenance, physical activity, and cell survival.

Mitohormesis The metabolic fitness of mitochondria to respond to cellular stress is a measure of mitochondria quality [51]. Mitohormesis [58–60] is a term that

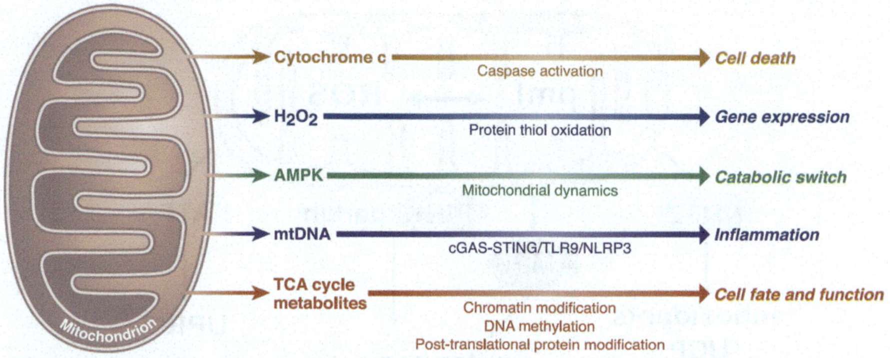

Fig. 7.28 Essential signaling functions of mitochondria: Mitochondria have evolved distinct strategies to integrate environmental cues and communicate their fitness to the rest of the cell to maintain cellular homeostasis. Release of cytochrome *c* to invoke caspase-dependent cell death, release of reactive oxygen species (ROS) to oxidize thiols within redox-regulated proteins, and induced gene expression, with activation of AMPK under energetic stress to control mitochondrial dynamics are three prominent mitochondrial-dependent signaling events. Additionally, release of mitochondrial DNA into the cytosol triggers inflammasome activation and pro-inflammatory responses through the cGAS–STING cytosolic DNA-sensing pathway. Signaling roles of TCA cycle metabolites are in part mediated by controlling chromatin modifications and DNA methylation, as well as post-translational protein modifications. (Scheme (Fig. 1) and its modified legend resourced with license (https://creativecommons.org/licenses/by/4.0/) from Martinez-Reyes and Chandel [52] (review))

describes an adaptive cellular response triggered by a mild mitochondrial stress that activates cytoprotective mechanisms to compensate for severe stress. This action can induce long-lasting, broad, metabolic, and molecular changes that can lead to continuing cell function by direct communication with the nucleus to trigger the adaptive cellular programs with signaling pathways emanating from both mitochondria and the nucleus. Such an interaction of mitochondria with the nucleus, as early endosymbionts (Sect. 4.2), are likely to have contributed to the increasing complexity of early eukaryotes over prokaryotes.

ROS Signaling There is now much evidence to support hypotheses that generation of low levels of ROS provide protection for mitochondria and its cell [36] whereas high levels of ROS oxidize and destroy cell structures as depicted in Fig. 7.29 [36, 61]. This is opposite to the long held, free-radical theory of aging and the long held protective role of antioxidants. Indeed, low amounts of ROS act as signaling molecules, as presented in Sects. 7.2 and 8.4, are also known to activate gene expression programs [62]. This signaling function of mitochondria may have been selected by nature from the inception of the early prokaryotes engulfing bacteria to initiate a symbiotic relationship. Because chemiosmosis results in some *e*- leak, the ROS generated, e.g., H_2O_2, by early eukaryotes could be utilized as an early signaling molecule of cell stress.

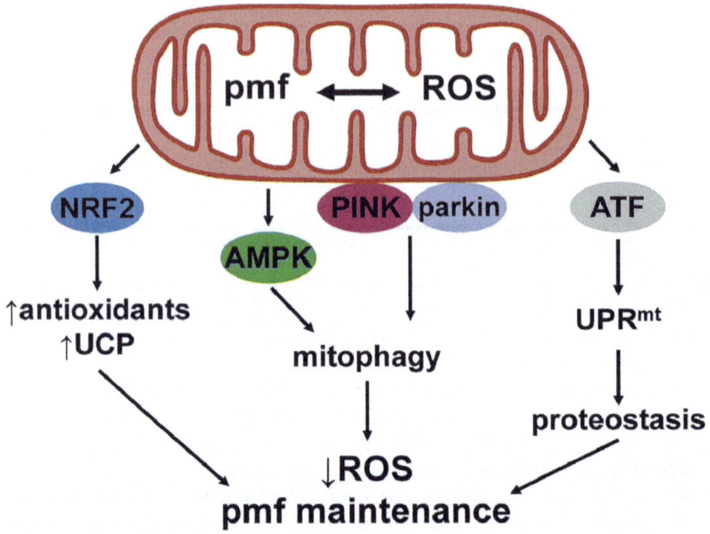

Fig. 7.29 Stress signaling regulated by the *pmf* and ROS: ROS signaling paradigms center around mitochondria and can impact the *pmf*. Nuclear factor (erythroid-like 2) related factor (NRF2) is activated by ROS and leads to upregulation of UCP3 activity (along with antioxidant defenses), decreasing the *pmf*, and potentially regulating ROS production. The mitochondrial unfolded protein response (UPRmt) is activated by ROS and impacts proteostasis and cellular health downstream of activating transcription factors (ATFs). Localized collapse of the *pmf* activates mitophagy through PINK and parkin proteins, stimulating mitochondrial turnover, preserving the *pmf* by degrading depolarized mitochondria. Changes in metabolic flux and *pmf* can activate the metabolic sensor 5′ AMP activated protein kinase (AMPK) to activate mitophagy. These signaling networks integrate the *pmf* and ROS production to maintain homeostasis. The interactions shown do not represent every interaction between the outlined pathways but highlight how the *pmf* and ROS coordinate them. (Diagram (Fig. 3) and modified legend resourced with permission from copyright owner of Berry et al. [61] (review))

It is also known that mild release of ROS induces an antioxidant response mediated by Nrf2 (Fig. 7.29) [61, 63], as well as also triggering mitochondrial biogenesis and expression of OxPhos genes [64]. Another example is regulation of the activity of ATP-synthase by its physiological inhibitor, the ATP-ase inhibitory factor 1 (IF1), which has a functional role in cellular homeostasis [60]. Figure 7.30 demonstrates that in liposomes containing F_0F_1-ATP synthase, and treated with valinomycin to allow K^+ uptake, the rates of H^+ uptake (A) and release (B) are reduced when the liposomes are incubated with increasing concentrations of IF1 [60]. Overexpression of IF1 causes a reprogramming of energy metabolism that is exhibited by enhanced glycolysis and activation of reactive ROS-dependent signaling pathways that promote cell survival. ATP-synthase modulates signaling pathways that activate the mitohormesis response, namely ATP, ROS, and TOR. ROS as signaling molecules during cardiac stress are also discussed in Sects. 7.3 and 8.2.

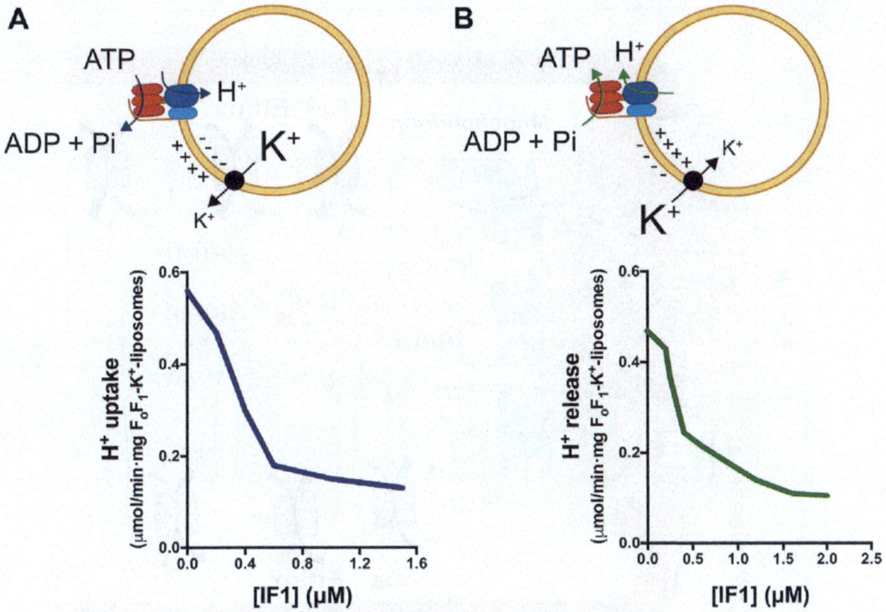

Fig. 7.30 ATPase inhibitory factor 1: (IF1) inhibits H^+ translocation through the H^+-ATP synthase in both synthetic and hydrolytic modes. (**a**) H^+ uptake is induced by valinomycin mediated K+ release from F_oF_1-K^+ liposomes with the H^+-ATP synthase functioning in the hydrolytic mode. (**b**) H^+ release is induced by valinomycin-mediated K^+ uptake in F_oF_1-K^+ liposomes with the H^+-ATP synthase functioning in the synthetic mode. Rates of H^+ uptake (**a**) and H^+ release (**b**) are reduced when the liposomes are incubated with increasing concentrations of isolated IF1. Black circle valinomycin. (Graphic diagram (Fig. 3) adapted from Ref. [84] and resourced with its legend with license (https://creativecommons.org/licenses/by/4.0/) from Esparza-Molto et al. [60] (review))

MAM Cardiac mitochondria-associated membranes (MAM), also known as mitochondria endoplasmic—sarcoplasmic reticular (SR) contact sites (MERCs), are specialized regions in which the interaction at the tight SR-mitochondrial contact sites allows regulation of multiple important cellular processes including Ca^{2+} homeostasis and regulation, lipid metabolism, unfolded protein response, ROS production, mitochondrial dynamics, autophagy, apoptosis, and inflammation [65]. The structure of MAM and their MAM-associated proteins consists of apposition of the SR membrane and the OMM with approximately 1000 associated proteins that drape around the SR and mitochondria and provide the contact sites for communication.

Signaling by Ca^{2+} between the cardiac SR and mitochondria is intense (Figs. 7.4 and 7.31) [66]. Tethering proteins, ranging from Ca^{2+} channels to apoptotic proteins, constitute the molecular bridges bringing the ER and MAM together as depicted in Fig. 7.31 [66]. These proteins are classified into (a) MAM-localized proteins that are only located at the MAM, (b) MAM-enriched proteins that can also be detected

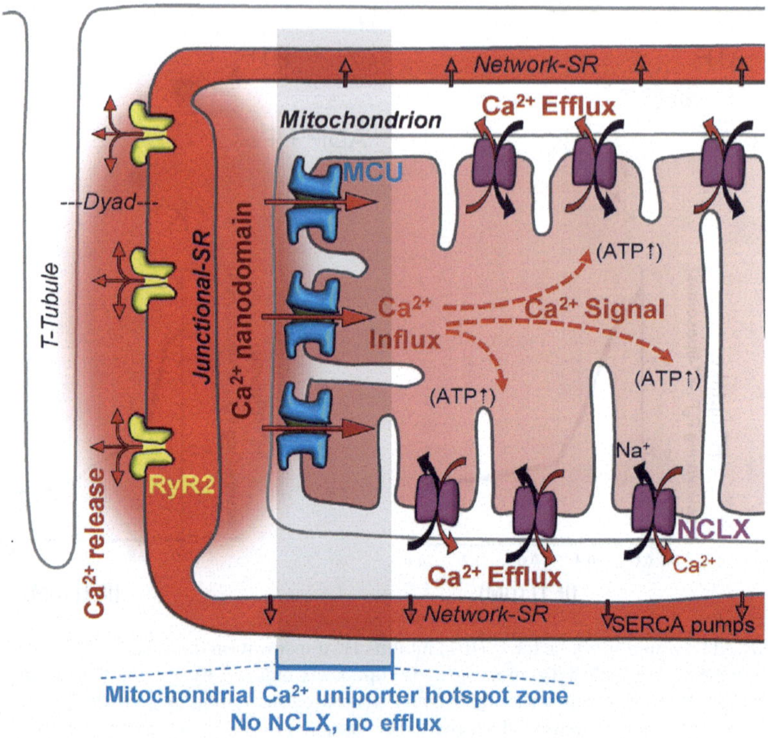

Fig. 7.31 Ca^{2+} signaling between cardiac cell sarcoplasmic reticulum (SR) and mitochondria: Ca^{2+} signals control mitochondrial fuel generation. In heart mitochondria, Ca^{2+} uptake via the mitochondrial Ca^{2+} uniporter (MCU) and extrusion via the Na^{+}-Ca^{2+} exchanger (NCLX = NCE) are spatially separated. Ca^{2+} is released from the SR via ryanodine receptor (RyR2) channels into transverse (T)-tubules during contraction (systole). Extruded Ca^{2+} is taken up by the sarcoendoplasmic reticular Ca^{2+} ATPase (SERCA) during cardiac relaxation (diastole). The most Ca^{2+}-exposed area is an uptake hotspot (deep red), but it lacks Ca^{2+} extrusion to optimize signaling efficiency and minimize the energy expense of controlling mitochondrial function by Ca^{2+}. (Diagram (graphic abstract) and modified legend resourced with license (https://creativecommons.org/licenses/BY-NC-ND/4.0/) from De La Fuente et al. [66])

in other cell regions, and c) MAM-associated proteins that are transiently found in MAM in a condition-dependent manner. Proteins in MAM include GPR75 (increases MAM formation and mCa^{2+} uptake), IP_3Rs (regulate cCa^{2+}), VDAC (facilitates mCa^{2+} uptake), and Mfn2 (a fusion stimulating factor) [65]. MAM and associated proteins are highly involved in regulating mitochondrial quality control, i.e., fusion, fission, mitophagy, and apoptosis [40], as discussed above in Sect. 7.3.

References

1. Brown D. The discovery of water channels (aquaporins). Ann Nutr Metab. 2017;70(suppl 1):37–42. https://doi.org/10.1159/000463061. PubMed PMID: 28614812.
2. Ledoux J, Werner ME, Brayden JE, Nelson MT. Calcium-activated potassium channels and the regulation of vascular tone. Physiology (Bethesda). 2006;21:69–78. https://doi.org/10.1152/physiol.00040.2005. PubMed PMID: 16443824.
3. Kopec W, Rothberg BS, de Groot BL. Molecular mechanism of a potassium channel gating through activation gate-selectivity filter coupling. Nat Commun. 2019;10(1):5366. Epub 20191126. https://doi.org/10.1038/s41467-019-13227-w. PubMed PMID: 31772184; PMCID: PMC6879586.
4. Hendriks K, Oster C, Lange A. Structural plasticity of the selectivity filter in cation channels. Front Physiol. 2021;12:792958. Epub 20211207. https://doi.org/10.3389/fphys.2021.792958. PubMed PMID: 34950061; PMCID: PMC8689586.
5. Ashcroft F, Gadsby D, Miller C, editors. Introduction. The blurred boundary between channels and transporters. Philos Trans R Soc Lond B Biol Sci; 2009 Jan 27 PMC2592634. Available from: https://www.ncbi.nlm.nih.gov/pubmed/18957372.
6. Tsujii M, Tanudjaja E, Uozumi N. Diverse physiological functions of cation proton antiporters across bacteria and plant cells. Int J Mol Sci. 2020;21(12). Epub 20200626. https://doi.org/10.3390/ijms21124566. PubMed PMID: 32604959; PMCID: PMC7349511.
7. Csanady L, Mindell JA, editors. The twain shall meet: channels, transporters and things between. EMBO Rep; 2008 OctPMC2572114. Available from: https://www.ncbi.nlm.nih.gov/pubmed/18772896.
8. Carafoli E, Krebs J. Why calcium? How calcium became the best communicator. J Biol Chem. 2016;291(40):20849–57. Epub 20160726. https://doi.org/10.1074/jbc.R116.735894. PubMed PMID: 27462077; PMCID: PMC5076498.
9. Bonora M, Patergnani S, Rimessi A, De Marchi E, Suski JM, Bononi A, Giorgi C, Marchi S, Missiroli S, Poletti F, Wieckowski MR, Pinton P. ATP synthesis and storage. Purinergic Signal. 2012;8(3):343–57. Epub 20120412. https://doi.org/10.1007/s11302-012-9305-8. PubMed PMID: 22528680; PMCID: PMC3360099.
10. Denton RM, McCormack JG, Edgell NJ. Role of calcium ions in the regulation of intramitochondrial metabolism. Effects of Na^+, Mg^{2+} and ruthenium red on the Ca^{2+}-stimulated oxidation of oxoglutarate and on pyruvate dehydrogenase activity in intact rat heart mitochondria. Biochem J. 1980;190(1):107–17. https://doi.org/10.1042/bj1900107. PubMed PMID: 6160850; PMCID: PMC1162068.
11. Viola HM, Hool LC. How does calcium regulate mitochondrial energetics in the heart? - new insights. Heart Lung Circ. 2014;23(7):602–9. Epub 20140228. https://doi.org/10.1016/j.hlc.2014.02.009. PubMed PMID: 24657282.
12. Gherardi G, Monticelli H, Rizzuto R, Mammucari C. The mitochondrial Ca^{2+} uptake and the fine-tuning of aerobic metabolism. Front Physiol. 2020;11:554904. Epub 20201007. https://doi.org/10.3389/fphys.2020.554904. PubMed PMID: 33117189; PMCID: PMC7575740.
13. Denton RM, Randle PJ, Martin BR. Stimulation by calcium ions of pyruvate dehydrogenase phosphate phosphatase. Biochem J. 1972;128(1):161–3. https://doi.org/10.1042/bj1280161. PubMed PMID: 4343661; PMCID: PMC1173580.
14. Arnold PK, Finley LWS. Regulation and function of the mammalian tricarboxylic acid cycle. J Biol Chem. 2023;299(2):102838. Epub 20221226. https://doi.org/10.1016/j.jbc.2022.102838. PubMed PMID: 36581208; PMCID: PMC9871338.
15. Gunter TE, Sheu SS. Characteristics and possible functions of mitochondrial Ca^{2+} transport mechanisms. Biochim Biophys Acta. 2009;1787(11):1291–308. Epub 20090106. https://doi.org/10.1016/j.bbabio.2008.12.011. PubMed PMID: 19161975; PMCID: PMC2730425.
16. Mason FE, Pronto JRD, Alhussini K, Maack C, Voigt N. Cellular and mitochondrial mechanisms of atrial fibrillation. Basic Res Cardiol. 2020;115(6):72. Epub 20201130. https://doi.org/10.1007/s00395-020-00827-7. PubMed PMID: 33258071; PMCID: PMC7704501.

17. Traaseth N, Elfering S, Solien J, Haynes V, Giulivi C. Role of calcium signaling in the activation of mitochondrial nitric oxide synthase and citric acid cycle. Biochim Biophys Acta. 2004;1658(1–2):64–71. https://doi.org/10.1016/j.bbabio.2004.04.015. PubMed PMID: 15282176.
18. McCormack JG, Denton RM. Role of calcium ions in the regulation of intramitochondrial metabolism. Properties of the Ca^{2+}-sensitive dehydrogenases within intact uncoupled mitochondria from the white and brown adipose tissue of the rat. Biochem J. 1980;190(1):95–105. https://doi.org/10.1042/bj1900095. PubMed PMID: 6778477; PMCID: PMC1162067.
19. Wan B, LaNoue KF, Cheung JY, Scaduto RC Jr. Regulation of citric acid cycle by calcium. J Biol Chem. 1989;264(23):13430–9. PubMed PMID: 2503501.
20. Haumann J, Dash RK, Stowe DF, Boelens AD, Beard DA, Camara AK. Mitochondrial free [Ca^{2+}] increases during ATP/ADP antiport and ADP phosphorylation: exploration of mechanisms. Biophys J. 2010;99(4):997–1006. https://doi.org/10.1016/j.bpj.2010.04.069. PubMed PMID: 20712982; PMCID: PMC2920628.
21. Boelens AD, Pradhan RK, Blomeyer CA, Camara AK, Dash RK, Stowe DF. Extra-matrix Mg^{2+} limits Ca^{2+} uptake and modulates Ca^{2+} uptake-independent respiration and redox state in cardiac isolated mitochondria. J Bioenerg Biomembr. 2013;45(3):203–18. Epub 20130303. https://doi.org/10.1007/s10863-013-9500-5. PubMed PMID: 23456198; PMCID: PMC3670670.
22. Laskowski M, Augustynek B, Kulawiak B, Koprowski P, Bednarczyk P, Jarmuszkiewicz W, Szewczyk A. What do we not know about mitochondrial potassium channels? Biochim Biophys Acta. 2016;1857(8):1247–57. Epub 20160304. https://doi.org/10.1016/j.bbabio.2016.03.007. PubMed PMID: 26951942.
23. O'Rourke B. Mitochondrial ion channels. Annu Rev Physiol. 2007;69:19–49. https://doi.org/10.1146/annurev.physiol.69.031905.163804. PubMed PMID: 17059356; PMCID: PMC2712118.
24. Heinen A, Camara AK, Aldakkak M, Rhodes SS, Riess ML, Stowe DF. Mitochondrial Ca^{2+}-induced K^+ influx increases respiration and enhances ROS production while maintaining membrane potential. Am J Physiol Cell Physiol. 2007;292(1):C148–56. Epub 20060726. https://doi.org/10.1152/ajpcell.00215.2006. PubMed PMID: 16870831.
25. Stowe DF, Aldakkak M, Camara AK, Riess ML, Heinen A, Varadarajan SG, Jiang MT. Cardiac mitochondrial preconditioning by Big Ca^{2+}-sensitive K^+ channel opening requires superoxide radical generation. Am J Physiol Heart Circ Physiol. 2006;290(1):H434–40. Epub 20050826. https://doi.org/10.1152/ajpheart.00763.2005. PubMed PMID: 16126810.
26. Riess ML, Camara AK, Heinen A, Eells JT, Henry MM, Stowe DF. K_{ATP} channel openers have opposite effects on mitochondrial respiration under different energetic conditions. J Cardiovasc Pharmacol. 2008;51(5):483–91. https://doi.org/10.1097/FJC.0b013e31816bf4a4. PubMed PMID: 18437094; PMCID: PMC3010203.
27. Yang M, Camara AKS, Aldakkak M, Kwok WM, Stowe DF. Identity and function of a cardiac mitochondrial small conductance Ca^{2+}-activated K^+ channel splice variant. Biochim Biophys Acta Bioenerg. 2017;1858(6):442–58. Epub 20170322. https://doi.org/10.1016/j.bbabio.2017.03.005. PubMed PMID: 28342809; PMCID: PMC5749404.
28. Stowe DF, Yang M, Heisner JS, Camara AKS. Endogenous and agonist-induced opening of mitochondrial big versus small Ca^{2+}-sensitive K^+ channels on cardiac cell and mitochondrial protection. J Cardiovasc Pharmacol. 2017;70(5):314–28. https://doi.org/10.1097/FJC.0000000000000524. PubMed PMID: 28777255; PMCID: PMC5726766.
29. Stowe DF, Gadicherla AK, Zhou Y, Aldakkak M, Cheng Q, Kwok WM, Jiang MT, Heisner JS, Yang M, Camara AK. Protection against cardiac injury by small Ca^{2+}-sensitive K^+ channels identified in guinea pig cardiac inner mitochondrial membrane. Biochim Biophys Acta. 2013;1828(2):427–42. Epub 20120908. https://doi.org/10.1016/j.bbamem.2012.08.031. PubMed PMID: 22982251; PMCID: PMC3534888.

30. Szteyn K, Singh H. BK_{Ca} channels as targets for cardioprotection. Antioxidants (Basel). 2020;9(8). Epub 20200817. https://doi.org/10.3390/antiox9080760. PubMed PMID: 32824463; PMCID: PMC7463653.
31. Heinen A, Aldakkak M, Stowe DF, Rhodes SS, Riess ML, Varadarajan SG, Camara AK. Reverse electron flow-induced ROS production is attenuated by activation of mitochondrial Ca^{2+}-sensitive K^+ channels. Am J Physiol Heart Circ Physiol. 2007;293(3):H1400–7. Epub 20070518. https://doi.org/10.1152/ajpheart.00198.2007. PubMed PMID: 17513497.
32. Kevin LG, Camara AK, Riess ML, Novalija E, Stowe DF. Ischemic preconditioning alters real-time measure of O_2 radicals in intact hearts with ischemia and reperfusion. Am J Physiol Heart Circ Physiol. 2003;284(2):H566–74. Epub 20021031. https://doi.org/10.1152/ajpheart.00711.2002. PubMed PMID: 12414448.
33. Kevin LG, Novalija E, Riess ML, Camara AKS, Rhodes SS, Stowe DF. Sevoflurane exposure generates superoxide but leads to decreased superoxide during ischemia and reperfusion in isolated hearts. Anesth Analg. 2003;96(4):949–55. https://doi.org/10.1213/01.ANE.0000052515.25465.35. PubMed PMID: 12651639.
34. Chavda V, Lu B. Reverse electron transport at mitochondrial complex I in ischemic stroke, aging, and age-related diseases. Antioxidants (Basel). 2023;12(4). Epub 20230406. https://doi.org/10.3390/antiox12040895. PubMed PMID: 37107270; PMCID: PMC10135819.
35. Lindsay DP, Camara AK, Stowe DF, Lubbe R, Aldakkak M. Differential effects of buffer pH on Ca^{2+} induced ROS emission with inhibited mitochondrial complexes I and III. Front Physiol. 2015;6:58. Epub 20150310. https://doi.org/10.3389/fphys.2015.00058. PubMed PMID: 25805998; PMCID: PMC4354303.
36. Stowe DF, Camara AK. Mitochondrial reactive oxygen species production in excitable cells: modulators of mitochondrial and cell function. Antioxid Redox Signal. 2009;11(6):1373–414. https://doi.org/10.1089/ars.2008.2331. PubMed PMID: 19187004; PMCID: PMC2842133.
37. Malas KM, Lambert DS, Heisner JS, Camara AKS, Stowe DF. Time and charge/pH-dependent activation of K^+ channel-mediated K^+ influx and K^+/H^+ exchange in guinea pig heart isolated mitochondria; role in bioenergetic stability. Biochim Biophys Acta Bioenerg. 2022;1863(8):148908. Epub 20220809. https://doi.org/10.1016/j.bbabio.2022.148908. PubMed PMID: 35961396.
38. Haumann J, Camara AKS, Gadicherla AK, Navarro CD, Boelens AD, Blomeyer CA, Dash RK, Boswell MR, Kwok WM, Stowe DF. Slow Ca^{2+} efflux by Ca^{2+}/H^+ exchange in cardiac mitochondria Is modulated by Ca^{2+} re-uptake via MCU, extra-mitochondrial pH, and H^+ pumping by F_OF_1-ATPase. Front Physiol. 2018;9:1914. Epub 20190204. https://doi.org/10.3389/fphys.2018.01914. PubMed PMID: 30804812; PMCID: PMC6378946.
39. Natarajan GK, Glait L, Mishra J, Stowe DF, Camara AKS, Kwok WM. Total matrix Ca^{2+} modulates Ca^{2+} efflux via the Ca^{2+}/H^+ exchanger in cardiac mitochondria. Front Physiol. 2020;11:510600. Epub 20200916. https://doi.org/10.3389/fphys.2020.510600. PubMed PMID: 33041851; PMCID: PMC7526510.
40. Scheffer DDL, Garcia AA, Lee L, Mochly-Rosen D, Ferreira JCB. Mitochondrial fusion, fission, and mitophagy in cardiac diseases: challenges and therapeutic opportunities. Antioxid Redox Signal. 2022;36(13–15):844–63. Epub 20220418. https://doi.org/10.1089/ars.2021.0145. PubMed PMID: 35044229; PMCID: PMC9125524.
41. Kuznetsov AV, Margreiter R. Heterogeneity of mitochondria and mitochondrial function within cells as another level of mitochondrial complexity. Int J Mol Sci. 2009;10(4):1911–29. Epub 20090424. https://doi.org/10.3390/ijms10041911. PubMed PMID: 19468346; PMCID: PMC2680654.
42. Sorriento D, Di Vaia E, Iaccarino G. Physical exercise: a novel tool to protect mitochondrial health. Front Physiol. 2021;12:660068. Epub 20210427. 10.3389/fphys.2021.660068. PubMed PMID: 33986694; PMCID: PMC8110831.
43. Picca A, Mankowski RT, Burman JL, Donisi L, Kim JS, Marzetti E, Leeuwenburgh C. Mitochondrial quality control mechanisms as molecular targets in cardiac ageing. Nat Rev

Cardiol. 2018;15(9):543–54. https://doi.org/10.1038/s41569-018-0059-z. PubMed PMID: 30042431; PMCID: PMC6283278.
44. Coronado M, Fajardo G, Nguyen K, Zhao M, Kooiker K, Jung G, Hu DQ, Reddy S, Sandoval E, Stotland A, Gottlieb RA, Bernstein D. Physiological mitochondrial fragmentation is a normal cardiac adaptation to increased energy demand. Circ Res. 2018;122(2):282–95. Epub 20171212. https://doi.org/10.1161/CIRCRESAHA.117.310725. PubMed PMID: 29233845; PMCID: PMC5775047.
45. Scarpulla RC. Nuclear control of respiratory chain expression by nuclear respiratory factors and PGC-1-related coactivator. Ann N Y Acad Sci. 2008;1147:321–34. https://doi.org/10.1196/annals.1427.006. PubMed PMID: 19076454; PMCID: PMC2853241.
46. Slavin MB, Memme JM, Oliveira AN, Moradi N, Hood DA. Regulatory networks coordinating mitochondrial quality control in skeletal muscle. Am J Physiol Cell Physiol. 2022;322(5):C913-C26. Epub 20220330. https://doi.org/10.1152/ajpcell.00065.2022. PubMed PMID: 35353634.
47. Hernandez-Resendiz S, Prunier F, Girao H, Dorn G, Hausenloy DJ, Action E-CC. Targeting mitochondrial fusion and fission proteins for cardioprotection. J Cell Mol Med. 2020;24(12):6571–85. Epub 20200514. https://doi.org/10.1111/jcmm.15384. PubMed PMID: 32406208; PMCID: PMC7299693.
48. Hill BG, Benavides GA, Lancaster JR Jr, Ballinger S, Dell'Italia L, Jianhua Z, Darley-Usmar VM. Integration of cellular bioenergetics with mitochondrial quality control and autophagy. Biol Chem. 2012;393(12):1485–512. https://doi.org/10.1515/hsz-2012-0198. PubMed PMID: 23092819; PMCID: PMC3594552.
49. Palikaras K, Lionaki E, Tavernarakis N. Mechanisms of mitophagy in cellular homeostasis, physiology and pathology. Nat Cell Biol. 2018;20(9):1013–22. Epub 20180828. https://doi.org/10.1038/s41556-018-0176-2. PubMed PMID: 30154567.
50. Kinnally KW, Peixoto PM, Ryu SY, Dejean LM. Is mPTP the gatekeeper for necrosis, apoptosis, or both? Biochim Biophys Acta. 2011;1813(4):616–22. Epub 20101001. https://doi.org/10.1016/j.bbamcr.2010.09.013. PubMed PMID: 20888866; PMCID: PMC3050112.
51. Chen L, Zhou M, Li H, Liu D, Liao P, Zong Y, Zhang C, Zou W, Gao J. Mitochondrial heterogeneity in diseases. Signal Transduct Target Ther. 2023;8(1):311. Epub 20230823. https://doi.org/10.1038/s41392-023-01546-w. PubMed PMID: 37607925; PMCID: PMC10444818.
52. Martinez-Reyes I, Chandel NS. Mitochondrial TCA cycle metabolites control physiology and disease. Nat Commun. 2020;11(1):102. Epub 20200103. https://doi.org/10.1038/s41467-019-13668-3. PubMed PMID: 31900386; PMCID: PMC6941980.
53. Huttemann M, Helling S, Sanderson TH, Sinkler C, Samavati L, Mahapatra G, Varughese A, Lu G, Liu J, Ramzan R, Vogt S, Grossman LI, Doan JW, Marcus K, Lee I. Regulation of mitochondrial respiration and apoptosis through cell signaling: cytochrome c oxidase and cytochrome c in ischemia/reperfusion injury and inflammation. Biochim Biophys Acta. 2012;1817(4):598–609. Epub 20110713. https://doi.org/10.1016/j.bbabio.2011.07.001. PubMed PMID: 21771582; PMCID: PMC3229836.
54. Chen Q, Camara AK, Stowe DF, Hoppel CL, Lesnefsky EJ. Modulation of electron transport protects cardiac mitochondria and decreases myocardial injury during ischemia and reperfusion. Am J Physiol Cell Physiol. 2007;292(1):C137–47. Epub 20060913. https://doi.org/10.1152/ajpcell.00270.2006. PubMed PMID: 16971498.
55. Gustafsson AB, Gottlieb RA. Heart mitochondria: gates of life and death. Cardiovasc Res. 2008;77(2):334–43. Epub 20070821. https://doi.org/10.1093/cvr/cvm005. PubMed PMID: 18006487.
56. Martinou JC, Desagher S, Antonsson B. Cytochrome c release from mitochondria: all or nothing. Nat Cell Biol. 2000;2(3):E41–3. https://doi.org/10.1038/35004069. PubMed PMID: 10707095.
57. Bennett CF, Latorre-Muro P, Puigserver P. Mechanisms of mitochondrial respiratory adaptation. Nat Rev Mol Cell Biol. 2022;23(12):817–35. Epub 20220708. https://doi.org/10.1038/s41580-022-00506-6. PubMed PMID: 35804199; PMCID: PMC9926497.

58. Ristow M. Unraveling the truth about antioxidants: mitohormesis explains ROS-induced health benefits. Nat Med. 2014;20(7):709–11. https://doi.org/10.1038/nm.3624. PubMed PMID: 24999941.
59. Yun J, Finkel T. Mitohormesis. Cell Metab. 2014;19(5):757–66. Epub 20140220. https://doi.org/10.1016/j.cmet.2014.01.011. PubMed PMID: 24561260; PMCID: PMC4016106.
60. Esparza-Molto PB, Nuevo-Tapioles C, Cuezva JM. Regulation of the H^+-ATP synthase by IF1: a role in mitohormesis. Cell Mol Life Sci. 2017;74(12):2151–66. Epub 20170206. https://doi.org/10.1007/s00018-017-2462-8. PubMed PMID: 28168445; PMCID: PMC5425498.
61. Berry BJ, Trewin AJ, Amitrano AM, Kim M, Wojtovich AP. Use the protonmotive force: mitochondrial uncoupling and reactive oxygen species. J Mol Biol. 2018;430(21):3873–91. Epub 20180404. https://doi.org/10.1016/j.jmb.2018.03.025. PubMed PMID: 29626541; PMCID: PMC6170739.
62. Chandel NS. Evolution of mitochondria as signaling organelles. Cell Metab. 2015;22(2):204–6. Epub 20150611. https://doi.org/10.1016/j.cmet.2015.05.013. PubMed PMID: 26073494.
63. Nguyen T, Nioi P, Pickett CB. The Nrf2-antioxidant response element signaling pathway and its activation by oxidative stress. J Biol Chem. 2009;284(20):13291–5. Epub 20090130. https://doi.org/10.1074/jbc.R900010200. PubMed PMID: 19182219; PMCID: PMC2679427.
64. Chae S, Ahn BY, Byun K, Cho YM, Yu MH, Lee B, Hwang D, Park KS. A systems approach for decoding mitochondrial retrograde signaling pathways. Sci Signal. 2013;6(264):rs4. Epub 20130226. https://doi.org/10.1126/scisignal.2003266. PubMed PMID: 23443683.
65. Luan Y, Luan Y, Yuan RX, Feng Q, Chen X, Yang Y. Structure and function of mitochondria-associated endoplasmic reticulum membranes (MAMs) and their role in cardiovascular diseases. Oxid Med Cell Longev. 2021;2021:4578809. Epub 20210711. https://doi.org/10.1155/2021/4578809. PubMed PMID: 34336092; PMCID: PMC8289621.
66. De La Fuente S, Lambert JP, Nichtova Z, Fernandez Sanz C, Elrod JW, Sheu SS, Csordas G. Spatial separation of mitochondrial calcium uptake and extrusion for energy-efficient mitochondrial calcium signaling in the heart. Cell Rep. 2018;24(12):3099–107 e4. https://doi.org/10.1016/j.celrep.2018.08.040. PubMed PMID: 30231993; PMCID: PMC6226263.

Part III
Modulation of Bioenergetics in Cardiac Myocyte Mitochondria

Overview of Part III: The mitochondria of cardiac cells furnish the chemical energy to power the heart to deliver oxygen and nutrients and remove carbon dioxide and waste products. All invertebrates with mesoderm and all vertebrates have hearts, or at least a heart vascular tube with an incomplete circulation. Some heart genes are conserved in species from fruit flies to humans indicating an ancient blueprint for evolutionary changes, and over time, modifications in individual gene expression and regulatory interactions with other genes. Locomotion and strenuous exercise require enhanced cardiac power to increase heart rate and contractility accompanied by changes in action potential duration, coronary flow, myoplasmic Ca^{2+} concentration, actin-myosin cross bridge cycling, ATP generation, and O_2 consumption. Heart rate and contractility are largely regulated by the autonomic nervous system through release of catecholamines (sympathetic system), acetylcholine (parasympathetic system), and by blood borne factors. Coronary flow is mediated mostly by local factors like nitric oxide generated in the endothelial lining of the vascular bed. The five to sixfold increase in cardiac power for ATP consumption is met by rapid increases in glycolysis, citric acid cycling, and faster forward electron flow during oxidative phosphorylation. Energy is stored as creatine phosphate and the ADP and ATP are recycled between the cytosol and mitochondria via adenine nucleotide translocase. The capacity for high metabolic reserves, when needed, is a selective advantage of mitochondrial adaptions of myocytes of warm-blooded mammals and birds for fitness and survival.

Chapter 8
Regulation of Bioenergetics in Cardiac Myocytes

Abstract Cardiac mitochondria power the mechanical twin pump that delivers nutrients, O_2, hormones, and other factors to all cells of the body while extruding waste products and carbon dioxide. The heart itself is fed by its own coronary circulatory system. The sinoatrial node pacemaker self-initiates an action potential that spreads throughout the heart to cause muscle contraction and relaxation. A depolarization wave transiently opens voltage-gated Na^+ channels and then induces Ca^{2+} release via L-type Ca^{2+} channels in the sarcoplasmic reticulum, which primarily results in a polarized state long enough to increase myofibrillar Ca^{2+} for contraction that propels blood ejection. Action potential repolarization occurs with closure of L-type Ca^{2+} channels and the influx of K^+ so that myocytes lengthen and relax. Na^+ entry stops with depolarization and Na^+ is extruded by Na^+-K^+-ATPase. The excitation-contraction coupling of the action potential with cell Ca^{2+} influx and efflux is mediated by fast Ca^{2+} release via ryanodine receptors in the sarcoplasmic reticulum with reuptake and extrusion by Ca^{2+}-ATPases. The binding of Ca^{2+} to troponin C causes the myosin filament to slide over the actin filament which shortens the muscle cells in unison. The rise and fall in both cell Ca^{2+} and in the generated left ventricular pressure can be measured together in isolated hearts. Cardiac mitochondria dynamically buffer large amounts of Ca^{2+} as calcium phosphates. The buffering system capacity is assessed by measuring the change in ionized mitochondrial Ca^{2+} and the Ca^{2+} retention capacity and is dynamically responsive to energy availability. Excess mitochondrial ionized Ca^{2+} is slowly removed, mostly by mitochondrial Na^+-Ca^{2+} exchange. Studies suggest that during mitochondrial stress due to cardiac ischemia, mitochondrial Ca^{2+}-H^+ and K^+-H^+ exchangers remove excess Ca^{2+} and K^+ while inducing proton influx (H^+ leak); this stimulates respiration and reduces release of superoxide radicals and hydrogen peroxide. Nitric oxide, whether produced in the cell or in mitochondria, can react with excess superoxide to form the toxin peroxynitrite that can damage lipids, proteins, and nucleotides, both within mitochondria and within the cell.

Keywords Actin-myosin filaments · Antioxidants · Bioenergetics · Ca^{2+}-induced Ca^{2+} release · Cardiac cell action potential · Cardiac diastole · Cardiac systole · Cardioprotection · Cell Na^+-Ca^{2+} exchange · Cell Na^+/H^+ exchange · Cell

D. F. Stowe, *Evolution of Bioenergetics from Elements to Life*,
https://doi.org/10.1007/978-3-031-76668-8_8

membrane potential · Cell Ca^{2+} L-type Ca^{2+} channels · Cell Ca^{2+}-ATPase · Cell K^+ channels · Cell Na^+-K^+-ATPase · Cell voltage gradient · Contraction/relaxation · Depolarization · Electron transport system · Endoplasmic reticulum · Excitation contraction coupling · Ischemia reperfusion injury · Ion homeostasis · K^+ equilibrium potential · Lusiotropic effect · Matrix phosphate buffering system · Mitochondrial-associated membranes · Mitochondrial Ca^{2+} buffering capacity · Mitochondrial Ca^{2+}-H^+ exchanger · Mitochondrial Ca^{2+} uniporter · Mitochondrial Na^+-Ca^{2+} exchanger · Mitochondrial Na^+-H^+ exchanger · Mitochondrial membrane potential · Myofilaments · Na^+ equilibrium potential · Negative inotropic drugs · Nitric oxide · Nitric oxide synthases · Oxidative stress · O_2 consumption · Peroxynitrite · Positive inotropic drugs · Preconditioning · Proton motive force · Reactive oxygen species · Redox couples · Reduction potential · Refractory period · Repolarization · Ryanodine receptors · Sarcomeres · Sarcoplasmic reticulum · Sinoatrial node · Subsarcolemmal cisternae · Superoxide radical · Superoxide dismutases · Tropomyosin · Troponin · Voltage dependent anion channel · Voltage-gated Na^+ channels

8.1 Electrophysiology of Ion Flux in Cardiac Muscle Cells

The heart functions as a mechanical pump that requires time for blood to enter and to be ejected into the circulation. Heart rate is inversely proportional to the size of the animal. At 60 beats/min, the bioelectric signals for contraction must have a 1 s delay from one activation signal to the next. This function depends on a selectively permeable membrane and ion gradients created by energy requiring cation pumps. In cardiac muscle cells stimulated by the automatic, sinoatrial node pacemaker [1, 2] down through the Purkinje fibers, the cell membrane voltage gradient fluctuates (action potential, AP) from about −85 to +40 mV [3] with each heartbeat [4]. At faster or slower heart rates, the delay (refractory period) must be adjusted shorter or longer, respectively. This is done by variations in cation and anion flow in and out of the cell during four designated phases. This is a complicated mechanism aided in understanding by computational modeling [3].

Cardiac Action Potential The resting phase 4 (cardiac diastole) is due primarily to a high K^+ permeability because of open K^+ channels; thus, the cell membrane potential approaches the K^+ equilibrium potential of approximately −85 mV. An AP arriving from a nearby cell will itself slightly depolarize a cell and if a threshold of about −70 mV is attained, this triggers voltage-gated Na^+ channels to regeneratively and rapidly open (phase 0) causing the cell membrane gradient to reverse quickly to approximately +40 mV.

When Na^+ enters via the Na^+ channels, the cells depolarize rapidly in <2 ms from negative to positive membrane gradients that stops further Na^+ entry because Na^+ attains its equilibrium potential with rapid inactivation of the inner Na^+ gate (phase 1). Phase 1 is also terminated by opening of a transient outward K^+ channel and reduced conductance of the inward rectifier K^+ channel. Na^+ is extruded by the constant pumping of 3 Na^+ out for 2 K^+ in; the required energy is furnished by the Na^+/K^+-ATP dependent pump, which during the resting phase maintains the cell membrane potential close to the K^+ equilibrium potential. Otherwise, this pump has no effect on the AP. Once activated, the AP spreads passively throughout regions of ventricular cardiomyocytes through gap junction connections among cells.

The depolarized plateau phase 2 is largely a result of a period of cell Ca^{2+} entry via L-type Ca^{2+} channels that triggers release of Ca^{2+} from large stores in the SR; this is called Ca^{2+}-induced Ca^{2+} release (CICR) and is presented in the schematic depiction of Fig. 8.1 [5] of a cardiac myocyte with an AP, SR, cell, and mCa^{2+} channels and exchangers [4–6]. Ca^{2+} influx also activates Cl^- channels (I_{to2}) to allow Cl^- to enter the cell, which contributes to a slight depolarization. In addition, I_{to1} K^+ channels allow K^+ efflux making the membrane slightly less positive. In all, phase 2 equilibration is largely maintained by increases in Ca^{2+} and Cl^- influx and K^+ outflux. Each Ca^{2+} ion is more slowly removed in exchange for 3 Na^+ ions entering (cytosolic Na^+/Ca^{2+} exchange, cNCE).

Phase 3 (rapid repolarization) is marked by net outward positive charges by closure of L-type Ca^{2+} channels and maintained opening of slow delayed rectifier K^+ channels (I_{ks}). As the membrane potential drops, K^+ channels (rapid delayed rectifier, I_{kr}, and inwardly rectifying I_{k1}) open to let K^+ rapidly flow out so that

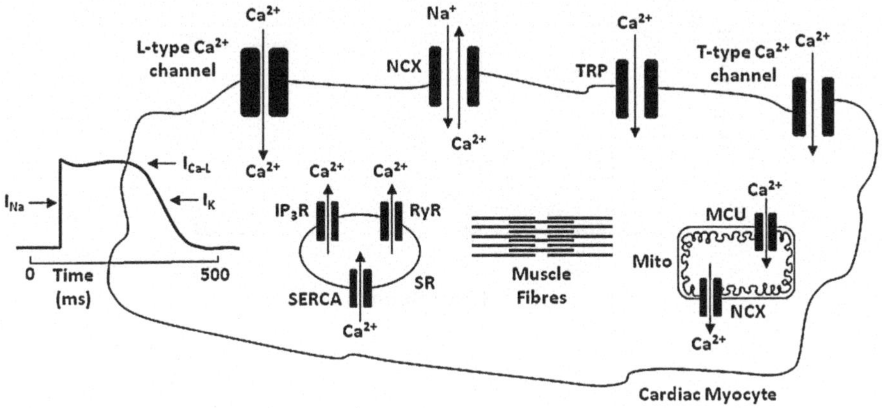

Fig. 8.1 Ca^{2+} and contraction in the heart: The ventricular action potential (AP) is triggered by depolarization of the plasma membrane resulting from Na^+ influx through Na^+ channels (I_{Na}). Ca^{2+} influx through the L-type Ca^{2+} channel ($I_{Ca\text{-}L}$) is responsible for the plateau phase of the AP and repolarization occurs due to K^+ efflux via K^+ channels (I_K). Ca^{2+} influx through $I_{Ca\text{-}L}$ triggers Ca^{2+} release from the sarcoplasmic reticulum (SR) via inositol triphosphate receptors (IP_3R) and ryanodine receptors (RyR), i.e., Ca^{2+}-induced Ca^{2+} release (CICR). Elevated intracellular Ca^{2+} activates muscle fibers and contraction. Subsequent relaxation occurs due to uptake of Ca^{2+} by the SR via sarco/endoplasmic reticulum Ca^{2+}-ATPase (SERCA). The remaining Ca^{2+} is extruded via the Na^+/Ca^{2+} exchanger (NCE) or is taken up by mitochondria via the mitochondrial Ca^{2+} uniporter (MCU). Transient receptor potential channels (TRP) and T-type Ca^{2+} channels also contribute to Ca^{2+} influx under conditions of stress and cardiac development, respectively. (Diagram (Fig. 1) and its abridged legend resourced with permission from copyright owner of Viola and Hool [5] (review))

repolarization occurs which sets the stage for phase 4 and the next cycle [4]. Figure 8.2 [4] is a cartoon of one cardiac cycle that simultaneously shows left atrial and ventricular pressure, aortic pressure, ventricular volume, heart sounds, and electrocardiogram (ECG). Figure 8.3 [5] depicts a cardiomyocyte showing the influence of cell Ca^{2+} on mitochondrial function. Conformational changes in the L-type Ca^{2+}

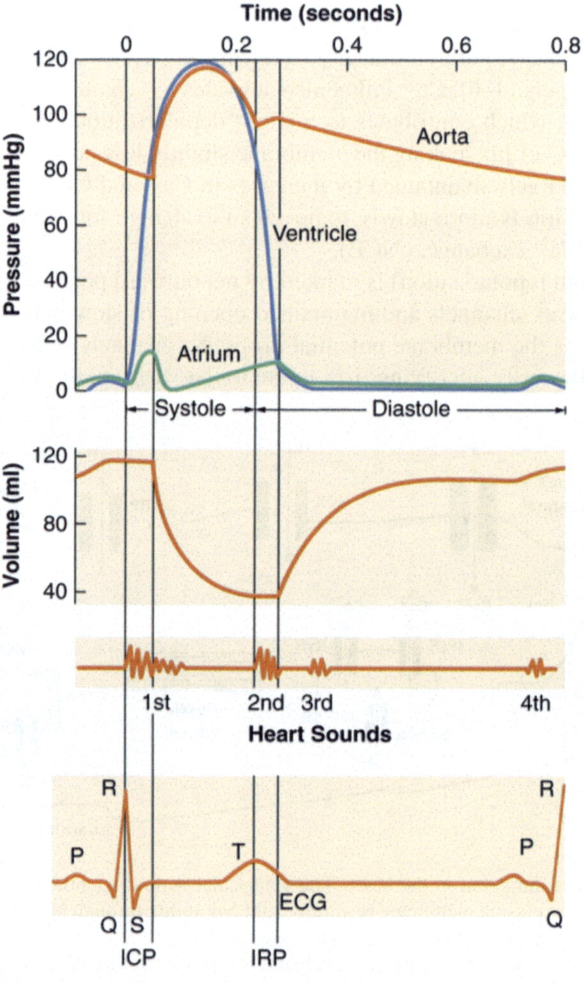

Fig. 8.2 Mechanical and electrical events of the cardiac cycle: Shown simultaneously are the left ventricular (LV) volume curve and the heart sounds. Note the LV isovolumic contraction period (ICP) and the relaxation period (IRP) during which there is no change in LV volume because the aortic and mitral valves are closed. The LV decreases in volume as it ejects its contents into the aorta. During the first third of systolic ejection (the rapid ejection period), the curve of emptying is steep. *ECG* electrocardiogram. (Diagram (Figs. 3–7) with legend resourced with permission from copyright owner of Pagel and Stowe [4] (book chapter))

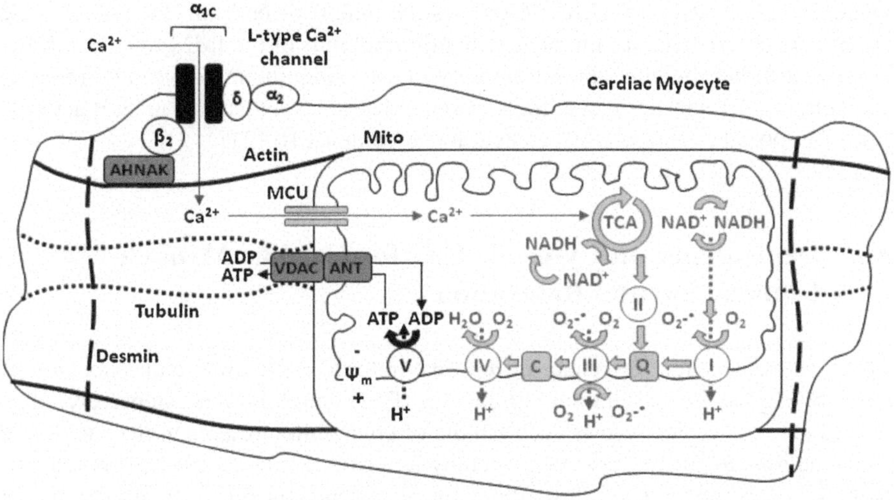

Fig. 8.3 Ca^{2+}-independent regulation of mitochondrial energetics: Cardiac L-type Ca^{2+} channels comprise of α_1C, α_2d and β_2 subunits. The β_2 subunit is tethered to F-actin via AHNAK. The mitochondrial voltage dependent anion channel (VDAC) also associates with cytoskeletal proteins. The L-type Ca^{2+} channel regulates $\Delta\Psi_m$ via cytoskeletal proteins when conformational changes in the channel occur during activation and inactivation due to transmission of movement from the β_2 subunit of the channel to mitochondria. Mitochondrial VDAC appears to play a role in this mechanism by responding to movement transmitted from the L-type Ca^{2+} channel via the cytoskeleton resulting in alterations in $\Delta\Psi_m$ in a Ca^{2+}-independent manner. Mechanism for Ca^{2+}-independent regulation of mitochondrial energetics is shown in black. Ca^{2+}-dependent regulation of mitochondrial energetics is shown in grey. (Diagram (Fig. 3) and its abridged legend resourced with permission from copyright owner of Viola and Hool [5] (review))

channel may also alter $\Delta\Psi_m$ via cytoskeletal proteins and VDAC, which are independent of changes in cell [Ca^{2+}] (Fig. 8.3) [5].

It is evident that the sarcolemmal Na^+, K^+, Cl^-, and especially Ca^{2+} channels, are of paramount importance in the electrophysiology of contraction/relaxation of cardiac myocytes. For homeostasis, cNCE and Na^+-H^+ exchange (cNHE) are expressed in the sarcolemmal membrane of myocytes where they couple to extrude Ca^{2+} (NCE) with Na^+ entry, and to extrude Na^+ (NHE) with H^+ entry [7–13]. Other potential exchangers in the sarcolemma, like the Ca^{2+}-H^+ exchanger (CHE) and the K^+-H^+ exchanger (KHE), may not exist or may be physiologically inactive because of the presence of the sarcolemmal Na^+/K^+-ATPase pump and Ca^{2+}-ATPases.

It is interesting that mitochondria do not express any Na^+ channels [9], but rather move Na^+ only via mitochondrial ion exchangers, i.e., mNCE [14–17] and mNHE [18, 19]. For example, as Ca^{2+} enters the mitochondrial matrix via the MCU channels within the IMM, the removal of one Ca^{2+} is only accomplished electrogenically by the uptake of three Na^+ by mNCE [14, 20]. In turn, the Na^+ is removed as H^+ enters via mNHE, so ionic homeostasis by the mitochondrial exchangers is essentially governed by the IMM proton gradient. Although Na^+ plays an important role in Ca^{2+} and H^+ homeostasis in mitochondria, Na^+ may also act as a second messenger. Activation of mNCE promotes the import of Na^+ into the matrix. One report [21] shows that Na^+ interacts with phospholipids of the IMM to reduce IMM fluidity, and that this hinders the mobility of free ubiquinone between complex II and complex III, but not inside the supercomplexes. As a result of mNCE activity, Na^+ is proposed to enhance $O_2^{\bullet -}$ generation at complex III [21].

8.2 Dependence on Cytosolic Ca^{2+} for Cardiac Myocyte Contraction and Relaxation

The AP sets the stage for synchronized Ca^{2+}-induced contraction and relaxation of cardiomyocytes, i.e., excitation contraction (E-C) coupling. An extensively intertwined network of SR invests each bundle of contractile proteins in the sarcomere and functions as a Ca^{2+} reservoir, thereby assuring homogenous distribution and reuptake of activator Ca^{2+} throughout the myofilaments during contraction and relaxation, respectively [4, 22].

Excitation-Contraction Coupling The subsarcolemmal cisternae of the SR are specialized structures located immediately adjacent to, but not continuous with, the sarcolemmal and transverse tubular membranes. The SR contains large numbers of ryanodine receptors (RyR) that function as the primary Ca^{2+} release channel for the SR [4, 23, 24]. Ca^{2+} entering the myocyte during the AP triggers CICR from the SR (Figs. 7.24 and 8.4) [23, 24, 26]. Figure 8.4 shows drawings of cardiomyocyte components, e.g., L-type Ca^{2+} channels, T-tubules, SR, RyR channels, Ca^{2+} exchangers, SERCA, and MCU, that convey Ca^{2+} as a regulator of muscle contraction and relaxation [23]. Perhaps as expected, other cations can modulate the activity of cCa^{2+} as the essential cation controlling contraction/relaxation. Inhibition of the Na^+/K^+-ATP dependent pump (Figs. 2.1 and 2.3) (e.g., by cardiac glycosides) increases cell (c) $c[Na^+]$, which increases $c[Ca^{2+}]$ via cNCE and enhances contractility (but not relaxation).

Mg^{2+} can interfere with E-C coupling by competing with Ca^{2+} at the Ca^{2+} activation site of the SR [27]. Cytosolic acidosis reduces the strength of cardiac contractility [28] by modulating the activity of osmo-regulators cNHE, and cNCE [8, 10, 11]. Excess K^+ can also play a role in decreasing excitability by attenuation of the Na^+/K^+-ATP dependent pump activity. It is important to reiterate that removal of cell cCa^{2+} by reuptake into the SR via SR Ca^{2+}-ATPase, and extrusion of cell Ca^{2+} by cell membrane Ca^{2+}-ATPase and cNCE, are equally important so that complete relaxation (lusiotropic effect) of the myofilaments is achieved during diastole.

Muscle Proteins Cytosolic free Ca^{2+} is a regulator of two muscle proteins, troponin and tropomyosin, that control muscle contraction by unblocking the binding of myosin to actin as depicted in Fig. 8.5 [4]. The left cartoon shows the mechanism for a contraction and relaxation cycle with ATP-mediated release of the actin head from myosin; the right cartoon shows the role of Ca^{2+} to bind troponin C and dis-

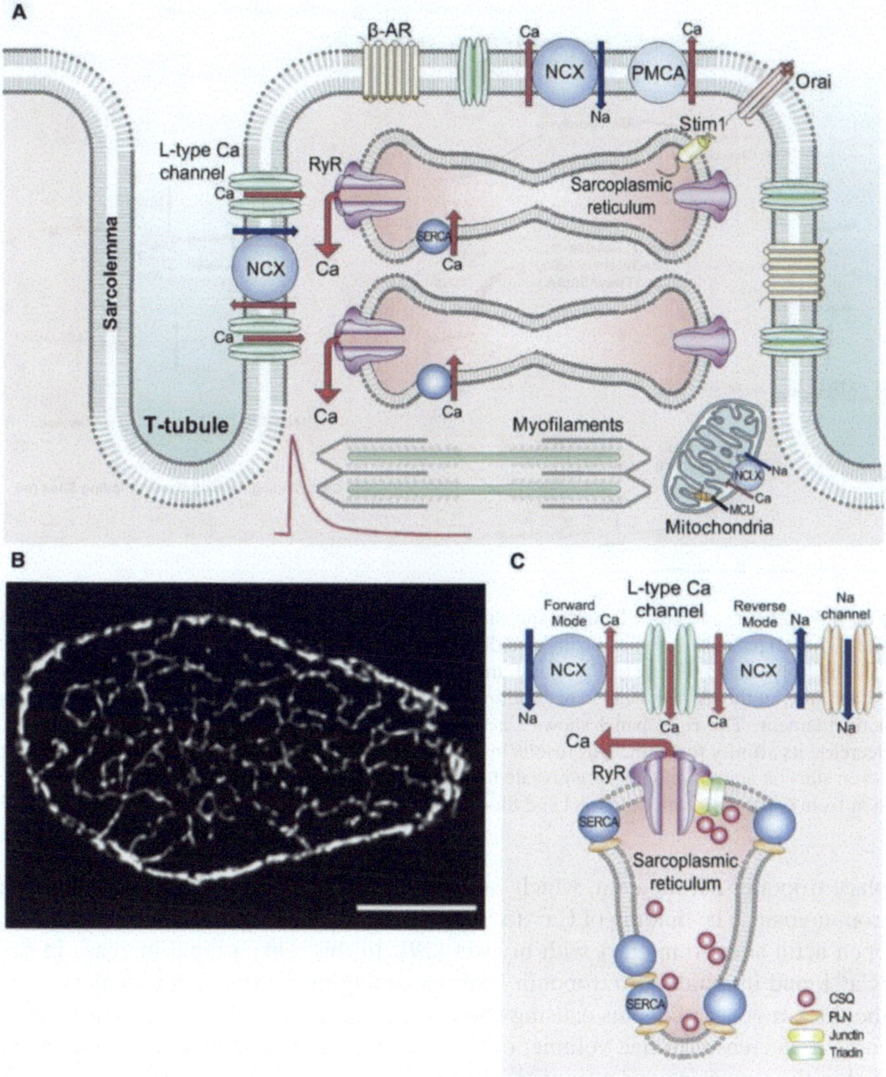

Fig. 8.4 Structures involved in Ca^{2+} cycling: (**a**) Schematic diagram. This shows surface membrane, transverse tubule, sarcoplasmic reticulum (SR), and mitochondria, as well as the various channels and transporters mentioned elsewhere. (**b**) High-resolution transverse section of a ventricular myocyte showing T-tubule network, as reprinted from text Ref. [25]. (**c**) Cartoon of dyad emphasizing the major proteins involved in Ca^{2+} cycling. *β-AR* beta adrenoceptor, *MCU* mitochondrial Ca uniporter, *NCX* sodium–calcium exchange, *NCLX* mitochondrial Na^+–Ca^{2+} exchange (mNCE), *PMCA* plasma membrane Ca^{2+}-ATPase, *RyR* ryanodine receptor, *SERCA* sarco/endoplasmic reticulum Ca-ATPase. (Diagrams, photomicrograph (Fig. 1) and modified legend resourced with permission from copyright owner of Eisner et al. [23] (review))

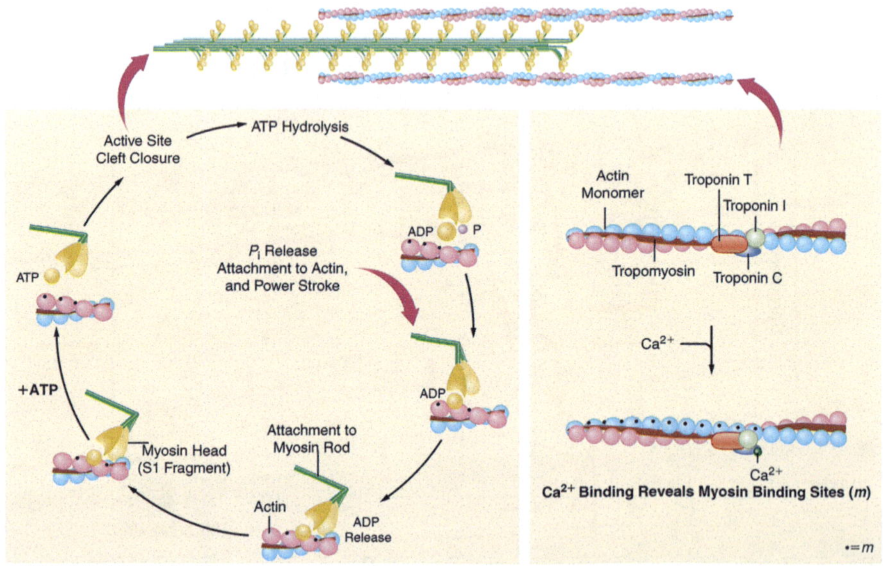

Fig. 8.5 Schema of actin filaments and their individual monomers and active myosin bindings sites: The left panel shows the myosin head dissociating from actin by binding with ATP. Subsequent ATP hydrolysis and release of P_i "cocks" the head group into a tension-generating configuration. Attachment of the myosin head to actin allows the head to apply tension to the myosin rod and the actin filament. The right panel shows Ca^{2+} binding to troponin C, which causes troponin I to decrease its affinity for actin. This results in a conformational shift in the tropomyosin position, so seven sites on actin monomers are revealed. (Cartoon (Fig. 8) with legend resourced with permission from copyright owner of Pagel and Stowe [4] (book chapter))

place troponin I from actin, which causes a conformational shift in the position of tropomyosin. The binding of Ca^{2+} to troponin on the tropomyosin molecule exposes open actin sites to interact with myosin [29]. In this way, a rapid increase in cell $[Ca^{2+}]_c$ and its binding to troponin leads to sliding of the thin actin filaments over the thick myosin filaments causing shortening of the cells, which together create a force to decrease cardiac volume; one-way valves causes development of pressure within the ventricles and atria with ejection of blood [4, 23, 24, 30]. At each pulse approximately 20 g of blood is accelerated from 0.25 to 0.35 ms over a 0.10 s duration [31]. This force, as in mass • acceleration, is dependent on troponin being bound by cell Ca^{2+}. Both the magnitude and duration of blood ejection depend on the amount and rate of Ca^{2+} binding to troponin and the strength of Ca^{2+} binding.

Cardiac Cell Ca^{2+} and Ventricular Pressure Cycling The subsequent rapid reuptake of SR Ca^{2+} and the fall in cell $[Ca^{2+}]$ by NCE releases Ca^{2+} from troponin and causes sarcomere relaxation. If cell systolic $[Ca^{2+}]$ is too low, or diastolic $[Ca^{2+}]$ is too high, contraction (left ventricular pressure (LVP) is greatly impaired. The rise

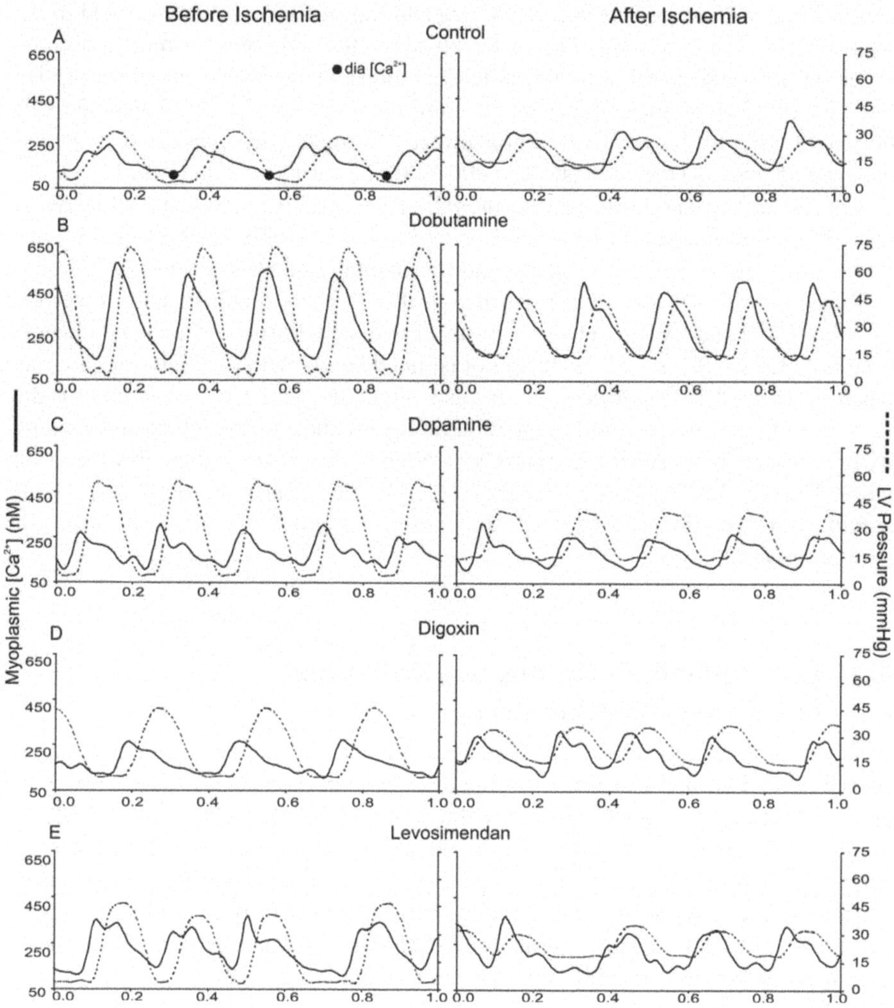

Fig. 8.6 Cardiac Ca^{2+}-activated ventricular pressure cycles with positive inotropic drugs: Sample time series plots of left ventricular (LV) pressure (LVP) and myoplasmic Ca^{2+} concentration ($[Ca^{2+}]$) from control (**a**), dobutamine (**b**), dopamine (**c**), digoxin (**d**), and levosimendan (**e**) before (left) and after (right) ischemia in guinea pig isolated hearts. Diastolic Ca^{2+} (dia $[Ca^{2+}]$) detection points are shown for control. Extracellular $CaCl_2$ was half-normal (1.2 mM) to allow for a full range of responses to the drugs. This accounts for the low control values of systolic LVP and $[Ca^{2+}]$. Note how the rise and fall of myocyte Ca^{2+} precedes the rise and fall in LVP. Phasic LVP was lower after ischemia in all groups, but phasic $[Ca^{2+}]$ was higher only in the control group (see Table 1 of text). (Data tracings (Fig. 1) and modified legend resourced with permission from copyright owner of Rhodes et al. [37])

and fall in cell [Ca^{2+}] occur before the rise and fall of LVP, as demonstrated in isolated hearts [7, 11, 32–36]. Figure 8.6 displays the effects of several ionotropic drugs on heart rate and on myoplasmic Ca^{2+} transients and LVP in guinea pig isolated hearts [32]. In these experiments, coronary perfusate [Ca^{2+}] was made half-normal in order to show the differential effects. Normally, myoplasmic [Ca^{2+}] varies between 75 and 100 nM at diastole to 400–600 nM during systole [7, 8, 11, 37, 38].

Cardiac ischemia retards the rise and fall of the cell Ca^{2+} transients, while delaying contraction-relaxation transients (systolic and diastolic LVP). Ischemia also slows and reduces cell Ca^{2+} release and the reuptake of SR Ca^{2+} and extrusion of cell Ca^{2+} by Ca^{2+}-ATPase. On reperfusion, these events result in both lower cell systolic [Ca^{2+}] and LVP, and by elevations in both cell diastolic [Ca^{2+}] and diastolic LVP [4, 23, 24, 30, 37, 38]. Perfusion of isolated hearts with positive (Fig. 8.6) and negative (Fig. 8.7) ionotropic drugs also alters the cell Ca^{2+}-LVP relationship (loops) and furnishes useful information on the mechanism of action and sensitivity of contraction/relaxation to cell [Ca^{2+}]. Note in these two figures how the orientation (phase shift) and area (shape) of LVP-Ca^{2+} transient loops display the simultaneous relationship between fluctuations in cell Ca^{2+} and LVP during the cardiac cycle [33, 39].

8.3 Ca^{2+} Uptake, Buffering, and Extrusion in Cardiac Mitochondria

matrix Ca^{2+} Uptake Cardiac mitochondria provide the energy of ATP for cyclic cardiac contraction and relaxation via the action of Na^+/K^+-ATP and Ca^{2+}-ATPase pumps. Ca^{2+} ions carry messages to virtually all important functional components of cells (Fig. 7.3) [25]. In cardiomyocytes, increases and decreases in cell Ca^{2+} levels and CICR drive the power stroke of actin-myosin cycling [40]. In cardiac myocytes, the mitochondrial associated membrane (MAM) [41, 42] acts as a conduit for rapid Ca^{2+} transfer between the SR sink and the mitochondrial matrix Ca^{2+} sink via the mitochondrial Ca^{2+} uniporter (MCU) within the IMM (Figs. 7.31 and 8.8) [25, 26]. As depicted in Fig. 8.8, the MCU complex has multiple regulating components [44]. The MCU provides fast Ca^{2+} uptake at the myofilaments with an overall balance in cytosolic [Ca^{2+}] and mCa^{2+} uptake when coupled with cNCE for Ca^{2+} extrusion.
There is an inverse relationship between mCa^{2+} uptake and MAM parameters, such as MAM distance, coupling expansion, and coupling number [41, 42]. In addition to the MCU, other reported Ca^{2+} uptake pathways are the rapid mode of uptake (RaM) [45], a pathway, not a channel, and the proposed mitochondrial ryanodine receptor (mRyR), which appears linked with SR RyR [27]. Whether the MCU associates with mRyR or is a separate pathway remains unclear [44].

As noted above and in Sects. 2.3, 7.2 and 8.2, Mg^{2+} can compete with Ca^{2+} to alter its regulation. In cardiac isolated mitochondria, adding $MgCl_2$ does not increase

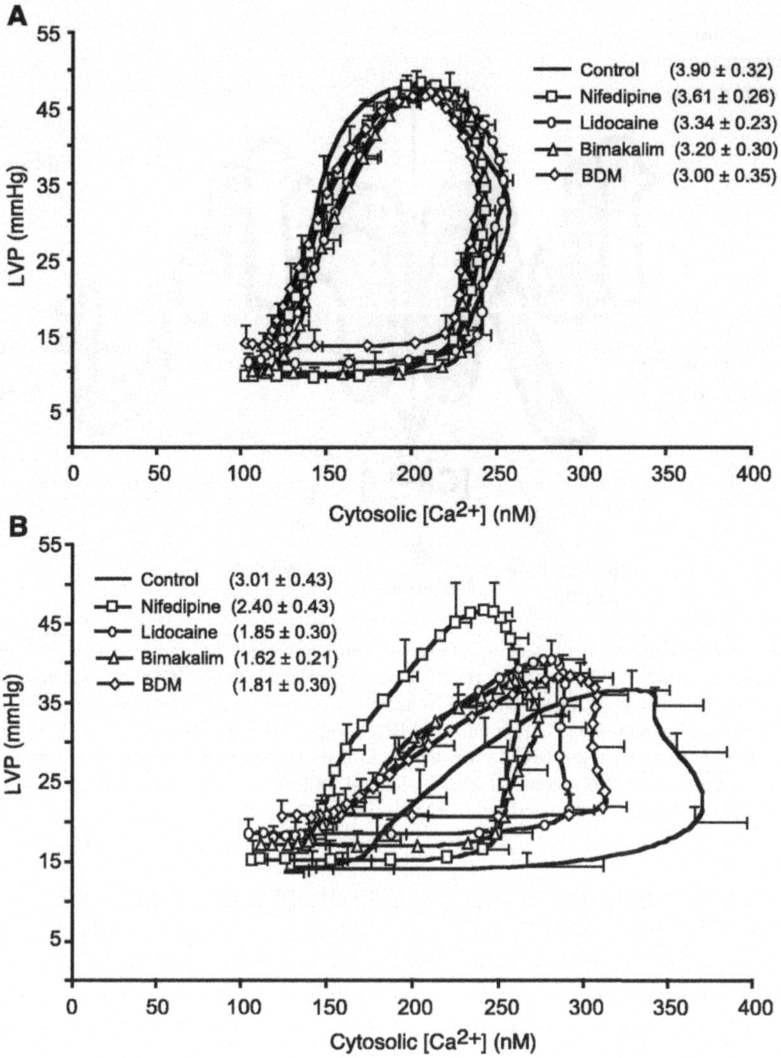

Fig. 8.7 Left ventricular pressure as a function of cytosolic Ca^{2+} with negative inotropic drugs: (**a**) Averaged [Ca^{2+}]-LVP loops (6–10 cardiac cycles in 8 guinea pig isolated hearts) during baseline conditions for four negative inotropic drug groups before drug perfusion, and control (no drug), before 30 min global ischemia. Loop area was acquired by integrating [Ca^{2+}]-LVP over 6–10 beats for 8 hearts/group. Mean loop area (103 nM • mmHg) before and after ischemia for each treatment is shown next to the drug name; there were no differences in loop area. (**b**) Averaged [Ca^{2+}]-LVP loops at 30 min of reperfusion (no drugs) after hearts were treated with bimakalim (Bim), 2, 3, butanedione monoxime (BDM), nifedipine (Nif), or lidocaine (Lid) before ischemia. The effects of negative inotropic drugs after IR injury compared to drug controls before IR injury were increased diastolic and systolic [Ca^{2+}] and increased diastolic LVP with decreased systolic LVP. Loop areas after ischemia were similar in the control and Nif groups, lower and similar in the Lid, Bim, and BDM groups, and not different between the Nif and Lid groups ($P < 0.05$). See Table 1 in text for significance. (Data tracings (Fig. 3) and legend resourced with permission from copyright owner of Camara et al. [39])

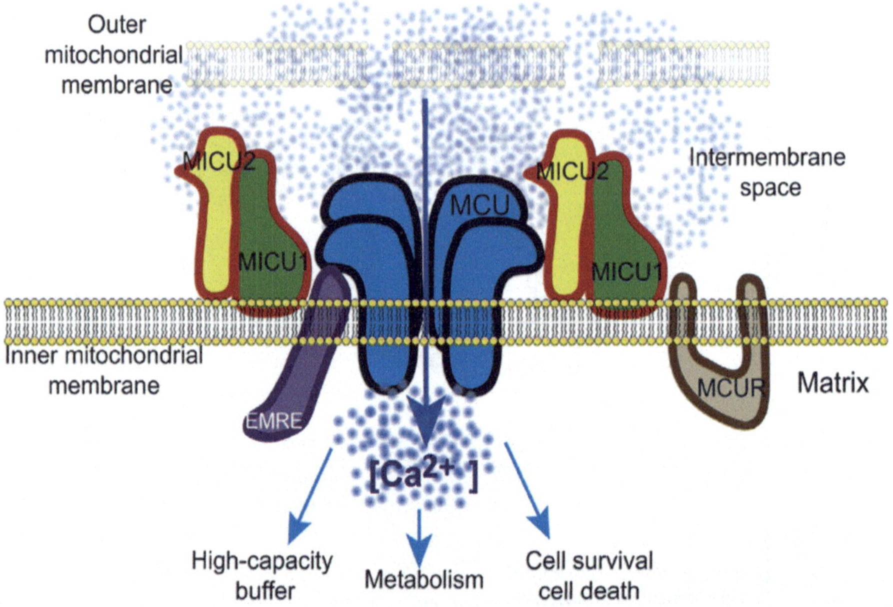

Fig. 8.8 Graphic depiction of a general view of the mitochondrial Ca^{2+} uniporter (MCU) complex: The channel-forming MCU is a tetrameric 40-KDa protein (a 33-KDA MCUb protein incorporates into the tetrameric MCU channel, reducing its activity). Accessory proteins (MICUs, MCU receptor (MCUR), and essential MCU regulator (EMRE)) variously associated with the IMM regulate the gating and activity of the MCU channel. Blue dots represent Ca^{2+} ions. (Cartoon (Fig. 4) adapted from Ref. [43] and resourced with its legend with license (https://creativecommons.org/licenses/BY/4.0/) from Carafoli and Krebs [25] (review))

m[Mg^{2+}] but it alters bioenergetics by a direct effect to decrease mCa^{2+} uptake. Moreover, at a given m[Ca^{2+}], state 3 respiration is incrementally attenuated, and state 4 respiration is enhanced, by higher external e[Mg^{2+}]. Thus, an increase in e[Mg^{2+}] without a change in m[Mg^{2+}] can modulate bioenergetic function independently of MCU-mediated mCa^{2+} [46].

matrix Ca^{2+} Buffering Cardiac mitochondria buffer large amounts of Ca^{2+} that can enter from the cytosol when the $\Delta\Psi_m$ is highly charged [14, 47–51]. The rise and fall in cytosolic Ca^{2+} transients observed in isolated beating hearts are not visible in the mitochondrial compartment likely because of the large and dynamic matrix Ca^{2+} buffering system capacity (mβ_{Ca}). Free (ionized) m[Ca^{2+}] is only approximately 100–200 nM in part because once the free Ca^{2+} is bound, it is removed from the cation pool, but resting m[Ca^{2+}] can increase five to sixfold over time, e.g., during IR injury in isolated hearts, as mitochondrial Ca^{2+} buffering capacity (mβCa) is exceeded [43, 52–54]. If mβ_{Ca} is overwhelmed, and the excess Ca^{2+} cannot be removed, mitochondrial function will fail because mCa^{2+} overload will depolarize

$\Delta\Psi_m$ despite an increase in respiration that cannot pump enough protons to repolarize $\Delta\Psi_m$.

In guinea pig, cardiac isolated mitochondria, adding $CaCl_2$ when $\Delta\Psi_m$ is fully charged causes a rapid uptake of mCa^{2+} via the MCU, followed by slower uptake as Ca^{2+} becomes sequestered (buffered) primarily by mP_I (Sects. 1.4 and 2.3) [14, 46, 49]. Phosphorylation of ADP enhances free $[Ca^{2+}]_m$ (in nM), likely by way of decreasing Ca phosphate precipitation in the matrix; the increase in free nM $[Ca^{2+}]_m$ by ADP is blocked if ANT or complex V is first inhibited by carboxyatractyloside (CATR) or oligomycin (OMN), respectively (Fig. 8.9) [55]. In rat isolated mitochondria, $m\beta_{Ca}$ increases if ADP and OMN are present in the physiological buffer solution before testing for mCa^{2+} buffering, as summarized in Fig. 8.10 [51]. These two studies suggest that removing phosphate from matrix stores to make ATP from ADP, reduces $m\beta_{Ca}$, but that blocking ADP to form ATP enhances $m\beta_{Ca}$.

Assessing mCa^{2+} Buffering Capacity ($m\beta_{Ca}$) Mitochondrial $m\beta_{Ca}$ is estimated in cardiac isolated mitochondria in part by measuring the Ca^{2+} retention capacity (CRC). CRC is assessed by adding $CaCl_2$ to a mitochondrial sample while measuring extra-matrix $e[Ca^{2+}]$ and $m[Ca^{2+}]$ using various Ca^{2+} fluorescent dyes and spectrophotometry [14, 46, 48, 49, 51]. Na^+ is absent in the buffer and drugs so mNHE and mNCE are inactivated; mNCE can also be blocked pharmacologically. The tolerance for added Ca^{2+} before mitochondrial $\Delta\Psi_m$ collapse indicates CRC and the greater the CRC attained, the greater is the initial residual Ca^{2+} buffering capacity, indicating a higher capacity to retain Ca^{2+} as a precipitate. This technique utilizes a physiological respiration buffer containing a substrate, usually pyruvic acid, glutamate + malate, or succinate, in which mitochondria are treated with a number of "pulses" of $CaCl_2$ until additional Ca^{2+} cannot be taken up by the matrix. When $e[Ca^{2+}]$ (Fig. 8.11) and $m[Ca^{2+}]$ (Fig. 8.12) are both measured in isolated mitochondria [48], $CaCl_2$ pulses first show rapid and initial complete uptake of Ca^{2+} from the mitochondria buffer to the matrix where it precipitates and disappears as a cation. Repeated boluses of the same concentration of $CaCl_2$, or increasing concentrations of $CaCl_2$, are met with less Ca^{2+} uptake when the matrix phosphate buffering system cannot accept more Ca^{2+}. Ru360 can be used to prevent the expelled Ca^{2+} from reentering the matrix via the MCU [48]. Assessment of the timing and dynamics of Ca^{2+} disappearance (buffering) and reappearance, as during added NaCl, is used to calculate $m\beta_{Ca}$.

$m\beta_{Ca}$ and mNa^+-Ca^{2+} Exchange In the absence of extra-matrix (e) Na^+, or in the presence of an inhibitor of NCE, mNCE cannot occur so mCa^{2+} cannot be expelled in exchange for Na^+. In the matrix, mCa^{2+} is shown to enter and remain constant but that is because most of it is buffered and so cannot be detected. Adding the MCU inhibitor ruthenium 360 (Ru360) after the $CaCl_2$ pulse prevents additional mCa^{2+} uptake via the MCU [48]. The difference between $e[Ca^{2+}]$ and $m[Ca^{2+}]$ represents the amount of Ca^{2+} that becomes buffered. Uptake of excess Ca^{2+} coincides temporally with reduced and eventual collapse of $\Delta\Psi_m$, ΔpH_m, and $NADH/NAD^+$ redox

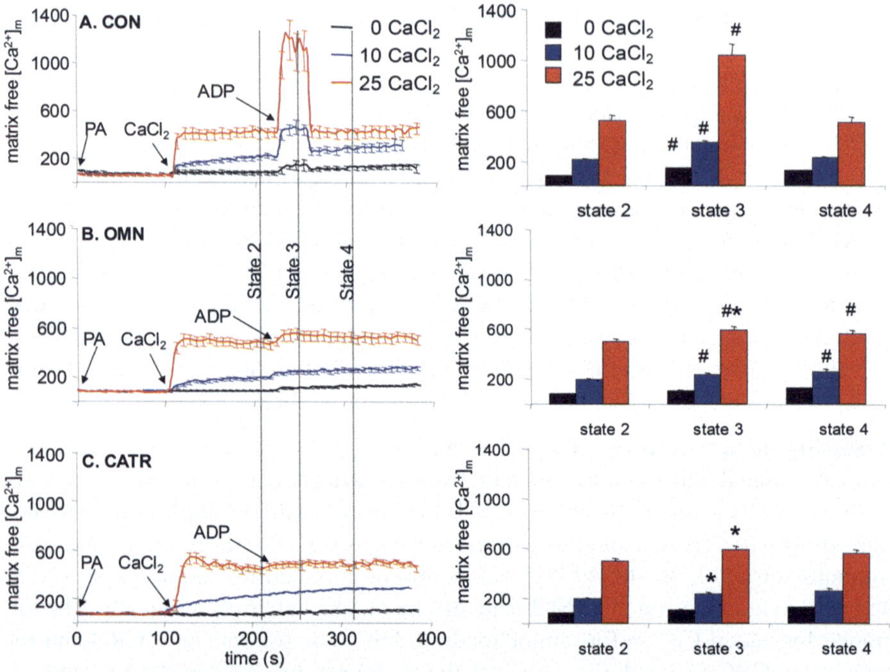

Fig. 8.9 Effect of ADP on $[Ca^{2+}]_m$ during inhibited ADP phosphorylation and blocked ADP/ATP transport: (**a**) Adding 10 and 25 μM $CaCl_2$ in energized mitochondria (after pyruvic acid, PA) caused abrupt, graded increases in $Ca^{2+}]_m$ to steady-state values. Adding ADP caused an abrupt but transient rise in nM $[Ca^{2+}]_m$ above the values established with added $CaCl_2$. (**b**) Adding ADP after $CaCl_2$ and blocking F_oF_1-ATP ase with oligomycin (OMN) caused a significant increase in nM $[Ca^{2+}]_m$ after addition of both 10 and 25 μM $CaCl_2$. (**c**) Adding ADP after $CaCl_2$ when the ADP/ATP carrier (ANT) was blocked with carboxy-atractyloside (CATR) inhibited the ADP-induced increase in nM $[Ca^{2+}]_m$. The buffer contained 40 μM EGTA. The results indicate that ADP likely reduces Ca-phosphate precipitation when phosphorylating ADP. For $P < 0.05$, #indicates state 3 versus states 2 and 4 respiration; *indicates the OMN and CATR versus CON groups at the same respiration state and $CaCl_2$. (Data plots (Fig. 6) and its modified legend resourced with permission from copyright owner of Haumann et al. [55])

state [51], and with release of cytochrome *c*. A low CRC (only a few $CaCl_2$ pulses before Ca^{2+} overload and loss of $\Delta\Psi_m$) indicates exceeded $m\beta_{Ca}$. A high CRC (many $CaCl_2$ pulses before Ca^{2+} overload) indicates there was less Ca^{2+} already buffered after a treatment. A high CRC together with a high respiratory control index (RCI) is a simple indicator of efficient bioenergetics that represent a well coupled OxPhos apparatus and an effective mCa^{2+} sequestering system [14, 46, 48, 49, 51, 56].

Measurements of both $e[Ca^{2+}]$ release and mCa^{2+} uptake in cardiac isolated mitochondria after $CaCl_2$ pulses at different e[NaCl], and after adding Ru360, show [14]

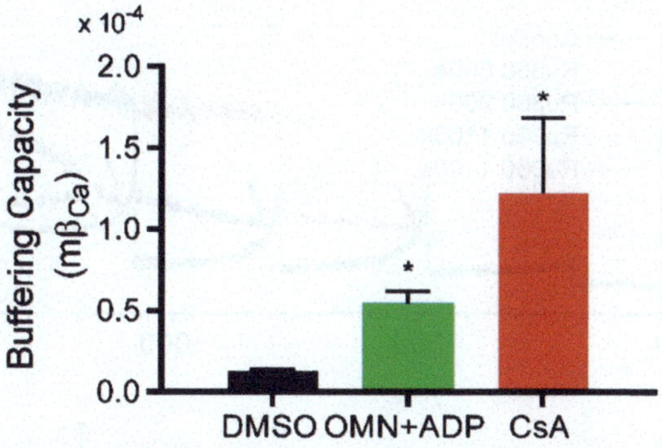

Fig. 8.10 Calcium buffering capacity ($m\beta_{Ca}$) in guinea pig isolated mitochondria: Cyclosporin A (CsA), an inhibitor of cyclophilin D and attenuator of mitochondrial permeability transition pore (mPTP) opening, and OMM + ADP, enhance mitochondrial Ca^{2+} buffering capacity, $m\beta_{Ca}$, when these drugs are added before adding $CaCl_2$ pulses. $m\beta_{Ca}$ was calculated from trend fits of extra-mitochondrial (e) e[Ca^{2+}] and m[Ca^{2+}] for DMSO-(control), CsA, and OMN + ADP-treated mitochondria as described by Eqs. (2) to (4) (see text). $m\beta_{Ca}$ for each treatment was calculated from three to five experiments each for e[Ca^{2+}] and m[Ca^{2+}] and averaged. Error bars represent mean + SEM (*$P < 0.01$ compared with DMSO. Oligomycin (OMN) is an inhibitor of F_oF_1-ATP synthase/ase. (Summary data plot (Fig. 4) and modified legend resourced with license (https://creativecommons.org/licenses/BY/4.0/) from Mishra et al. [51])

that the Na^+- induced Ca^{2+} release is an electrogenic exchange of these ions. The experiments exclude that it is an electroneutral exchange by showing that mCa^{2+} handling between the extra-matrix and matrix compartments is not reciprocal; this indicates that the matrix Ca^{2+} buffering system, $m\beta_{Ca}$ is dynamic, rather than static. Figure 8.13 demonstrates that challenging isolated mitochondria with $CaCl_2$ pulses causes Ca^{2+} ions to leave the respiration buffer and enter ther matrix where most of it is precipitated and thus removed from the Ca^{2+} ion pool by the Ca^{2+} buffering system. Na^+ can enter mitochondria only via Na^+ ion exchangers. This causes Ca^{2+} to be expelled from the matrix because of enhanced mNCE activity [14]. Mitochondrial Na^+ influx and efflux via mNHE is normally impeded by the high Δp_m, that is, a large *pmf* [18].

The difference between e[Ca^{2+}] and m[Ca^{2+}] (Fig. 8.13c) over a range of mNCE activities exposes the un-ionized calcium that is being buffered. A similar experimental mitochondrial model [49], when applied to a computational model [50], demonstrates that the general Ca^{2+} binding properties of the matrix Ca^{2+} buffering system are not only dynamic, but also independent of the mode of mCa^{2+} uptake, i.e., through the fast or slow mode of uptake. Results from the above experimental studies have led to an integrated model of Ca^{2+} handling and mitochondrial bioenergetics. Figure 8.14 depicts models of Ca^{2+} channels and exchangers as related to overall bioenergetic function, Ca^{2+} sequestration systems that are dependent on the mechanism of Ca^{2+} uptake, and modulation of Ca^{2+} uptake as modified by cytosolic Mg^{2+} [47].

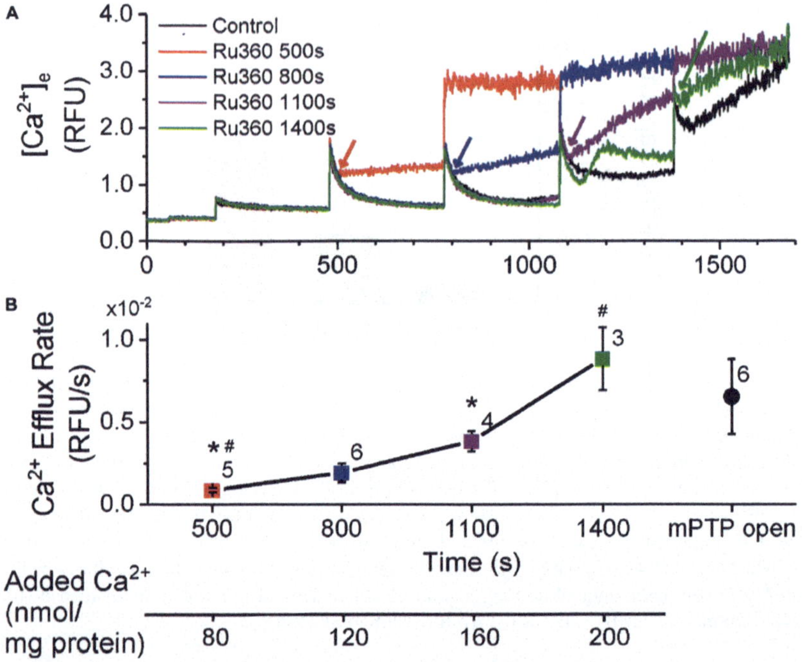

Fig. 8.11 Ca^{2+}-efflux increases with an increase in total matrix Ca^{2+} in rat heart isolated mitochondria: (**a**) Extra-matrix Ca^{2+} measured as relative fluorescence changes in Na^{+}-free experimental buffer loaded with the Ca^{2+}-sensitive, ratiometric dye Fura-4F. 20 μM $CaCl_2$ pulses were added first at 180 s and at every 300 s thereafter, in the absence (control, black trace) or the presence of MCU-inhibitor Ru360 (1 μM) added at different times as follows: red trace—500 s; blue trace—800 s; purple trace—1110 s; green trace—1400 s. Times of addition of Ru360 are indicated by color-matched arrows. (**b**) Efflux rates obtained from the slope values (m) were calculated from fits ($y = m \bullet x + c$) to the linear phase of Ca^{2+} efflux from 5 s to 1 min after addition of 1 μM Ru360 in each trace in (**a**). Symbol colors correspond to the trace colors and conditions as specified in (**a**). The efflux rate for Control (black circle) was obtained by fitting the equation ($y = e^{-(x-t)/\tau} + m \bullet x + c$) to changes in Ca^{2+} when mitochondrial permeability transition pore (mPTP) opens (1380 s onward in control in **a**). Numbers besides the *symbols denote the number of experiments for each data point. The scale below the X-axis shows the cumulative Ca^{2+} added before the corresponding time-point of efflux rate calculation. The physiological buffer contained approximately 40 μM EGTA # indicates significant difference ($P < 0.05$) between indicated pairs. (Data plot (Fig. 1) and modified legend resourced with license (https://creativecommons.org/licenses/BY/4.0/) from Natarajan et al. [48])

The computational model (Fig. 8.14) [47] also validates the experimental studies demonstrating that the Ca^{2+} buffering system, $m\beta_{Ca}$, is composed of at least two classes of Ca^{2+} binding; the first class representing prototypical mCa^{2+} binding, and the second class encompassing complex binding events associated with formation of various amorphous calcium phosphates [14, 48–50]. Two such phosphates formed are tricalcium phosphate, $Ca_3(PO^4)_2$, and hydroxyapatite, $[Ca_3(PO^4)_2]_3Ca$

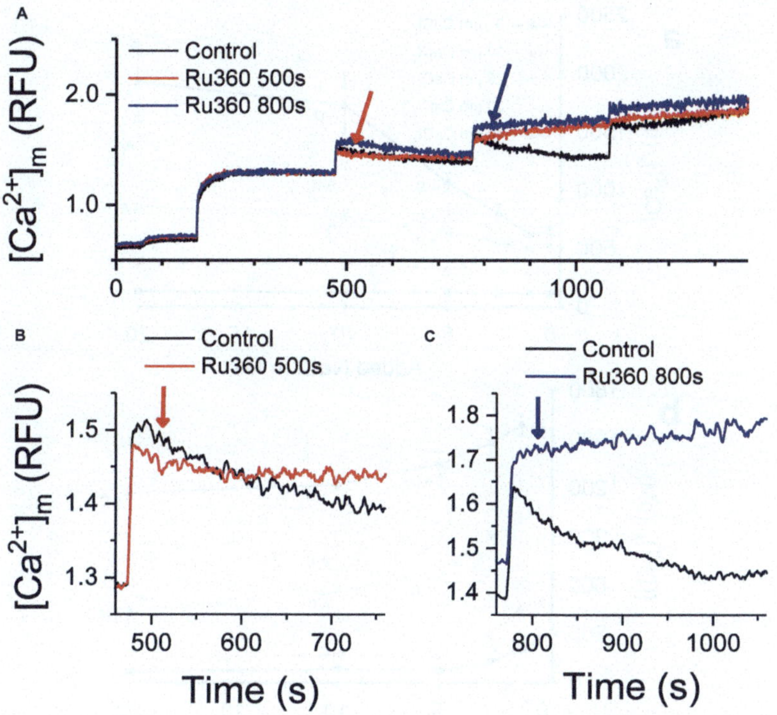

Fig. 8.12 Inhibition of Ca^{2+} uptake regulates Ca^{2+} buffering and intra-matrix Ca^{2+}: (**a**) Changes in intra-matrix Ca^{2+} in mitochondria isolated from rat hearts in response to 20 μM $CaCl_2$ boluses in the absence (control, black) or presence of 1 μM RU360, MCU inhibitor, applied at 500 s (red) or 800 s (blue). (**b**, **c**) are expanded graphs of (**a**) to highlight the differences in the kinetics of Ca^{2+} buffering and steady state mCa^{2+} in response to application of Ru360. (Data plot (Fig. 2) and modified legend resourced with license (https://creativecommons.org/licenses/BY/4.0/) from Natarajan et al. 48)

$(OH)_2$ [40]. In addition, an increase in extra-matrix $e[Mg^{2+}]$ hinders mCa^{2+} sequestration and attenuates Na^+-dependent Ca^{2+} extrusion [49].

Ca^{2+} buffers in the matrix include a low affinity buffer with a capacity of approximately 40 nM Ca^{2+}/mg mitochondrial protein, and a high affinity buffer with a capacity of 1.2 nM Ca^{2+}/mg mitochondrial protein [40]. Ca^{2+} also binds to the IMM at a capacity of about 80 nM/mg mitochondrial protein. The net effect of these buffers is the reason that free $m[Ca^{2+}]$ is only about 100–200 nM [57], which is between $1 \bullet 10^{-4}$ and $3 \bullet 10^{-4}$ times less than the total estimated bound $m[Ca_nPO^4{}_n]$ [40].

Factors Affecting Matrix Ca^{2+} Buffering It is now known that many external and internal factors can affect mCa^{2+} buffering dynamics and capacity. An example, in addition to ADP and OMN (Fig. 8.10), is cyclosporin A (CsA). Cyclophilin D (Cyp D) is a mitochondrial matrix peptidyl-prolyl cis-trans isomerase (PPIase) that is translocated to the IMM during high mCa^{2+} conditions. Cyp D is inhibited by CsA and CsA markedly increases $m\beta_{Ca}$ by up to tenfold without dissipating ΔpH_m and

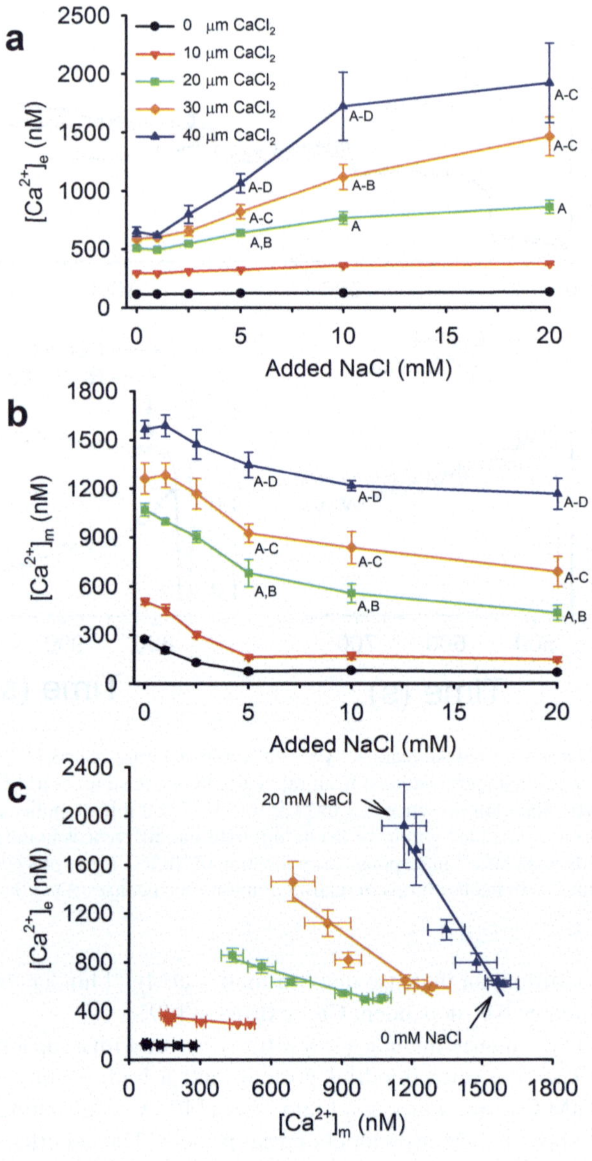

Fig. 8.13 Dynamic buffering of Ca^{2+} during mitochondrial Ca^{2+} uptake and Na^{+}-Ca^{2+} exchange (mNCE): mCa^{2+} buffering was determined from the final extra-matrix free [Ca^{2+}] and matrix free [Ca^{2+}] measured after adding $CaCl_2$ and or NaCl to guinea pig cardiac isolated mitochondria. Panels (**a**) and (**b**) show extra-matrix free [Ca^{2+}] (e[Ca^{2+}]) and matrix free [Ca^{2+}] (m[Ca^{2+}]) measured at the end of each experiment (see Figs. 2 and 3 of text) as a function of e[NaCl]. Panel (**c**) explores the correlation between e[Ca^{2+}] and m[Ca^{2+}] with respect to added $CaCl_2$. Panel c demonstrates that in the absence of mNCE, e[Ca^{2+}] is absorbed into the matrix and m[Ca^{2+}] rises with increases of boluses of added $CaCl_2$; in the presence of mNCE, matrix Ca^{2+} is extruded so e[Ca^{2+}]

$\Delta\Psi_m$, or reducing the NADH/NAD$^+$ redox state (Fig. 8.10) [51, 57]. CSA enhances mβ_{Ca} twice as much as OMN + ADO. This CsA-induced effect to greatly enhance mβ_{Ca} is mP$_i$-dependent because in a PO_4^{3-} free buffer condition, the CsA-mediated increase in mβ_{Ca} capacity is inhibited, whereas in the presence of CsA, addition of P$_i$ just before collapse of mCa^{2+} uptake under the P$_i$-depleted condition reinstates mβ_{Ca} [51].

Figure 8.15 [48] presents data from isolated mitochondria treated with $CaCl_2$ pulses; the peak in e[Ca^{2+}] with a $CaCl_2$ pulse is followed by Ca^{2+} uptake into the matrix and its disappearance from the extra-matrix buffer. Ru360 block any further Ca^{2+} uptake after a $CaCl_2$ pulse so e[Ca^{2+}] remains static. Ca^{2+} efflux rates were obtained from the slope values (m) calculated from fits ($y = m \bullet x + c$) to the linear phase of Ca^{2+} efflux from 5 s to 1 min after adding Ru360. Under these conditions, ADP, and less so CsA, decrease the Ca^{2+}efflux rates and increase the CRC as Ca^{2+} entering after more $CaCl_2$ pulses continues to be buffered. It is not yet clear if CsA acts directly on the buffering system or indirectly through PPIase activity. What does appear clear is that inhibition of mitochondrial permeability transition pore (MPTP) opening by CsA is largely due to its role to promote matrix Ca^{2+} sequestration and prevent overloading of mCa^{2+} that dissipates Δp_m (Fig. 8.15) [48, 51, 57].

Overall, mitochondrial Ca^{2+} buffering capacity, mβ_{Ca}, is complex, dynamic, and modulated by many factors [51]. The studies described above demonstrate that mitochondria exhibit fast and slow modes of mCa^{2+} uptake, with the slow, but not fast, mCa^{2+} uptake mode attenuated by an increase in e[Mg^{2+}], and a dynamic mβ_{Ca} that depends on the substrate utilized, and on other factors (buffer Na^+, Mg^{2+}, ADP and phosphate (P$_i$) levels, and inhibitors such as RuR, Ru360, CATR, OMN, and CsA that alter either the mP$_i$ pool or enhance the matrix binding of Ca^{2+}. The utility of a dynamic, responsive cardiac myocyte mCa^{2+} buffering system attests to selective adaptions over time in mammals and birds that permit tight control of m[Ca^{2+}] during abrupt changes, e.g., from resting to running (flying) activity that requires a higher supply and utilization of energy.

Matrix Ca^{2+} Extrusion The Na^+ and H^+ cations have large and essential roles in expelling the Ca^{2+} that enters the mitochondrial matrix when there are voltage and or concentration gradients that permit Ca^{2+} influx. Because there are no known

Fig. 8.13 (continued) rises more, with a lesser increase in m[Ca^{2+}]. The difference is represented by the slopes with greater Ca^{2+} dynamic buffering at low Ca^{2+} loading and less dynamic buffering at high Ca^{2+} loading. Each $CaCl_2$ group consists of data from each of the six [NaCl] levels (as indicated by the arrows), which were best fitted with linear trendlines. Data were obtained at a mitochondrial protein concentration of 0.5 mg/ml and are presented as mean ± SEM for both x and y-axes, where applicable. Significant differences ($P < 0.05$) in [Ca^{2+}]e or [Ca^{2+}]m for a given NaCl (5, 10 and 20 μM): 'A' vs. 0 μM $CaCl_2$; 'B' vs. 10 μM $CaCl_2$; 'C' vs. 20 μM $CaCl_2$; 'D' vs. 30 μM $CaCl_2$. (Data plots (Fig. 4) and modified legend resourced with permission from copyright owner of Blomeyer et al. [14])

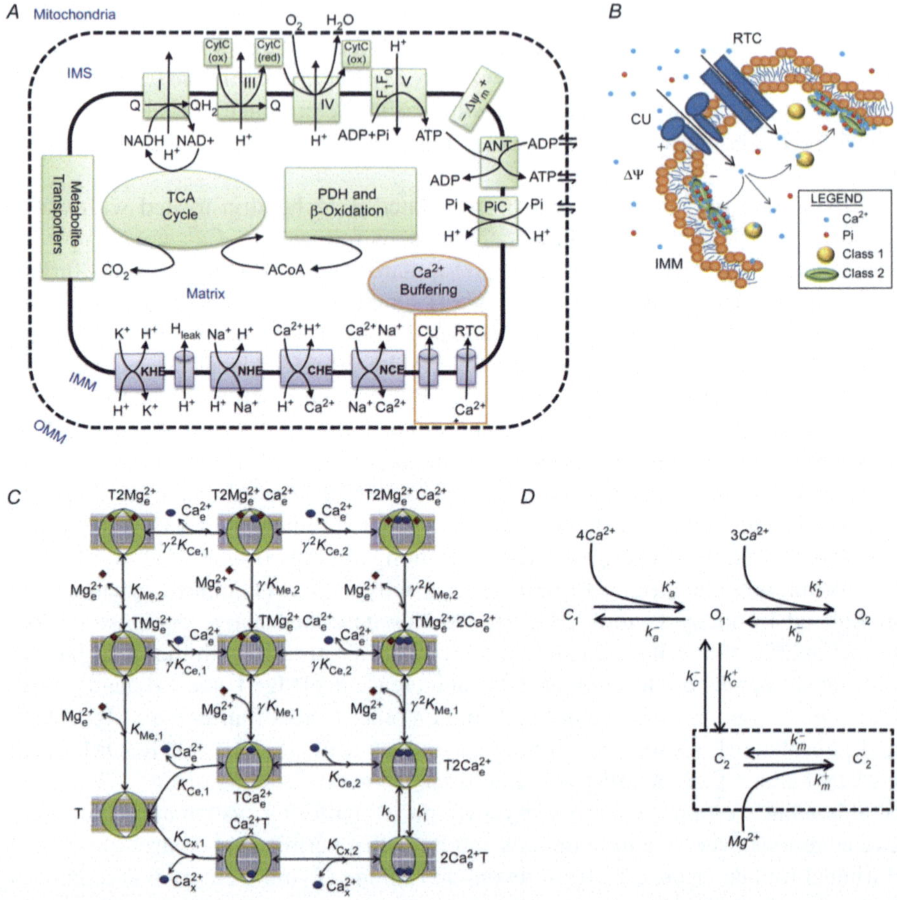

Fig. 8.14 Integrated model of mitochondrial bioenergetics and Ca^{2+} handling: (**a**) Schematic diagram of the integrated model of mitochondrial bioenergetics and Ca^{2+} handling of Bazil et al. [50] (see text) with the revised MCU (CU) model and an additional RTC model (highlighted with an orange box). (**b**) Mitochondrial Ca^{2+} sequestration system depicting how Ca^{2+} is differentially buffered depending on how it is taken up by mitochondria (see arrows). (**c**) Modified MCU model for Mg^{2+} inhibition of the MCU occurring only at the cytosolic side. (**d**) Keizer–Levine model of cardiac RyR modified to include modulation of the ryanodine receptor (RyR)-type channel (RTC) to account for the effects by cytosolic Mg^{2+}. The dotted box represents Mg^{2+} binding with the C_2 state, which is assumed to be in rapid equilibrium. (Diagrams (Fig. 1) and modified legend resourced with permission from copyright owner of Tewari et al. [47])

mitochondrial Na^+ channels or Na^+ pumps in the IMM, mNa^+ influx via mNCE is responsible for ejecting the excess mCa^{2+} (Fig. 8.13) in order to restore resting m[Ca^{2+}] [12, 14–16, 20, 58, 59]. If cell [Ca^{2+}] is elevated due to cardiomyocyte injury, mNCE is essential to efflux the excess Ca^{2+} from mitochondria that would otherwise incapacitate mitochondrial function. However, unlike the rapid uptake

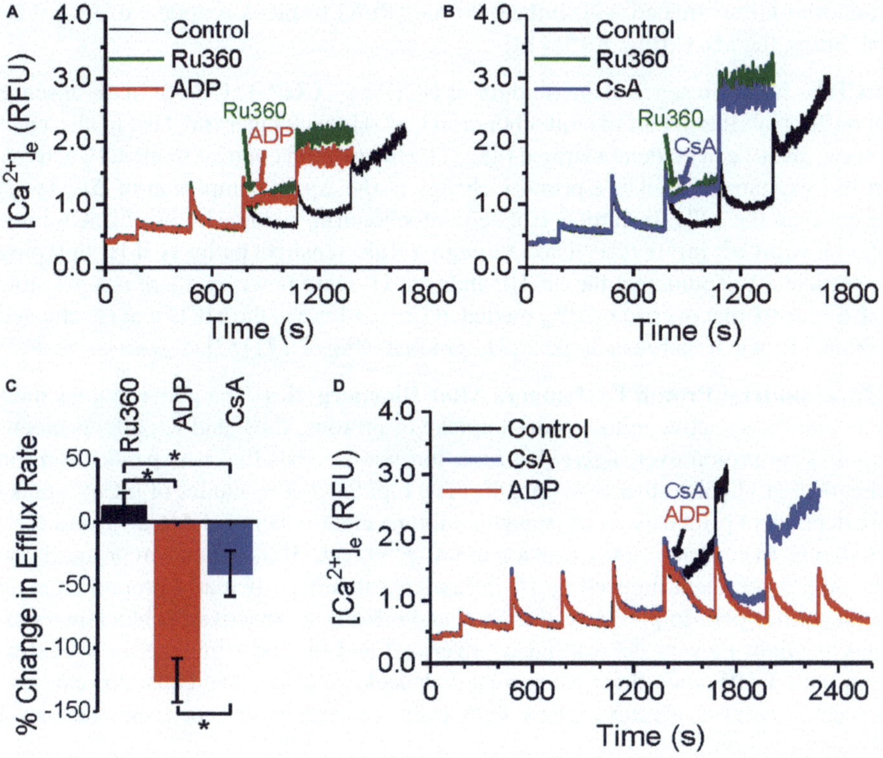

Fig. 8.15 Effect of ADP and cyclosporin A (CsA), an inhibitor of cyclophilin D (Cyp D), on Ca^{2+}-efflux and mitochondrial permeability transition pore (mPTP) opening: (**a**–**c**) Representative raw traces of extra-matrix Ca^{2+} measured as relative fluorescence changes in Na^+-free experimental buffer loaded with the Ca^{2+}-sensitive ratiometric dye Fura-4F. 20 μM $CaCl_2$ pulses were given at 180 and 480 s, followed by 1 μM Ru360 at 800 s (green trace), and (**a**) 250 μM ADP at 860 s or (**b**) 0.5 μM CsA at 860 s. (**c**) % change in the efflux rate before and after addition of 1 μM Ru360 (n = 9), 250 μM ADP (n = 6) or 0.5 μM CsA (n = 4). * indicates significant difference ($p < 0.05$) between indicated pairs. (**d**) Rescue of mPTP opening in Control (black trace) after addition of 250 μM ADP (red trace) or 0.5 μM CsA (blue trace). (Data plots (Fig. 3) and modified legend resourced with license (https://creativecommons.org/licenses/BY/4.0/) from Natarajan et al. [48])

and sequestration of mCa^{2+}, the extrusion of Ca^{2+} by mCHE is orders of magnitude slower.

mNCE The cytosolic or sarcoplasmic cNCE selectively exchanges only Na^+ with Ca^{2+}; the mitochondrial variant of NCE (mNCE) is sometimes called NCLX because it can exchange Na^+ or Li^+ with Ca^{2+} if present in the buffer [15]. The effect of Li^+ (Sect. 2.4) to efflux Ca^{2+} is slower and half that of Na^+ and is isotope independent for Li^+ [60]. Thermodynamically balanced modeling of mNCE proposes it is an electrogenic 3:1 exchanger, i.e., $3Na^+$ are exchanged for every Ca^{2+} [20]. The stoichiometry of the mNCE is supported by experimental data in which $e[Ca^{2+}]$ and $m[Ca^{2+}]$ were measured in the presence of $CaCl_2$ and NaCl added to isolated mitochondrial buffer treated with ruthenium red (RuR) to block reuptake of mCa^{2+} by inhibiting the MCU (Fig. 8.13) [14].

mCHE In the presence of an inhibitor of NCE (e.g., CGP-37157), or in the absence of buffer Na^+, the role of the mitochondrial Ca^{2+}-H^+ exchanger (mCHE) to also eject excess mCa^{2+} can be demonstrated [48, 57]. However, in contrast to mNCE, activity of this exchanger invokes a primary change in the ΔpH_m component of Δp_m (*pmf*) rather than the $\Delta\Psi_m$. In cardiac isolated mitochondria, a single bolus of $CaCl_2$ initiates slow mCa^{2+} influx (Ca^{2+} leak) through a RuR-sensitive pathway if H^+ pumping is sufficient to counteract the mCHE-induced H^+ leak; however, there is a net slow mCa^{2+} efflux that overrides $\Delta\Psi_m$-mediated Ca^{2+} influx via the MCU that is activated via mCHE when there is a large ΔpH_m gradient (Fig. 7.17) [57].

Mitochondrial Proton Exchangers Alter Bioenergetics The above studies indicate that the selective mitochondrial uptake of protons, Ca^{2+}, and K^+, affects bioenergetics by proton exchangers that reset ion homeostasis that is dependent on ion chemical gradients rather than on $\Delta\Psi_m$. The rapid and slow modes of mCa^{2+} uptake are dependent primarily on the trans-membrane electrochemical $\Delta[Ca^{2+}]_m$ and $\Delta\Psi_m$ gradients. In contrast, slow extrusion of mCa^{2+} via mCHE is dependent primarily on the $\Delta[H^+]_m$ gradient induced by H^+ influx/leak; this is particularly prominent at an acidic buffer pH_e. Importantly, if mNCE and mNHE are inactivated, blocking complex V might prevent delayed mCa^{2+} overload and instead stimulate mCa^{2+} extrusion via mCHE when there is an inward H^+ leak [57]. In intact cells, this can also serve to preserve glycolysis and TCA cycle-generated ATP, i.e., substrate level phosphorylation.

Such a passive homeostatic balance of $m[Ca^{2+}]$ may occur during cardiac injury when there is mCa^{2+} loading accompanied by declines in NADH redox state, pH_m and $\Delta\Psi_m$. One of the above studies [57] demonstrates that the differences in the rate and magnitude of mCa^{2+} influx/efflux in partially depolarized mitochondria, in the presence or absence of complex V activity, can be ascribed to the underlying changes in both *pmf* components, ΔpH_m and $\Delta\Psi_m$, after rapid mCa^{2+} loading [57].

Another related study in cardiac isolated mitochondria demonstrates that Ca^{2+} efflux via the mCHE is modulated not only by the ΔpH_m gradient but also by the total (free + bound) matrix Ca^{2+} [48]. This conclusion is based on results in which the presence of ADP and CsA increase CRC (enhanced $m\beta_{Ca}$) and decrease the Ca^{2+} efflux rate (reduced mCHE activity). A mechanism is proposed whereby activation of CHE is sensitive to changes in both total (bound) mCa^{2+} and the free $[mCa^{2+}]$ [48]. The important role of mKHE activity on inducing a proton leak and stimulating mitochondrial respiration [61] is discussed in Sect. 7.2.

8.4 Roles of Free Radicals $O_2^{\bullet-}$ and $NO^\bullet$ on Inducing Cardiac Cell Protection and Injury

Superoxide This free radical anion ($O_2^{\bullet-}$) is generated (Figs. 5.16, 5.17, 5.18 and 6.2) in mitochondria because of the leak of *e*- by the ETS and the proximity of O_2 (Sects. 5.2, 5.3 and 6.1). $O_2^{\bullet-}$ release has consequences not only for mitochondrial respiration but also for cardiomyocyte and overall cardiac function [62]. Because of its very short half-life of 10^{-6} s at 37 °C, the presence of hemoglobin, and the rapid conversion of $O_2^{\bullet-}$ to other reactants, $O_2^{\bullet-}$ release cannot be satisfactorily measured in vivo. However, the release of $O_2^{\bullet-}$ over time can be assessed in buffer perfused isolated hearts using the fluorescence probe dihydroethidium (DHE, ETH) (Fig. 6.4) [8, 63–67]. Although O_2 delivery is impeded during cardiac ischemia, these studies show that $O_2^{\bullet-}$ levels nevertheless rise slowly during ischemia. Importantly, during reperfusion, there is a burst of $O_2^{\bullet-}$ release that may or may not return to pre-ischemia levels depending on application of experimental therapeutics before IR injury.

Antioxidant systems are necessary to keep excess ROS in check. The roles of SODs and other antioxidant systems are additionally discussed in Sects. 2.3, 2.4, 5.2 and 6.1. In one example of reduced antioxidant efficacy, it was found that 17 °C hypothermia of isolated hearts increases $O_2^{\bullet-}$ emission by decreasing dismutation of $O_2^{\bullet-}$ as a result of the cold-induced reduction of enzyme activity [68]. This and the deceased activity of other antioxidant systems allow for more $O_2^{\bullet-}$ and its toxic reactants to persist [69]. Mitochondrial DNA, proteins, and lipid membranes are all subject to attack by excess ROS. ROS can damage TCA-cycle enzymes, especially aconitase and α-KG, and the respiratory complexes, mostly complex I, but also complexes III and IV. Phospholipids, especially mitochondrial cardiolipin, are major targets of excess ROS. Transition metal radicals like $(Fe = O)^{2+}$ can initiate lipid peroxidation cascades, e.g., on C-C bonds of polyunsaturated fatty acids (PUFAs) causing damage to membrane function [69, 70].

Pre-ischemia Oxidative Stress Reduces Cardiac IR Injury It is now generally held that a small amount of $O_2^{\bullet-}$ release modifies various signaling pathways that lead to cell protection [62, 69], particularly if there is sufficient MnSOD activity to convert most of the $O_2^{\bullet-}$ to H_2O_2. There are multiple experimental ways to protect the heart before an ischemic episode as demonstrated by less damage and function during ischemia and reperfusion (IR). As introduced in Sect. 5.2, cardiac preconditioning is a brief coronary infusion of a drug (Fig. 7.9) [63, 67] or brief ischemia (Fig. 6.4) [64, 71] followed by later and longer-term IR. Experimental studies indicate that induced release of small amounts of $O_2^{\bullet-}$ induces cardioprotection [69]. Ischemic preconditioning [64], sevoflurane (general anesthetic) preconditioning (Fig. 5.12) [54, 63, 72], and the complex I inhibitor amobarbital [66] each afford protection against subsequent IR injury, assessed by smaller infarct size, improved contractility and relaxation, reduced matrix Ca^{2+} loading, a higher NADH/NAD^+ ratio, and, importantly, reduced release of $O_2^{\bullet-}$ on reperfusion. This manner of cardioprotection is dependent on controlled, limited release of $O_2^{\bullet-}$, attenuating effects of drugs on respiratory complexes [73], high NADH/NAD$^+$ redox status [74], and activation of protective mitochondrial and cellular signaling pathways [62, 75].

Protective ROS Release Here are several other examples for evidence of protective ROS. Exogenous, brief opening of BK_{Ca} and SK_{Ca} channels protect against IR injury in an $O_2^{\bullet-}$ dependent manner. There is reduced release of ROS during IR and loss of protection on reperfusion if the $O_2^{\bullet-}$ dismutator Mn-TBAP (Fig. 8.16), or K_{Ca} channel blockers like paxilline and NS8693 (Fig. 7.12), are given before the K^+ channel openers [65, 67, 76]. The results presented in Fig. 8.16 show that the reduction in $O_2^{\bullet-}$ levels during ischemia and reperfusion in isolated hearts is largely obviated if TBAP is given before ischemia to bracket the effect of SK_{Ca} channel opener DCEB [67]; this suggests that DCEB reduces cardiac injury via a mechanism that initiates a small release of $O_2^{\bullet-}$ as also discussed in Sects. 5.2 and 7.2. It is likely that the underlying mechanism of many of these protective effects is due to effects of compounds on the ETS respiratory enzymes that transiently impede rapid *e*- transfer, and or to the NADH-linked free radical scavenging systems of mitochondria and cytosol [69]. It should be noted that therapeutically treating the heart on reperfusion only after ischemia is much less successful than prophylactically preparing the heart before ischemia. Thus, "post-conditioning" immediately after ischemia

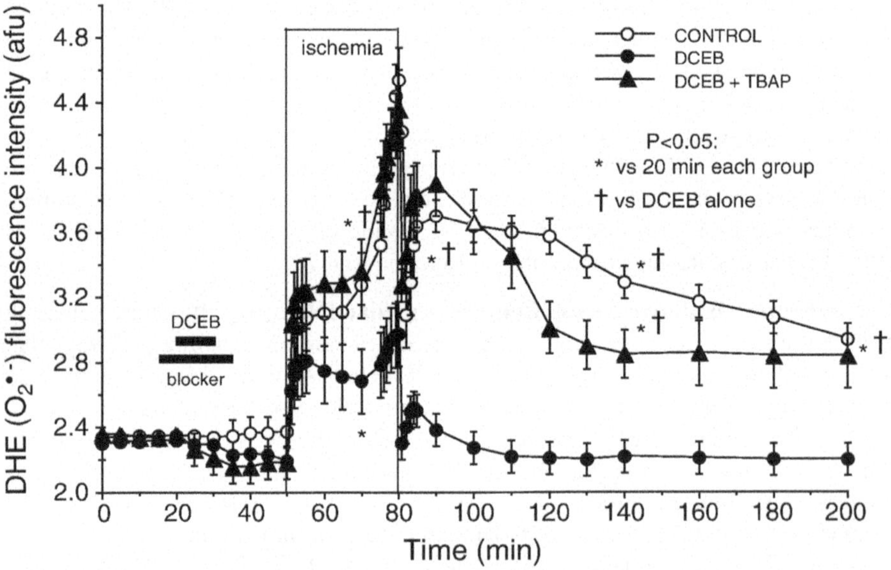

Fig. 8.16 Superoxide ($O_2^{\bullet-}$) measurements in isolated hearts: $O_2^{\bullet-}$ was assessed by dihydroethidium (DHE) fluorescence after preconditioning with the SK_{Ca} channel activator DCEB. Note the increases in these signals during ischemia and the slow decline during reperfusion. DCEB attenuated the increase in these signals during ischemia and reperfusion and this was reversed by the mitochondrial superoxide dismutase mimetic MnTBAP, (Mn(III)tetrakis(4-benzoic acid)porphyrin chloride). (Data plot (Fig. 4a) and legend resourced with permission from copyright owner of Stowe et al. [67])

would be a more practical method to reduce reperfusion injury. New approaches are needed in this area of heart research.

Signaling H_2O_2 It is probable that H_2O_2, after is conversion from $O_2^{\bullet-}$, is the most relevant molecule responsible for ROS signaling effects (Fig. 6.6), including cardioprotection, because of its stability, permeability, and longer half-life. Figure 8.17 [77] depicts a scheme for formation of H_2O_2 via SODs and its inactivation by linked scavenging systems in and out of the matrix, and ultimately by catalase to dismutate H_2O_2 ($2H_2O_2 \rightarrow 2H_2O + O_2$). Up to a limit, mitochondrial and cell antioxidant systems are capable of neutralizing excess ROS [69]. Recall that cellular antioxidant defenses depend on the reduction potential of the *e*- carriers and the reducing capacity of linked redox couples in the matrix ($NADH/NAD^+$ and $FADH_2/FAD$) and cyto-

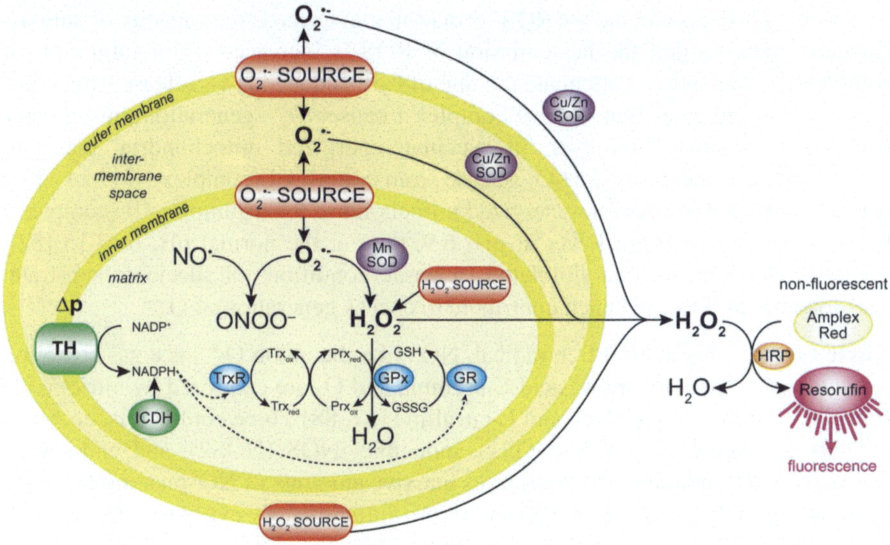

Fig. 8.17 Reactive oxygen species production by isolated mitochondria: Production of $O_2^{\bullet-}$ within the mitochondrial matrix, IMS, and OMM leads to the formation of H_2O_2 from SOD-catalyzed dismutation. Some $O_2^{\bullet-}$ can react directly with nitric oxide ($NO^{\bullet}$) to form peroxynitrite ($ONOO^-$). There are also sources outside mitochondria that produce H_2O_2 directly. The H_2O_2 efflux from mitochondria can be measured following reaction with a non-fluorescent substrate such as Amplex Red in conjunction with horseradish peroxidase (HRP) to form a fluorescent product, resorufin. Within mitochondria, H_2O_2 is degraded by glutathione peroxidases (GPx) or peroxiredoxins (Prx) which depend on glutathione (GSH) and thioredoxin-2 (Trx) for their reduction respectively. Glutathione disulfide (GSSG) is reduced back to GSH by glutathione reductase (GR). Trx is reduced by thioredoxin reductase-2 (TrxR). Both enzymes receive reducing equivalents from the NADPH and NADH pools, which are kept reduced by the Δp_m-dependent transhydrogenase (TH), and by isocitrate dehydrogenase (ICDH). Note that many mitochondrial preparations, particularly those from the liver, contain large amounts of catalase (converts H_2O_2 to H_2O) contamination. The effects of such extramitochondrial H_2O_2 sinks are not indicated here as they are usually accounted for by appropriate H_2O_2 calibration curves in the presence of mitochondria. ox, oxidized; red. (Cartoon with scheme (Fig. 2) and its modified legend resourced with license (https://creativecommons.org/licenses/BY-NC/4.0/) from Murphy [77] (review))

plasm [69]. These include the intra- (MnSOD) (Figs. 2.6 and 8.17) and extra-matrix (CuSOD) (Figs. 6.3 and 8.17) SODs [78], glutathione (GSH) [79], and thioredoxin ($TRXSH_2$) [78] systems.

Experiments in cardiac isolated mitochondria that utilize H_2O_2 measurements (Figs. 6.7, 6.8 and 7.10) aid in understanding normal and abnormal *e*- transfer and *e*- leaks causing release of $O_2^{\bullet-}$ along with assessment of antioxidant activity by MnSOD. Ca^{2+} loading of cardiac isolated mitochondria affects the release of H_2O_2 by causing partial $\Delta\Psi_m$ depolarization. The $FADH_2$-linked substrate succinate accumulates to become the major substrate utilized by mitochondria during cardiac ischemia [80]. Under conditions that mimic the mitochondrial environment during late ischemia (high external e[Ca^{2+}] with succinate as the only substrate), isolated mitochondrial experiments show that complex III becomes a major source of H_2O_2 release [81].

Severe cell stress can induce ROS formation that exceeds the capacity of antioxidant enzymes, so that the net emission of ROS is increased. Prior inhibition of complex I *e*- transfer can attenuate the rate of Ca^{2+}-induced H_2O_2 release from complex III; this indicates that RET at complex I causes $O_2^{\bullet-}$ generation during early simulated ischemia. Moreover, in succinate-energized mitochondria, the high e[Ca^{2+}]-induced increases in H_2O_2 release from complex I (complex III blocked by antimycin A) and in pyruvate-energized mitochondria from complex III (complex I blocked by rotenone) are lower at pH_e 6.9, than at the normal pH_e of 7.15 [82]. These results indicate that simulated ischemic conditions of succinate substrate, Ca^{2+} loading, and low pH each contribute to excess generation of $O_2^{\bullet-}$.

Nitric Oxide This neutral free radical, $NO^{\bullet}$, like the anion $O_2^{\bullet-}$, is a gas signaling molecule [83–86]. The amino acid L-arginine and O_2 are catalyzed by nitric oxide synthases (NOS) to make $NO^{\bullet}$ and L-citrulline [87, 88] Three common isoenzymes of NOS are neuronal nNOS (NOS1), inducible iNOS (NOS2), and endothelial eNOS (NOS3). Inducible iNOS releases massive amounts of $NO^{\bullet}$ on stimulation by bacterial lipopolysaccharide (LPS) and or pro-inflammatory cytokines [69, 89, 90]; nNOS and eNOS constitutively release $NO^{\bullet}$ and levels of $NO^{\bullet}$ increase with activation of calmodulin and increased Ca^{2+} [83–86]. The reaction requires NADPH and each NOS contains Fe heme and tetrahydrobiopterin [85]. During catalysis, seven *e*- from a N atom and eight *e*- from an O atom are covalently shared to form the molecule $N \equiv O^{\bullet}$. Like all free radicals, $NO^{\bullet}$ can only interact with other free radicals with unpaired orbital *e*-, or it can react with transition metals, which have more than one oxidation state, e.g., Fe or Cu (Sects. 2.3, 5.2 and 5.3).

Although $NO^{\bullet}$ is a free radical like $O_2^{\bullet-}$, it is not charged and so it, like H_2O_2, can easily passes through membranes. $NO^{\bullet}$ gas and nitro-containing compounds were found in the 1970's to act on soluble guanylyl cyclase (sGC) coupled receptors to generate the second messenger cyclic (c) cGMP [91]. $NO^{\bullet}$ attaches to the heme group in sGC to convert GTP to cGMP. This is the major function of eNOS. $NO^{\bullet}$ was independently reported as a signaling molecule in 1987 [92, 93] by its identification as the endothelium-dependent relaxing factor (EDRF), which had been found to increase cGMP levels that produced vasodilation [94].

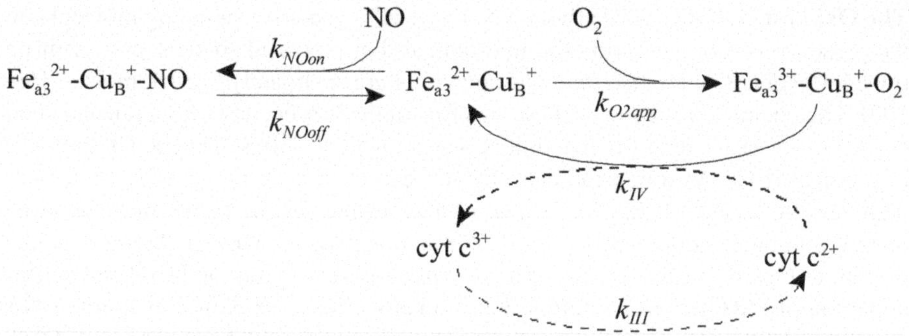

Fig. 8.18 Models to describe inhibition of the respiratory chain by $NO^{\bullet}$ in competition with O_2: Shown are a minimal model (continuous-line-depicted reactions, model 1) and an extended model (continuous- and dashed-line depicted reactions, model 2). $k_{NO\bullet on}$ and kO_2app refer to rate constants describing the combination of $NO^{\bullet}$ and O_2 with the reduced cytochrome a_3–Cu_B site (Fe_{a3}^{2+}–Cu_B^{+}) of COX, respectively. $k_{NO\bullet off}$ refers to the release of $NO^{\bullet}$ from the reduced cytochrome a_3–Cu_B site. k_{IV}, first-order rate constant in model 1 and second-order rate constant in model 2, refers to the reduction of the cyt a_3–Cu_B site entailing the arrival to COX of *e*- from the other mitochondrial complexes, via cytochrome c (cyt *c*). k_{III}, a first order rate constant, refers to the reduction of cytochrome *c* in complex III. Fe_{a3}^{3+}–Cu_B^{+}–O_2 species aggregates the intermediate forms appearing in the COX catalytic cycle. Mass action kinetics were assumed, and simulations were carried out with GEPASI (Ref. [46]) and PLAS (Ref. [47]). (Diagram (scheme 1) and its modified legend resourced with permission of owner (copyright 2024, National Academy of Sciences) of Antunes et al. [97])

$NO^{\bullet}$ is the active product of nitroglycerine, amyl nitrite, and Na-nitroprusside, three common vasodilatory drugs. $NO^{\bullet}$ increases cGMP to promote vasodilation [95] (Sect. 9.3). $NO^{\bullet}$ can also cause protein S-nitrosylation, which is the post-translational, covalent attachment of a $NO^{\bullet}$ group to a cysteine sulfur site in a protein that effectively acts as a signaling pathway for $NO^{\bullet}$ by changing the function of the protein [96] $NO^{\bullet}$ also competes with O_2 at cytochrome *c* oxidase (see also Sect. 6.1). Figure 8.18 [97] shows putative models to describe attenuation of respiratory chain function by $NO^{\bullet}$ at the Fe-Cu centers of cytochrome a_3 of complex IV (COX). In dynamite factories in Sweden 175 years ago, workers who inhaled fumes of nitroglycerine experienced headaches, and those who normally suffered chest pains (angina pectoralis) had relief while working. Only in the last 40–50 years was it recognized that the potent vasodilatory effect of these drugs was actually produced by $NO^{\bullet}$.

Measuring $NO^{\bullet}$ $NO^{\bullet}$ can be measured in the coronary effluent of physiological-buffer perfused, isolated hearts by a $NO^{\bullet}$ sensitive Clark type electrode [98–100]. These studies show a correlation between coronary flow rate and $NO^{\bullet}$ and L-citrulline release with bradykinin and serotonin, which are endothelium-dependent vasodilators, and with Na nitroprusside, but no correlation between coronary flow and $NO^{\bullet}$ with direct vascular dilating drugs [98–100]. The coronary vasodilatory effect of $NO^{\bullet}$ [95] is evidenced by blocking NOS with N^g L-arginine methyl ester (L-NAME) or other NOS inhibitors [99, 101].

The Oxidant $ONOO^-$ Even though $NO^•$ is itself a powerful signaling molecule, in the presence of $O_2^{•-}$, it forms the non-radical, but powerful oxidant peroxynitrite ($ONOO^-$) (Fig. 8.17) with a rate constant (k) of approximately 7 × 10^9 moles/s [69, 102]. The chemical reaction of these two free radicals proceeds with a rate constant (k) 3.5 times faster than the reaction between $O_2^{•-}$ with MnSOD (2 × 10^9 moles/s) to produce H_2O_2. As a non-radical, $ONOO^-$ can react with many more molecules than can $NO^•$. $ONOO^-$ itself is stable at an alkaline pH, as in the mitochondrial matrix, and fairly nonreactive, but $ONOO^-$ is readily protonated in the more acidic cytosol to ONOOH (peroxynitrous acid), which is very cytotoxic [103] and causes depletion of –SH groups and other anti-oxidants effects, oxidation of lipids, DNA strand breakage, as well as nitration of some amino acids and de-amination of DNA bases, especially guanine [104].

$ONOO^-$ Induced Damage to Proteins Important to mitochondrial function, two transport carriers, VDAC in the OMM, and ANT in the IMM (Fig. 6.18), can sustain damage and dysfunction due to nitration of their tyrosine residues by $ONOO^-$ [105, 106]. Thus the cellular effects of $ONOO^-$ are considered a combined action of $NO^•$ and $O_2^{•-}$ that are dependent on the relative concentrations of each reactant. In isolated hearts the deleterious effects of increased $O_2^{•-}$ and $ONOO^-$ on cardiac dysfunction during reperfusion after ischemia can be reduced by dismutation of $O_2^{•-}$ by Mn-SOD, by blocking L-arginine's conversion to L-citrulline by NOS with L-NAME, by glutathione (GSH), an antioxidant, and by catalase, which converts H_2O_2 to H_2O [52, 107–109].

Mitochondrial NOS It is probable that $NO^•$ is made in cardiac mitochondria by a mitochondrial nitric oxide synthase (mNOS) as well as in the cytosol because $ONOO^-$ is a product found in stressed isolated mitochondria [90, 110, 111]. As noted above, $NO^•$ levels can also be assessed using a $NO^•$ sensor, Clark-type electrode [98–100], or indirectly by measuring O_2 respiration and H_2O_2 release in the presence of L-arginine, NOS inhibitors, and Mn-SOD [112]. $ONOO^-$ will readily convert the amino acid L-tyrosine (Tyr) to fluorescent di-Tyr [84]. $ONOO^-$ levels can be assessed using fluorescence spectrophotometry in isolated hearts [68, 105, 107, 109], coronary effluent [68], and in isolated mitochondria [90], where $ONOO^-$ is indirectly measured from the level of fluorescent di-Tyr [84] in the presence of Tyr, which is proportionally oxidized by $ONOO^-$.

Figure 8.19 [90] summarizes results from data in isolated mitochondria showing changes in the production of mitochondrial $ONOO^-$. $ONOO^-$ was assessed by di-Tyr fluorescence; mitochondrial stress was stimulated by using only the substrate succinate with elevated buffer Ca^{2+}, and by menadione, a quinone redox cycler. Other drugs were used to verify that $NO^•$ in the presence of $O_2^{•-}$ was responsible for the changes. In the same study [90], western blotting and immunogold labeling with electron microscopy were utilized with specific antibodies to several NOS enzymes. The results indicate that induced release of $ONOO^-$ into the mitochondrial buffer is dependent both on $O_2^{•-}$ released from mitochondria and $NO^•$ derived from a mCa^{2+}-inducible nNOS isoform, possibly attached to mitochondria, and a mNOS isoform

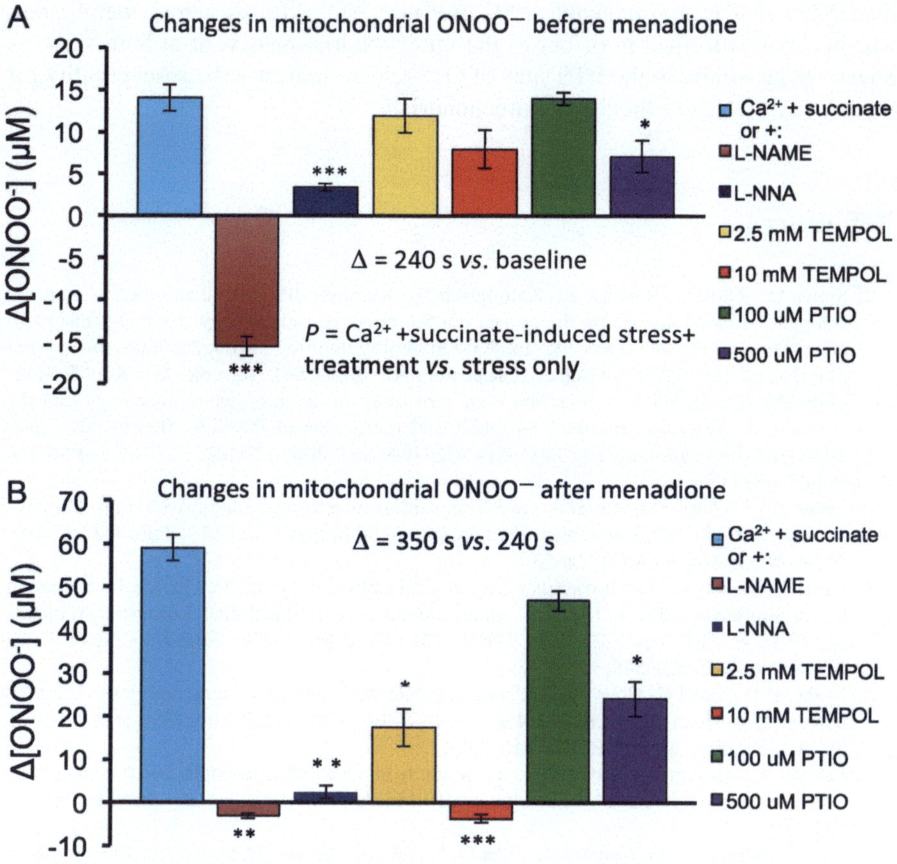

Fig. 8.19 Production of $ONOO^-$ during stress in guinea pig isolated cardiac mitochondria: Stress was induced by adding $CaCl_2$ to promote excess Ca^{2+} uptake before adding complex II substrate succinate. If cardiac mitochondrial contain nitric oxide synthase (mNOS), the generated $NO^{\bullet}$ interacts with $O_2^{\bullet -}$ to form $ONOO^-$, which in the presence of tyrosine, converts $ONOO^-$ to dityrosine (diTyr) in proportion to the $ONOO^-$ present. Menadione, a quinone redox cycler, was used to generate $O_2^{\bullet -}$. Exposing mitochondria to excess $CaCl_2$ before succinate resulted in an increase in diTyr (and amplex red fluorescence (H_2O_2) signals, indicating that mitochondrial oxidant stress, induced by elevated mCa^{2+} and succinate, increased mitochondrial $ONOO^-$ production via $NO^{\bullet}$ and $O_2^{\bullet -}$. Changes in mitochondrial $ONOO^-$ production dependent on NOS were evidenced by using NOS inhibitors L-NAME/L-NNA, TEMPOL, a superoxide dismutase (SOD) mimetic, and PTIO, a potent global $NO^{\bullet}$ scavenger. L-NAME and L-NNA decreased succinate and menadione-mediated $ONOO^-$ production, PTIO decreased production of $ONOO^-$, and TEMPOL decreased $ONOO^-$ levels by converting more $O_2^{\bullet -}$ to H_2O_2. Mitochondria undergoing oxidative stress induced by excess $CaCl_2$ plus with menadione (added at 350 s) produced a greater increase in $ONOO^-$ (60.0 ± 2.9 μM) (**b**) compared to $CaCl_2$ plus succinate alone (**a**). Pre-treatment with L-NAME or L-NNA) significantly decreased menadione-mediated $ONOO^-$ production. These results indicate that $ONOO^-$ production is greatly attenuated during oxidative stress when mNOS is inhibited, if $NO^{\bullet}$ is scavenged, if more $O_2^{\bullet -}$ is converted to H_2O_2 by TEMPOL, and if menadione is present to enhance generation of $O_2^{\bullet -}$ within the ETS. (Summary data plots (Fig. 4) and modified legend resourced with permission from copyright owner of Gerdes et al. [90])

like iNOS that is non-inducible [90]. Because $NO^{\bullet}$ diffuses across membranes, whether $NO^{\bullet}$ is formed in or out of the mitochondrial matrix, or at both sites, its release in proximity to the ETS sites of $O_2^{\bullet-}$ release indicates the susceptibility for formation of $ONOO^-$ directly in mitochondria.

References

1. Stowe DF, Dujic Z, Bosnjak ZJ, Kalbfleisch JH, Kampine JP. Volatile anesthetics attenuate sympathomimetic actions on the guinea pig SA node. Anesthesiology. 1988;68(6):887–94. https://doi.org/10.1097/00000542-198806000-00009. PubMed PMID: 2897808.
2. MacDonald EA, Madl J, Greiner J, Ramadan AF, Wells SM, Torrente AG, Kohl P, Rog-Zielinska EA, Quinn TA. Sinoatrial node structure, mechanics, electrophysiology and the chronotropic response to stretch in rabbit and mouse. Front Physiol. 2020;11:809. Epub 20200722. https://doi.org/10.3389/fphys.2020.00809. PubMed PMID: 32774307; PMCID: PMC7388775.
3. Cooper J, Scharm M, Mirams GR. The cardiac electrophysiology Web Lab. Biophys J. 2016;110(2):292–300. https://doi.org/10.1016/j.bpj.2015.12.012. PubMed PMID: 26789753; PMCID: PMC4724653.
4. Pagel P, Stowe DF. Cardiovascular anatomy and physiology. In: Powlovich L, Nemergut EC, Collins SR, editors. Barash's clinical anesthesia. Philadelphia: Lippincott Williams & Wilkins; 2017. p. 277–300. ISBN 13. 978-1496337009. https://doi.org/10.1213/ANE.0000000000003548.
5. Viola HM, Hool LC. How does calcium regulate mitochondrial energetics in the heart?—new insights. Heart Lung Circ. 2014;23(7):602-9. Epub 20140228. https://doi.org/10.1016/j.hlc.2014.02.009. PubMed PMID: 24657282.
6. Fabiato A. Calcium-induced release of calcium from the cardiac sarcoplasmic reticulum. Am J Phys. 1983;245(1):C1–14. https://doi.org/10.1152/ajpcell.1983.245.1.C1. PubMed PMID: 6346892.
7. An J, Varadarajan SG, Camara A, Chen Q, Novalija E, Gross GJ, Stowe DF. Blocking Na^+/H^+ exchange reduces $[Na^+]_i$ and $[Ca^{2+}]_i$ load after ischemia and improves function in intact hearts. Am J Physiol Heart Circ Physiol. 2001;281(6):H2398–409. https://doi.org/10.1152/ajpheart.2001.281.6.H2398. PubMed PMID: 11709405.
8. Aldakkak M, Stowe DF, Heisner JS, Spence M, Camara AK. Enhanced Na^+/H^+ exchange during ischemia and reperfusion impairs mitochondrial bioenergetics and myocardial function. J Cardiovasc Pharmacol. 2008;52(3):236–44. https://doi.org/10.1097/FJC.0b013e3181831337. PubMed PMID: 18806604; PMCID: PMC2659015.
9. O'Rourke B. Mitochondrial ion channels. Annu Rev Physiol. 2007;69:19–49. https://doi.org/10.1146/annurev.physiol.69.031905.163804. PubMed PMID: 17059356; PMCID: PMC2712118.
10. Stowe DF, Heisner JS, An J, Camara A, Varadarajan SG, Novalija E, Chen Q, Schelling P. Inhibition of Na^+/H^+ isoform-1 exchange protects hearts perfused after 6-hour cardioplegic cold storage. J Heart Lung Transplant. 2002;21(3):374–82. https://doi.org/10.1016/s1053-2498(01)00383-7. PubMed PMID: 11897527.
11. Camara AK, An J, Chen Q, Novalija E, Varadarajan SG, Schelling P, Stowe DF. Na^+/H^+ exchange inhibition with cardioplegia reduces cytosolic $[Ca^{2+}]$ and myocardial damage after cold ischemia. J Cardiovasc Pharmacol. 2003;41(5):686–98. https://doi.org/10.1097/00005344-200305000-00004. PubMed PMID: 12717098.
12. Tsuruya Y, Bersohn MM, Li Z, Nicoll DA, Philipson KD. Molecular cloning and functional expression of the guinea pig cardiac Na^+-Ca^{2+} exchanger. Biochim Biophys Acta.

1994;1196(1):97–9. https://doi.org/10.1016/0005-2736(94)90301-8. PubMed PMID: 7986817.

13. Karmazyn M, Gan XT, Humphreys RA, Yoshida H, Kusumoto K. The myocardial Na^+-H^+ exchange: structure, regulation, and its role in heart disease. Circ Res. 1999;85(9):777–86. https://doi.org/10.1161/01.res.85.9.777. PubMed PMID: 10532945.
14. Blomeyer CA, Bazil JN, Stowe DF, Pradhan RK, Dash RK, Camara AK. Dynamic buffering of mitochondrial Ca^{2+} during Ca^{2+} uptake and Na^+-induced Ca^{2+} release. J Bioenerg Biomembr. 2013;45(3):189–202. Epub 20121207. https://doi.org/10.1007/s10863-012-9483-7. PubMed PMID: 23225099; PMCID: PMC4065551.
15. Palty R, Silverman WF, Hershfinkel M, Caporale T, Sensi SL, Parnis J, Nolte C, Fishman D, Shoshan-Barmatz V, Herrmann S, Khananshvili D, Sekler I. NCLX is an essential component of mitochondrial Na^+/Ca^{2+} exchange. Proc Natl Acad Sci USA. 2010;107(1):436–41. Epub 20091215. https://doi.org/10.1073/pnas.0908099107. PubMed PMID: 20018762; PMCID: PMC2806722.
16. Luongo TS, Lambert JP, Gross P, Nwokedi M, Lombardi AA, Shanmughapriya S, Carpenter AC, Kolmetzky D, Gao E, van Berlo JH, Tsai EJ, Molkentin JD, Chen X, Madesh M, Houser SR, Elrod JW. The mitochondrial Na^+/Ca^{2+} exchanger is essential for Ca^{2+} homeostasis and viability. Nature. 2017;545(7652):93–7. Epub 20170426. https://doi.org/10.1038/nature22082. PubMed PMID: 28445457; PMCID: PMC5731245.
17. Murphy E, Eisner DA. Regulation of intracellular and mitochondrial sodium in health and disease. Circ Res. 2009;104(3):292–303. https://doi.org/10.1161/CIRCRESAHA.108.189050. PubMed PMID: 19213964; PMCID: PMC2662399.
18. Alvarez BV, Villa-Abrille MC. Mitochondrial NHE1: a newly identified target to prevent heart disease. Front Physiol. 2013;4:152. Epub 20130628. https://doi.org/10.3389/fphys.2013.00152. PubMed PMID: 23825461; PMCID: PMC3695379.
19. Santo-Domingo J, Demaurex N. Perspectives on: SGP symposium on mitochondrial physiology and medicine: the renaissance of mitochondrial pH. J Gen Physiol. 2012;139(6):415–23. https://doi.org/10.1085/jgp.201110767. PubMed PMID: 22641636; PMCID: PMC3362525.
20. Dash RK, Beard DA. Analysis of cardiac mitochondrial Na^+-Ca^{2+} exchanger kinetics with a biophysical model of mitochondrial Ca^{2+} handling suggests a 3:1 stoichiometry. J Physiol. 2008;586(13):3267–85. Epub 20080508. https://doi.org/10.1113/jphysiol.2008.151977. PubMed PMID: 18467367; PMCID: PMC2538784.
21. Hernansanz-Agustin P, Choya-Foces C, Carregal-Romero S, Ramos E, Oliva T, Villa-Pina T, Moreno L, Izquierdo-Alvarez A, Cabrera-Garcia JD, Cortes A, Lechuga-Vieco AV, Jadiya P, Navarro E, Parada E, Palomino-Antolin A, Tello D, Acin-Perez R, Rodriguez-Aguilera JC, Navas P, Cogolludo A, Lopez-Montero I, Martinez-Del-Pozo A, Egea J, Lopez MG, Elrod JW, Ruiz-Cabello J, Bogdanova A, Enriquez JA, Martinez-Ruiz A. Na^+ controls hypoxic signalling by the mitochondrial respiratory chain. Nature. 2020;586(7828):287–91. Epub 20200729. https://doi.org/10.1038/s41586-020-2551-y. PubMed PMID: 32728214; PMCID: PMC7992277.
22. Kuo IY, Ehrlich BE. Signaling in muscle contraction. Cold Spring Harb Perspect Biol. 2015;7(2):a006023. Epub 20150202. https://doi.org/10.1101/cshperspect.a006023. PubMed PMID: 25646377; PMCID: PMC4315934.
23. Eisner DA, Caldwell JL, Kistamas K, Trafford AW. Calcium and excitation-contraction coupling in the heart. Circ Res. 2017;121(2):181–95. https://doi.org/10.1161/CIRCRESAHA.117.310230. PubMed PMID: 28684623; PMCID: PMC5497788.
24. Endoh M. Cardiac Ca^{2+} signaling and Ca^{2+} sensitizers. Circ J. 2008;72(12):1915–25. Epub 20081104. https://doi.org/10.1253/circj.cj-08-0838. PubMed PMID: 18981594.
25. Carafoli E, Krebs J. Why calcium? How calcium became the best communicator. J Biol Chem. 2016;291(40):20849–57. Epub 20160726. https://doi.org/10.1074/jbc.R116.735894. PubMed PMID: 27462077; PMCID: PMC5076498.
26. De La Fuente S, Lambert JP, Nichtova Z, Fernandez Sanz C, Elrod JW, Sheu SS, Csordas G. Spatial separation of mitochondrial calcium uptake and extrusion for energy-efficient

mitochondrial calcium signaling in the heart. Cell Rep. 2018;24(12):3099–107 e4. https://doi.org/10.1016/j.celrep.2018.08.040. PubMed PMID: 30231993; PMCID: PMC6226263.
27. Laver DR, Lamb GD. Inactivation of Ca^{2+} release channels (ryanodine receptors RyR1 and RyR2) with rapid steps in [Ca^{2+}] and voltage. Biophys J. 1998;74(5):2352–64. https://doi.org/10.1016/S0006-3495(98)77944-5. PubMed PMID: 9591662; PMCID: PMC1299578.
28. Poole-Wilson PA. Acidosis and contractility of heart muscle. Ciba Found Symp. 1982;87:58–76. https://doi.org/10.1002/9780470720691.ch4. PubMed PMID: 6804193.
29. Squire JM. Muscle contraction: sliding filament history, sarcomere dynamics and the two Huxleys. Glob Cardiol Sci Pract. 2016;2016(2):e201611. Epub 20160630. https://doi.org/10.21542/gcsp.2016.11. PubMed PMID: 29043260; PMCID: PMC5642817.
30. Katz AM, Lorell BH. Regulation of cardiac contraction and relaxation. Circulation. 2000;102(20 Suppl 4):IV69–74. https://doi.org/10.1161/01.cir.102.suppl_4.iv-69. PubMed PMID: 11080134.
31. Giancoli DC. Physics principles with applications. 7th ed. Prentice Hall; 2023. ISBN 978-0321625922.
32. Rhodes SS, Camara AK, Aldakkak M, Heisner JS, Stowe DF. Stretch-induced increase in cardiac contractility is independent of myocyte Ca^{2+} while block of stretch channels by streptomycin improves contractility after ischemic stunning. Physiol Rep. 2015;3(8) https://doi.org/10.14814/phy2.12486. PubMed PMID: 26290532; PMCID: PMC4562572.
33. Chen Q, Camara AK, Rhodes SS, Riess ML, Novalija E, Stowe DF. Cardiotonic drugs differentially alter cytosolic [Ca^{2+}] to left ventricular relationships before and after ischemia in isolated guinea pig hearts. Cardiovasc Res. 2003;59(4):912–25. https://doi.org/10.1016/s0008-6363(03)00524-8. PubMed PMID: 14553831.
34. Camara AK, Chen Q, An J, Novalija E, Riess ML, Rhodes SS, Stowe DF. Comparison of hyperkalemic cardioplegia with altered [$CaCl_2$] and [$MgCl_2$] on [Ca^{2+}]$_i$ transients and function after warm global ischemia in isolated hearts. J Cardiovasc Surg. 2004;45(1):1–13. PubMed PMID: 15041929.
35. Ryu SY, Beutner G, Dirksen RT, Kinnally KW, Sheu SS. Mitochondrial ryanodine receptors and other mitochondrial Ca^{2+} permeable channels. FEBS Lett. 2010;584(10):1948–55. Epub 20100121. https://doi.org/10.1016/j.febslet.2010.01.032. PubMed PMID: 20096690; PMCID: PMC2866836.
36. Gunter TE, Sheu SS. Characteristics and possible functions of mitochondrial Ca^{2+} transport mechanisms. Biochim Biophys Acta. 2009;1787(11):1291–308. Epub 20090106. https://doi.org/10.1016/j.bbabio.2008.12.011. PubMed PMID: 19161975; PMCID: PMC2730425.
37. Rhodes SS, Ropella KM, Camara AK, Chen Q, Riess ML, Pagel PS, Stowe DF. Ischemia-reperfusion injury changes the dynamics of Ca^{2+}-contraction coupling due to inotropic drugs in isolated hearts. J Appl Physiol (1985). 2006;100(3):940–50. Epub 20051110. https://doi.org/10.1152/japplphysiol.00285.2005. PubMed PMID: 16282437.
38. Rhodes SS, Camara AK, Ropella KM, Audi SH, Riess ML, Pagel PS, Stowe DF. Ischemia reperfusion dysfunction changes model-estimated kinetics of myofilament interaction due to inotropic drugs in isolated hearts. Biomed Eng Online. 2006;5:16. Epub 20060302. https://doi.org/10.1186/1475-925X-5-16. PubMed PMID: 16512898; PMCID: PMC1431537.
39. Camara AK, Chen Q, Rhodes SS, Riess ML, Stowe DF. Negative inotropic drugs alter indexes of cytosolic [Ca^{2+}]-left ventricular pressure relationships after ischemia. Am J Physiol Heart Circ Physiol. 2004;287(2):H667–80. Epub 20040401. https://doi.org/10.1152/ajpheart.01142.2003. PubMed PMID: 15059780.
40. Jafri MS, Dudycha SJ, O'Rourke B. Cardiac energy metabolism: models of cellular respiration. Annu Rev Biomed Eng. 2001;3:57–81. https://doi.org/10.1146/annurev.bioeng.3.1.57. PubMed PMID: 11447057.
41. Csordas G, Renken C, Varnai P, Walter L, Weaver D, Buttle KF, Balla T, Mannella CA, Hajnoczky G. Structural and functional features and significance of the physical linkage between ER and mitochondria. J Cell Biol. 2006;174(7):915–21. Epub 20060918. https://doi.org/10.1083/jcb.200604016. PubMed PMID: 16982799; PMCID: PMC2064383.

42. Csordas G, Varnai P, Golenar T, Roy S, Purkins G, Schneider TG, Balla T, Hajnoczky G. Imaging interorganelle contacts and local calcium dynamics at the ER-mitochondrial interface. Mol Cell. 2010;39(1):121–32. https://doi.org/10.1016/j.molcel.2010.06.029. PubMed PMID: 20603080; PMCID: PMC3178184.
43. Varadarajan SG, An J, Novalija E, Smart SC, Stowe DF. Changes in $[Na^+]_i$, compartmental $[Ca^{2+}]$, and NADH with dysfunction after global ischemia in intact hearts. Am J Physiol Heart Circ Physiol. 2001;280(1):H280–93. https://doi.org/10.1152/ajpheart.2001.280.1.H280. PubMed PMID: 11123243.
44. Mishra J, Jhun BS, Hurst S, O-Uchi J, Csordas G, Sheu SS. The mitochondrial Ca^{2+} uniporter: structure, function, and pharmacology. Handb Exp Pharmacol. 2017;240:129–56. https://doi.org/10.1007/164_2017_1. PubMed PMID: 28194521; PMCID: PMC5554456.
45. Sparagna GC, Gunter KK, Sheu SS, Gunter TE. Mitochondrial calcium uptake from physiological-type pulses of calcium. A description of the rapid uptake mode. J Biol Chem. 1995;270(46):27510–5. https://doi.org/10.1074/jbc.270.46.27510. PubMed PMID: 7499209.
46. Boelens AD, Pradhan RK, Blomeyer CA, Camara AK, Dash RK, Stowe DF. Extra-matrix Mg^{2+} limits Ca^{2+} uptake and modulates Ca^{2+} uptake-independent respiration and redox state in cardiac isolated mitochondria. J Bioenerg Biomembr. 2013;45(3):203–18. Epub 20130303. https://doi.org/10.1007/s10863-013-9500-5. PubMed PMID: 23456198; PMCID: PMC3670670.
47. Tewari SG, Camara AK, Stowe DF, Dash RK. Computational analysis of Ca^{2+} dynamics in isolated cardiac mitochondria predicts two distinct modes of Ca^{2+} uptake. J Physiol. 2014;592(9):1917–30. Epub 20140303. https://doi.org/10.1113/jphysiol.2013.268847. PubMed PMID: 24591571; PMCID: PMC4230769.
48. Natarajan GK, Glait L, Mishra J, Stowe DF, Camara AKS, Kwok WM. Total matrix Ca^{2+} modulates Ca^{2+} efflux via the Ca^{2+}/H^+ exchanger in cardiac mitochondria. Front Physiol. 2020;11:510600. Epub 20200916. https://doi.org/10.3389/fphys.2020.510600. PubMed PMID: 33041851; PMCID: PMC7526510.
49. Blomeyer CA, Bazil JN, Stowe DF, Dash RK, Camara AK. Mg^{2+} differentially regulates two modes of mitochondrial Ca^{2+} uptake in isolated cardiac mitochondria: implications for mitochondrial Ca^{2+} sequestration. J Bioenerg Biomembr. 2016;48(3):175–88. Epub 20160127. https://doi.org/10.1007/s10863-016-9644-1. PubMed PMID: 26815005; PMCID: PMC5098337.
50. Bazil JN, Blomeyer CA, Pradhan RK, Camara AK, Dash RK. Modeling the calcium sequestration system in isolated guinea pig cardiac mitochondria. J Bioenerg Biomembr. 2013;45(3):177–88. Epub 20121122. https://doi.org/10.1007/s10863-012-9488-2. PubMed PMID: 23180139; PMCID: PMC3615037.
51. Mishra J, Davani AJ, Natarajan GK, Kwok WM, Stowe DF, Camara AKS. Cyclosporin A increases mitochondrial buffering of calcium: an additional mechanism in delaying mitochondrial permeability transition pore opening. Cells. 2019;8(9). Epub 20190907. https://doi.org/10.3390/cells8091052. PubMed PMID: 31500337; PMCID: PMC6770067.
52. Camara AK, Aldakkak M, Heisner JS, Rhodes SS, Riess ML, An J, Heinen A, Stowe DF. ROS scavenging before 27 °C ischemia protects hearts and reduces mitochondrial ROS, Ca^{2+} overload, and changes in redox state. Am J Physiol Cell Physiol. 2007;292(6):C2021–31. Epub 20070207. https://doi.org/10.1152/ajpcell.00231.2006. PubMed PMID: 17287367.
53. Riess ML, Camara AK, Kevin LG, An J, Stowe DF. Reduced reactive O_2 species formation and preserved mitochondrial NADH and $[Ca^{2+}]$ levels during short-term 17 °C ischemia in intact hearts. Cardiovasc Res. 2004;61(3):580–90. https://doi.org/10.1016/j.cardiores.2003.09.016. PubMed PMID: 14962488.
54. Riess ML, Camara AK, Novalija E, Chen Q, Rhodes SS, Stowe DF. Anesthetic preconditioning attenuates mitochondrial Ca^{2+} overload during ischemia in Guinea pig intact hearts: reversal by 5-hydroxydecanoic acid. Anesth Analg. 2002;95(6):1540–6, table of contents. https://doi.org/10.1097/00000539-200212000-00013. PubMed PMID: 12456413.

55. Haumann J, Dash RK, Stowe DF, Boelens AD, Beard DA, Camara AK. Mitochondrial free [Ca^{2+}] increases during ATP/ADP antiport and ADP phosphorylation: exploration of mechanisms. Biophys J. 2010;99(4):997–1006. https://doi.org/10.1016/j.bpj.2010.04.069. PubMed PMID: 20712982; PMCID: PMC2920628.
56. Kim J, Zimmerman MA, Shin WY, Boettcher BT, Lee JS, Park JI, Ali M, Yang M, Mishra J, Hagen CE, McGraw JE, Mathison A, Woehlck HJ, Lomberk G, Camara AKS, Urrutia RA, Stowe DF, Hong JC. Effects of subnormothermic regulated hepatic reperfusion on mitochondrial and transcriptomic profiles in a porcine model. Ann Surg. 2023;277(2):e366–e75. Epub 20230110. https://doi.org/10.1097/SLA.0000000000005156. PubMed PMID: 34387201; PMCID: PMC8840998.
57. Haumann J, Camara AKS, Gadicherla AK, Navarro CD, Boelens AD, Blomeyer CA, Dash RK, Boswell MR, Kwok WM, Stowe DF. Slow Ca^{2+} efflux by Ca^{2+}/H^+ exchange in cardiac mitochondria Is modulated by Ca^{2+} re-uptake via MCU, extra-mitochondrial pH, and H^+ pumping by F_OF_1-ATPase. Front Physiol. 2018;9:1914. Epub 20190204. https://doi.org/10.3389/fphys.2018.01914. PubMed PMID: 30804812; PMCID: PMC6378946.
58. Bernardi P. Mitochondrial transport of cations: channels, exchangers, and permeability transition. Physiol Rev. 1999;79(4):1127–55. https://doi.org/10.1152/physrev.1999.79.4.1127. PubMed PMID: 10508231.
59. Griffiths EJ. Mitochondrial calcium transport in the heart: physiological and pathological roles. J Mol Cell Cardiol. 2009;46(6):789–803. Epub 20090312. https://doi.org/10.1016/j.yjmcc.2009.03.001. PubMed PMID: 19285504.
60. Bukhteeva I, Rahman FA, Kendall B, Duncan RE, Quadrilatero J, Pavlov EV, Gingras MJP, Leonenko Z. Effects of lithium isotopes on sodium/lithium co-transport and calcium efflux through the sodium/calcium/lithium exchanger in mitochondria. Front Physiol. 2024;15:1354091. Epub 20240409. https://doi.org/10.3389/fphys.2024.1354091. PubMed PMID: 38655027; PMCID: PMC11036541.
61. Malas KM, Lambert DS, Heisner JS, Camara AKS, Stowe DF. Time and charge/pH-dependent activation of K^+ channel-mediated K^+ influx and K^+/H^+ exchange in guinea pig heart isolated mitochondria; role in bioenergetic stability. Biochim Biophys Acta Bioenerg. 2022;1863(8):148908. Epub 20220809. https://doi.org/10.1016/j.bbabio.2022.148908. PubMed PMID: 35961396.
62. Mailloux RJ. Teaching the fundamentals of electron transfer reactions in mitochondria and the production and detection of reactive oxygen species. Redox Biol. 2015;4:381–98. Epub 20150207. https://doi.org/10.1016/j.redox.2015.02.001. PubMed PMID: 25744690; PMCID: PMC4348434.
63. Kevin LG, Novalija E, Riess ML, Camara AKS, Rhodes SS, Stowe DF. Sevoflurane exposure generates superoxide but leads to decreased superoxide during ischemia and reperfusion in isolated hearts. Anesth Analg. 2003;96(4):949–55. https://doi.org/10.1213/01.ANE.0000052515.25465.35. PubMed PMID: 12651639.
64. Kevin LG, Camara AK, Riess ML, Novalija E, Stowe DF. Ischemic preconditioning alters real-time measure of O_2 radicals in intact hearts with ischemia and reperfusion. Am J Physiol Heart Circ Physiol. 2003;284(2):H566–74. Epub 20021031. https://doi.org/10.1152/ajpheart.00711.2002. PubMed PMID: 12414448.
65. Stowe DF, Aldakkak M, Camara AK, Riess ML, Heinen A, Varadarajan SG, Jiang MT. Cardiac mitochondrial preconditioning by Big Ca^{2+}-sensitive K^+ channel opening requires superoxide radical generation. Am J Physiol Heart Circ Physiol. 2006;290(1):H434–40. Epub 20050826. https://doi.org/10.1152/ajpheart.00763.2005. PubMed PMID: 16126810.
66. Aldakkak M, Stowe DF, Chen Q, Lesnefsky EJ, Camara AK. Inhibited mitochondrial respiration by amobarbital during cardiac ischaemia improves redox state and reduces matrix Ca^{2+} overload and ROS release. Cardiovasc Res. 2008;77(2):406–15. https://doi.org/10.1016/j.cardiores.2007.08.008. PubMed PMID: 17900548.
67. Stowe DF, Gadicherla AK, Zhou Y, Aldakkak M, Cheng Q, Kwok WM, Jiang MT, Heisner JS, Yang M, Camara AK. Protection against cardiac injury by small Ca^{2+}-sensitive K^+ chan-

nels identified in guinea pig cardiac inner mitochondrial membrane. Biochim Biophys Acta. 2013;1828(2):427–42. Epub 20120908. https://doi.org/10.1016/j.bbamem.2012.08.031. PubMed PMID: 22982251; PMCID: PMC3534888.
68. Camara AK, Riess ML, Kevin LG, Novalija E, Stowe DF. Hypothermia augments reactive oxygen species detected in the guinea pig isolated perfused heart. Am J Physiol Heart Circ Physiol. 2004;286(4):H1289–99. Epub 20031126. https://doi.org/10.1152/ajpheart.00811.2003. PubMed PMID: 14644763.
69. Stowe DF, Camara AK. Mitochondrial reactive oxygen species production in excitable cells: modulators of mitochondrial and cell function. Antioxid Redox Signal. 2009;11(6):1373–414. https://doi.org/10.1089/ars.2008.2331. PubMed PMID: 19187004; PMCID: PMC2842133.
70. Ayala A, Munoz MF, Arguelles S. Lipid peroxidation: production, metabolism, and signaling mechanisms of malondialdehyde and 4-hydroxy-2-nonenal. Oxid Med Cell Longev. 2014;2014:360438. Epub 20140508. https://doi.org/10.1155/2014/360438. PubMed PMID: 24999379; PMCID: PMC4066722.
71. Murry CE, Jennings RB, Reimer KA. Preconditioning with ischemia: a delay of lethal cell injury in ischemic myocardium. Circulation. 1986;74(5):1124–36. https://doi.org/10.1161/01.cir.74.5.1124. PubMed PMID: 3769170.
72. Kevin LG, Novalija E, Stowe DF. Reactive oxygen species as mediators of cardiac injury and protection: the relevance to anesthesia practice. Anesth Analg. 2005;101(5):1275–87. https://doi.org/10.1213/01.ANE.0000180999.81013.D0. PubMed PMID: 16243980.
73. Agarwal B, Dash RK, Stowe DF, Bosnjak ZJ, Camara AK. Isoflurane modulates cardiac mitochondrial bioenergetics by selectively attenuating respiratory complexes. Biochim Biophys Acta. 2014;1837(3):354–65. Epub 20131217. https://doi.org/10.1016/j.bbabio.2013.11.006. PubMed PMID: 24355434; PMCID: PMC4084852.
74. Agarwal B, Camara AK, Stowe DF, Bosnjak ZJ, Dash RK. Enhanced charge-independent mitochondrial free Ca^{2+} and attenuated ADP-induced NADH oxidation by isoflurane: implications for cardioprotection. Biochim Biophys Acta. 2012;1817(3):453–65. Epub 20111202. https://doi.org/10.1016/j.bbabio.2011.11.011. PubMed PMID: 22155157; PMCID: PMC3269543.
75. Agarwal B, Stowe DF, Dash RK, Bosnjak ZJ, Camara AK. Mitochondrial targets for volatile anesthetics against cardiac ischemia-reperfusion injury. Front Physiol. 2014;5:341. Epub 20140916. https://doi.org/10.3389/fphys.2014.00341. PubMed PMID: 25278902; PMCID: PMC4165278.
76. Stowe DF, Yang M, Heisner JS, Camara AKS. Endogenous and agonist-induced opening of mitochondrial big versus small Ca^{2+}-sensitive K^+ channels on cardiac cell and mitochondrial protection. J Cardiovasc Pharmacol. 2017;70(5):314–28. https://doi.org/10.1097/FJC.0000000000000524. PubMed PMID: 28777255; PMCID: PMC5726766.
77. Murphy MP. How mitochondria produce reactive oxygen species. Biochem J. 2009;417(1):1–13. https://doi.org/10.1042/BJ20081386. PubMed PMID: 19061483; PMCID: PMC2605959.
78. Inarrea P, Moini H, Han D, Rettori D, Aguilo I, Alava MA, Iturralde M, Cadenas E. Mitochondrial respiratory chain and thioredoxin reductase regulate intermembrane Cu,Zn-superoxide dismutase activity: implications for mitochondrial energy metabolism and apoptosis. Biochem J. 2007;405(1):173–9. https://doi.org/10.1042/BJ20061809. PubMed PMID: 17394422; PMCID: PMC1925252.
79. Pai HV, Starke DW, Lesnefsky EJ, Hoppel CL, Mieyal JJ. What is the functional significance of the unique location of glutaredoxin 1 (GRx1) in the intermembrane space of mitochondria? Antioxid Redox Signal. 2007;9(11):2027–33. https://doi.org/10.1089/ars.2007.1642. PubMed PMID: 17845131.
80. Zhang J, Wang YT, Miller JH, Day MM, Munger JC, Brookes PS. Accumulation of succinate in cardiac ischemia primarily occurs via canonical krebs cycle activity. Cell Rep. 2018;23(9):2617–28. https://doi.org/10.1016/j.celrep.2018.04.104. PubMed PMID: 29847793; PMCID: PMC6002783.

81. Aldakkak M, Stowe DF, Dash RK, Camara AK. Mitochondrial handling of excess Ca^{2+} is substrate-dependent with implications for reactive oxygen species generation. Free Radic Biol Med. 2013;56:193–203. Epub 20120923. https://doi.org/10.1016/j.freeradbiomed.2012.09.020. PubMed PMID: 23010495; PMCID: PMC3542420.
82. Lindsay DP, Camara AK, Stowe DF, Lubbe R, Aldakkak M. Differential effects of buffer pH on Ca^{2+} induced ROS emission with inhibited mitochondrial complexes I and III. Front Physiol. 2015;6:58. Epub 20150310. https://doi.org/10.3389/fphys.2015.00058. PubMed PMID: 25805998; PMCID: PMC4354303.
83. Wink DA, Mitchell JB. Chemical biology of nitric oxide: insights into regulatory, cytotoxic, and cytoprotective mechanisms of nitric oxide. Free Radic Biol Med. 1998;25(4-5):434–56. https://doi.org/10.1016/s0891-5849(98)00092-6. PubMed PMID: 9741580.
84. Beckman JS, Koppenol WH. Nitric oxide, superoxide, and peroxynitrite: the good, the bad, and ugly. Am J Phys. 1996;271(5 Pt 1):C1424–37. https://doi.org/10.1152/ajpcell.1996.271.5.C1424. PubMed PMID: 8944624.
85. Pfeiffer S, Mayer B, Hemmens B. Nitric oxide: chemical puzzles posed by a biological messenger. Angew Chem Int Ed Engl. 1999;38(12):1714–31. https://doi.org/10.1002/(SICI)1521-3773(19990614)38:12<1714::AID-ANIE1714>3.0.CO;2-3. PubMed PMID: 29711207.
86. Moncada S, Palmer RM, Higgs EA. Nitric oxide: physiology, pathophysiology, and pharmacology. Pharmacol Rev. 1991;43(2):109–42. PubMed PMID: 1852778.
87. Moncada S, Palmer RM, Higgs EA. Biosynthesis of nitric oxide from L-arginine. A pathway for the regulation of cell function and communication. Biochem Pharmacol. 1989;38(11):1709–15. https://doi.org/10.1016/0006-2952(89)90403-6. PubMed PMID: 2567594.
88. Nathan C, Xie QW. Regulation of biosynthesis of nitric oxide. J Biol Chem. 1994;269(19):13725–8. PubMed PMID: 7514592.
89. Cinelli MA, Do HT, Miley GP, Silverman RB. Inducible nitric oxide synthase: regulation, structure, and inhibition. Med Res Rev. 2020;40(1):158–89. Epub 20190613. https://doi.org/10.1002/med.21599. PubMed PMID: 31192483; PMCID: PMC6908786.
90. Gerdes HJ, Yang M, Heisner JS, Camara AKS, Stowe DF. Modulation of peroxynitrite produced via mitochondrial nitric oxide synthesis during Ca^{2+} and succinate-induced oxidative stress in cardiac isolated mitochondria. Biochim Biophys Acta Bioenerg. 2020;1861(12):148290. Epub 20200820. https://doi.org/10.1016/j.bbabio.2020.148290. PubMed PMID: 32828729; PMCID: PMC7560995.
91. Murad F, Mittal CK, Arnold WP, Katsuki S, Kimura H. Guanylate cyclase: activation by azide, nitro compounds, nitric oxide, and hydroxyl radical and inhibition by hemoglobin and myoglobin. Adv Cyclic Nucleotide Res. 1978;9:145–58. PubMed PMID: 27076.
92. Palmer RM, Ferrige AG, Moncada S. Nitric oxide release accounts for the biological activity of endothelium-derived relaxing factor. Nature. 1987;327(6122):524–6. https://doi.org/10.1038/327524a0. PubMed PMID: 3495737.
93. Ignarro LJ, Buga GM, Wood KS, Byrns RE, Chaudhuri G. Endothelium-derived relaxing factor produced and released from artery and vein is nitric oxide. Proc Natl Acad Sci USA. 1987;84(24):9265–9. https://doi.org/10.1073/pnas.84.24.9265. PubMed PMID: 2827174; PMCID: PMC299734.
94. Furchgott RF. Endothelium-derived relaxing factor: discovery, early studies, and identification as nitric oxide. Biosci Rep. 1999;19(4):235–51. https://doi.org/10.1023/a:1020537506008. PubMed PMID: 10589989.
95. Goodwill AG, Dick GM, Kiel AM, Tune JD. Regulation of coronary blood flow. Compr Physiol. 2017;7(2):321–82. Epub 20170316. https://doi.org/10.1002/cphy.c160016. PubMed PMID: 28333376; PMCID: PMC5966026.
96. Horenberg AL, Houghton AM, Pandey S, Seshadri V, Guilford WH. S-nitrosylation of cytoskeletal proteins. Cytoskeleton (Hoboken). 2019;76(3):243–53. Epub 20190429. https://doi.org/10.1002/cm.21520. PubMed PMID: 30969482.

97. Antunes F, Boveris A, Cadenas E. On the mechanism and biology of cytochrome oxidase inhibition by nitric oxide. Proc Natl Acad Sci USA. 2004;101(48):16774–9. Epub 20041116. https://doi.org/10.1073/pnas.0405368101. PubMed PMID: 15546991; PMCID: PMC534717.
98. Novalija E, Fujita S, Kampine JP, Stowe DF. Sevoflurane mimics ischemic preconditioning effects on coronary flow and nitric oxide release in isolated hearts. Anesthesiology. 1999;91(3):701–12. https://doi.org/10.1097/00000542-199909000-00023. PubMed PMID: 10485782.
99. Fujita S, Roerig DL, Bosnjak ZJ, Stowe DF. Effects of vasodilators and perfusion pressure on coronary flow and simultaneous release of nitric oxide from guinea pig isolated hearts. Cardiovasc Res. 1998;38(3):655–67. https://doi.org/10.1016/s0008-6363(98)00051-0. PubMed PMID: 9747433.
100. Fujita S, Roerig DL, Chung WW, Bosnjak ZJ, Stowe DF. Volatile anesthetics do not alter bradykinin-induced release of nitric oxide or L-citrulline in crystalloid perfused guinea pighearts. Anesthesiology. 1998;89(2):421–33. https://doi.org/10.1097/00000542-199808000-00019. PubMed PMID: 9710401.
101. Stowe DF, Ebert TJ. Neural and endothelial control of the peripheral circulation—implications for anesthesia: Part II, Endothelium-mediated effects in the normal and diseased circulation. J Cardiothorac Vasc Anesth. 1996;10(1):159–71. https://doi.org/10.1016/s1053-0770(96)80191-1. PubMed PMID: 8634382.
102. Habib S, Ali A. Biochemistry of nitric oxide. Indian J Clin Biochem. 2011;26(1):3–17. Epub 20110203. https://doi.org/10.1007/s12291-011-0108-4. PubMed PMID: 22211007; PMCID: PMC3068772.
103. Ahmad R, Hussain A, Ahsan H. Peroxynitrite: cellular pathology and implications in autoimmunity. J Immunoassay Immunochem. 2019;40(2):123–38. Epub 20190307. https://doi.org/10.1080/15321819.2019.1583109. PubMed PMID: 30843753.
104. Halliwell B, Gutteridge JMC. Free radicals in biology and medicine. 5th ed. New York: Oxford University Press; 2015. ISBN 9780191802133. https://doi.org/10.1093/acprof:oso/9780198717478.001.0001.
105. Yang M, Camara AK, Wakim BT, Zhou Y, Gadicherla AK, Kwok WM, Stowe DF. Tyrosine nitration of voltage-dependent anion channels in cardiac ischemia-reperfusion: reduction by peroxynitrite scavenging. Biochim Biophys Acta. 2012;1817(11):2049–59. Epub 20120615. https://doi.org/10.1016/j.bbabio.2012.06.004. PubMed PMID: 22709907; PMCID: PMC3985433.
106. Yang M, Xu Y, Heisner JS, Sun J, Stowe DF, Kwok WM, Camara AKS. Peroxynitrite nitrates adenine nucleotide translocase and voltage-dependent anion channel 1 and alters their interactions and association with hexokinase II in mitochondria. Mitochondrion. 2019;46:380–92. Epub 20181101. https://doi.org/10.1016/j.mito.2018.10.002. PubMed PMID: 30391711; PMCID: PMC6487210.
107. Novalija E, Varadarajan SG, Camara AK, An J, Chen Q, Riess ML, Hogg N, Stowe DF. Anesthetic preconditioning: triggering role of reactive oxygen and nitrogen species in isolated hearts. Am J Physiol Heart Circ Physiol. 2002;283(1):H44–52. https://doi.org/10.1152/ajpheart.01056.2001. PubMed PMID: 12063273.
108. Riess ML, Kevin LG, McCormick J, Jiang MT, Rhodes SS, Stowe DF. Anesthetic preconditioning: the role of free radicals in sevoflurane-induced attenuation of mitochondrial electron transport in guinea pig isolated hearts. Anesth Analg. 2005;100(1):46–53. https://doi.org/10.1213/01.ANE.0000139346.76784.72. PubMed PMID: 15616050.
109. Novalija E, Hogg N, Kevin LG, Camara AK, Stowe DF. Ischemic preconditioning: triggering role of nitric oxide-derived oxidants in isolated hearts. J Cardiovasc Pharmacol. 2003;42(5):593–600. https://doi.org/10.1097/00005344-200311000-00003. PubMed PMID: 14576506.
110. Ghafourifar P, Sen CK. Mitochondrial nitric oxide synthase. Front Biosci. 2007;12:1072–8. Epub 20070101. https://doi.org/10.2741/2127. PubMed PMID: 17127362.

111. Zaobornyj T, Valdez LB. Heart mitochondrial nitric oxide synthase: a strategic enzyme in the regulation of cellular bioenergetics. Vitam Horm. 2014;96:29–58. https://doi.org/10.1016/B978-0-12-800254-4.00003-9. PubMed PMID: 25189383.
112. Valdez LB, Zaobornyj T, Boveris A. Functional activity of mitochondrial nitric oxide synthase. Methods Enzymol. 2005;396:444–55. https://doi.org/10.1016/S0076-6879(05)96037-0. PubMed PMID: 16291252.

Chapter 9
Evolution and Physiology of the Mammalian Heart

Abstract As multicellular organisms evolved and increased in size, increasingly complex respiratory and circulatory systems developed to bring in O_2 and nutrients and to take out carbon dioxide and waste products. From simple gas exchange via ectoderm in sponges to tracheae in insects, water animals developed gills, while land animals developed lungs. These changes were associated with natural selection of bilateral symmetry, segmentation, and ventral-lateral axes, with locomotion supported by either exoskeletons (invertebrates) or endoskeletons (vertebrates). With increasing size and anatomical complexity, a simple vascular tube developed first in three layered animals, triploblasts, from the lateral plate mesoderm. A pulsating pump moves blood through a simple atrium and ventricle toward gills supplied with branchial arteries in fish and tadpoles; in all mammals and birds, there is a right atrial-ventricular pulmonary circulatory system distinctly for the lungs, and a separate left atrial-ventricular circulatory system for all other organs. Genetic data indicate that some genes for heart development are conserved from insects to mammals. This supports the notion that Nature starts with a basic blueprint which is adapted for fitness and survival during evolution. Embryology, in part, lays out the developmental history of these gradual evolutionary changes. Heart rate is modulated up and down by the autonomic nervous system and contractility is enhanced primarily by the sympathetic nervous system with its cardiac receptors and by other neuromodulators. Coronary vascular tone, or flow, is regulated mostly by local vasodilatory and vasoconstrictive factor from the endothelium. An important vasodilatory factor is nitric oxide, which acts on guanylyl cyclase in vascular cells to convert GTP to GMP; this stimulates K^+ channel opening and induces vascular cell membrane hyperpolarization. The O_2 supplied to the heart is utilized in the process of making ATP to drive cellular cation pumps and maintain ion homeostasis, to maintain the proton motive force and coupling efficiency, and to power actin-myosin cross bridge cycling. ATP produced in the matrix is exchanged for ADP in the sarcoplasm via adenine nucleotide translocase and stored as creatine phosphate. The human heart processes approximately 30 kg of ATP per day. The hearts and mitochondria of mammals are adapted to increase cardiac output and mitochondrial O_2 consumption during exercise to over five or six-fold, and in birds to more than twice as much as in mammals.

D. F. Stowe, *Evolution of Bioenergetics from Elements to Life*,
https://doi.org/10.1007/978-3-031-76668-8_9

Keywords Adenine nucleotide translocase · ATP synthase · Atrial ostium · Autonomic nervous system · Bilateral symmetry · Bioelectricity · Bioenergetics · Blood vascular system · Branchial arches · Contraction/relaxation · Cardiac power · Cell Ca^{2+}-ATPase · Cell Na^{+}-K^{+}-ATPase · Cordates · Ductus arteriosus · Ectoderm · Endoskeleton · Exoskeleton · Embryogenesis · Endoderm · Endothelium-derived vasoconstrictors · Endothelium-derived vasodilators · Gastrovascular cavity · Gene protein-coupled receptors · Heart rate variability · Homology · Invertebrates · L-type Ca^{2+} channels · Mitochondrial power · Mesoderm · Metabolic reserve · Negative inotropic drugs · Neuromodulators · Nitric oxide · Nitric oxide synthases · O_2 consumption · Ontogeny · Orthologous genes · Phylogeny · Positive inotropic drugs · Potential gradient · Proton motive force · Ryanodine channels · Sarcoplasmic reticulum · Sinoatrial node · Triploblasts · Vagus nerve · Vasomotor center · Ventral-lateral axes · Ventricular septum · Ventricular stroke work · Vertebrates

9.1 Evolution and Embryology of the Animal Fluid Pump

The movement of O_2 and CO_2 in and out of unicellular organisms during respiration occurs simply by diffusion down their concentration gradients. Multicellular organisms first appeared about 1.5–1.4 Bya. Sponges and jelly fish exchange gases in the watery pores of their ectoderm and gastrovascular cavity, respectively [1]. Insects with high surface to volume ratios exchange gases at the thorax and abdomen via a network of tubes (tracheae) that open and close to the air (spiracles) based on metabolic activity [2]. Three Mya when the O_2 fraction on Earth's surface was elevated (Fig. 1.11) from 21% to 30% (O_2 tension 27 kPa (Sect. 1.3), insects grew up to 10 times larger than today, e.g., a dragonfly could have a 72 cm wingspan [3].

As animals became more complex 600–700 Mya [4], a basic body design was selected with the appearance of a new third layer called mesoderm, the adaption of bilateral symmetry with development of anterior-posterior and ventral-lateral axes in invertebrates (arthropods, cephalopods, molluscs, annelids, and worms) and invertebrate cordates. These animals developed a dorsal nerve with a mesodermal-derived notochord (tunicates, e.g., sea squirt). Some animals developed exoskeletons while others developed bony endoskeletons (all vertebrates). Vertebrate cordates include all bony fish, amphibians, reptiles, mammals, and birds. In addition to these developments, the beginning of a circulatory system was apparent by an anterior to posterior pulsating tube with segmented branchial arches containing arteries that were eventually used for gas exchange [1]. This was necessary because of the time distance limitations of simple diffusion with increasing body size and activity, and the increasing complexity of evolving, segmented organisms.

Early Vascular Tube Figure 9.1 shows a diagram in the evolution of animals that highlights the development of a blood vascular system and a pump to deliver the blood to the body (and later to lungs) [4]. The first evidence of a vascular system may be an ancestor of a three-layered animal (triploblast) from about 600 Mya [4]. The heart and circulatory system develop from the lateral plate mesodermal layer. This tube system carried over into the first vertebrates, fish, which have one atrium, one ventricle, and four bilateral aortic (branchial) arch arteries that supply the gills for gas exchange. In fish embryos, the first two branchial arches and their arteries become modified for head structures [1].

Starting with the same genetic information, the circulatory system of amphibians becomes metamorphosed from the gill system of tadpoles (like fish) into body and lung circulations. During metamorphosis the arteries of the 3rd arches become carotid arteries to the head, the 4th arch arteries supplied the body, and the 6th arch arteries partially supplied the lungs (the 5th arches disappear). With the split of one atrium into two atria as shown in Fig. 9.2 [1], but with an incomplete ventricular septum, blood from the body and lungs are partially separated in amphibians and most reptiles.

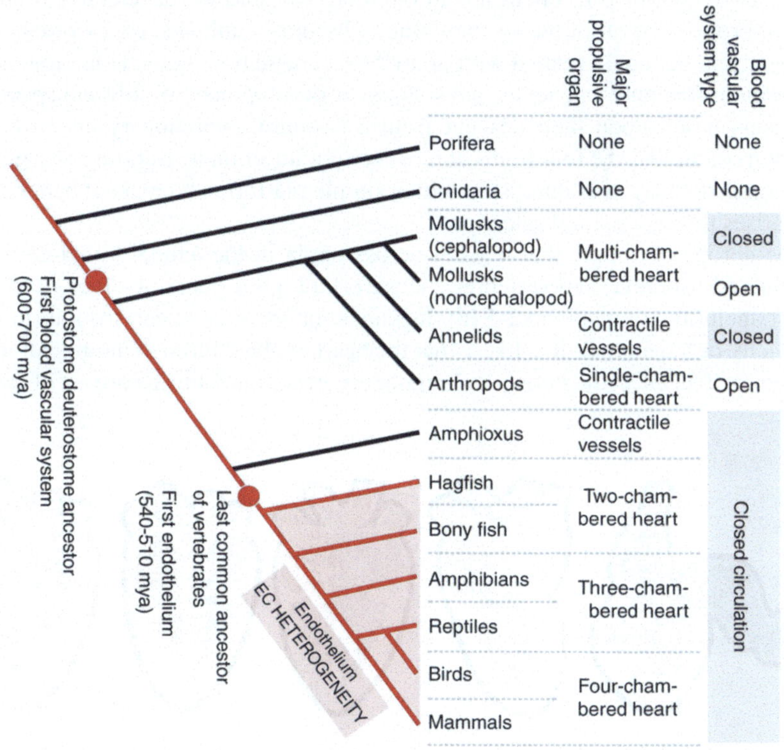

Fig. 9.1 Phylogenetic perspective of the blood vascular system: Blood vascular system types and major propulsive organs are shown for representative extant phyla. The phylogenetic tree is schematic only, and the timelines are not drawn to scale. *EC* endothelial cell, *Mya* million years ago. (Diagram (Fig. 7) and modified legend resourced with permission from copyright owner of Monahan-Early et al. [4])

Reptiles, which came after the amphibians, have the same evolutionary system except that the single ventricle is partially partitioned or completely partitioned into two ventricles (as in crocodiles) [1]. All mammals and birds are warm-blooded and have two-atrial, two ventricular hearts so that the systemic and pulmonary systems are completely separated (after the atrial ostium and ductus arteriosus close after birth in mammals). The left 4th arch artery became the aorta and the right 4th arch artery became the right subclavian artery. Vertebrates (except arctic ice fish) concentrate O_2 onto hemoglobin for transport to cells and tissues. Erythroblasts, when matured, lose all organelles, including the nucleus and mitochondria, to carry out the single task of hemoglobin packaging; so red blood "cells" are not actually cells anymore.

The dictum that ontogeny (embryogenesis) recapitulates phylogeny (evolution) has been mired in controversy for at least 150 years [5]; a newer phrase, "ontogeny produces phylogeny" was later proposed (Fig. 9.3) [5]. Is it possible that the natural selection of traits during embryogenesis also drives evolution? A basic tenet of evolution is that modifications within individual species are gradual and very slow if the species is adaptable to its environment but more variable and accelerated if certain traits improve survival of the species. But a phylum of animals, e.g., Cordata (vertebrates), can be highly related with up to 99% genetic homology. Homology is the similarity of structures, genetics, physiology, or development of different species of organisms based upon their descent from a common evolutionary ancestor. The common example is the four limbs of most vertebrate animals. Building on a genetic blueprint is the way of Nature; start with a simple platform and make it better as the need arises through natural selection.

One apropos example that reaches across phyla in the animal kingdom is that fruit flies (drosophila) and vertebrates share orthologous genes involved in making a heart, including tinman/Nkx2-5 [6]. Its knockout prevents a heart from forming in either a fly or a mammal. It is likely that the heart evolved through modification and expansion of an ancestral network of regulatory genes encoding a core set of cardiac

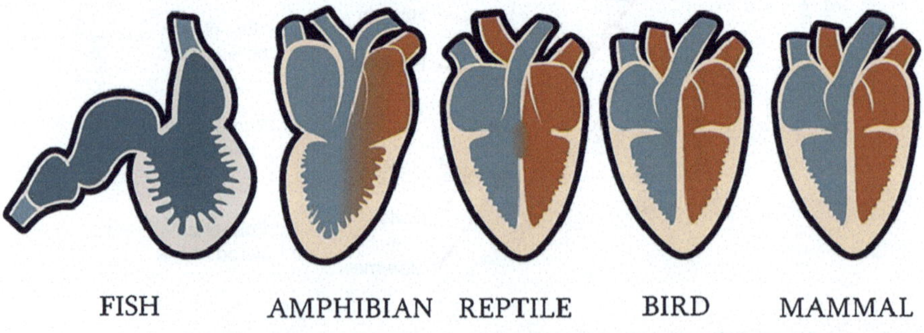

Fig. 9.2 Simplified evolution of the heart shape and chambers: Each vertebrate heart includes a closed circulation; Fish, 2 chambered; amphibian and reptiles (except crocodile), 3 chambered; mammals and birds, 4 chambered. Only the mammals and birds have true parallel circulatory systems-pulmonary and systemic. (Drawings (Fig. 3) and modified legend resourced with permission from copyright owner of Stephenson et al. [1])

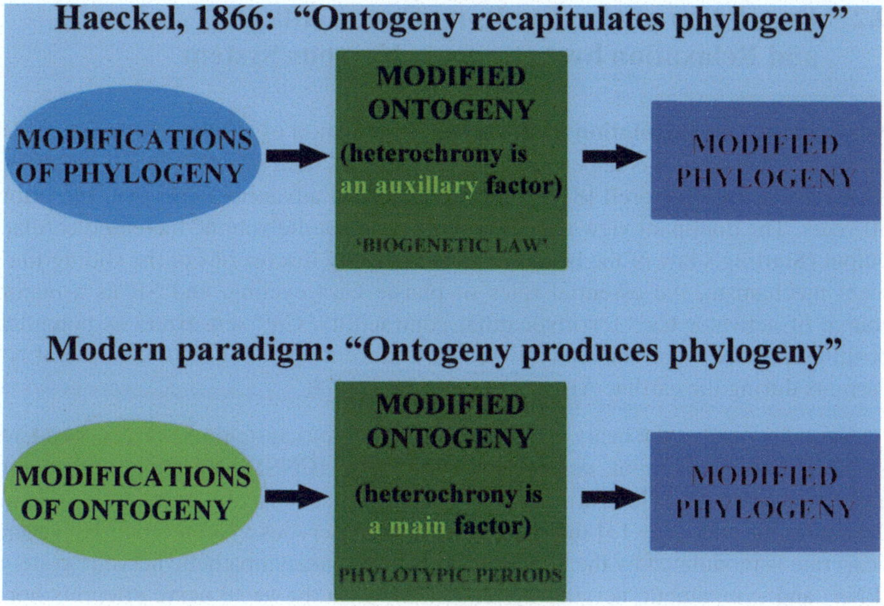

Fig. 9.3 Ontogeny and phylogeny: Scheme illustrating changes in the understanding of the "ontogeny and phylogeny" conceptual framework compared with the original view (text reference [68]) and the later paradigm (text reference [69]). (Scheme (Fig. 1) and modified legend resourced with license (https://creativecommons.org/licenses/BY-NC/4.0/) from Martynov et al. [5] (review))

transcription factors, including NK2, myocyte enhancer factor 2 (MEF2), GATA, Tbx and Hand, virtually all of which are present in all triploblasts [7].

Triploblasts consist of all animals that evolved to have all three primary germ layers; ectoderm, endoderm, and mesoderm, the last of which forms first during the gastrula stage of the embryo. Triploblasts include platyhelminthes, annelids, arthropods, molluscs, echinoderms, chordates, and all vertebrates. Complexity of the heart increased during evolution, likely due to expansion of the number of ancestral regulatory genes. During evolution, there was modification of the timing and pattern of expression of these genes, as well as changes in their regulatory interactions with each other and with other regulatory inputs [4].

Instead of creating a complete new model of a four chambered, valved heart with closed circulation and separation of pulmonary and systemic circulations in large active animals, embryogenesis and evolution both demonstrate that the basic heart tube plan was utilized, but then built upon, expanded, shortened, or altered to respond to evolutionary pressures for adaptions in which the newer models were better suited for the animal to survive (e.g., run) and replicate. Embryology lays out the model of the initial basic plan with ancient genes, as circulatory systems in all animals since fish follow the fish blueprint during early gestation, while newer gene modifications kick in during embryogenesis to reflect changes in the structure of the heart and circulation as useful adaptions selected naturally during evolution.

9.2 Regulation of Cardiac Rate, Contractility, and Relaxation by Autonomic Nervous System

Nervous System Regulation of the Heart Regulation of cardiac rate and muscle contraction/relaxation is dependent on many factors both intrinsic and extrinsic. A review by Katz and Lorell [8] in 2000 summarized advances made over the prior 50 years. The dominant view that increased end-diastolic volume increased cardiac output (Starling's law of the heart) was overtaken by discoveries of the sliding filament mechanism, the essential roles of phasic Ca^{2+} cycling, and SR as a major source of activator Ca^{2+} for myocardial contractility, Ca^{2+} sensitizers as potential positive inotropes, and the slow inward Ca^{2+} current operating with the several K^+ currents during the cardiac AP, as discussed in Sect. 8.

Neural control of the heart by the autonomic nervous system (ANS) is dependent on sensory afferent input, central nervous system (CNS) integration, and motor efferent outflow [9–11]. Heart rate is initiated spontaneously by pacemaker cells in the sinoatrial node [12, 13] that is independent of nervous system input. However, heart rate is modulated by the balance between the parasympathetic nervous system (PNS) and sympathetic nervous system (SNS) with the vagal nerve efferents normally exerting a slowing effect on heart rate at rest and the sympathetic nerve efferents having an effect to increase heart rate as well as contractility/relaxation, i.e., during exercise in which effects of epinephrine (adrenaline) at the β-adrenergic receptor triggers actions further downstream on effectors and pathways [14, 15]. Figure 9.4 [14] is a cartoon depicting how exercise stimulates cardiac rate and contraction/relaxation via L-type Ca^{2+}-induced activation of SR Ca^{2+} release via RyR channels, as well as β-adrenergic receptor activation of adenylyl cyclase (AC), stimulation of PKA, phosphorylation of L-type Ca^{2+} channels, and phosphorylation of the SERCA pump inhibitor phospholamban.

The major integration center for afferent signaling, primarily mediated by the vagus nerve (12th cranial nerve), is the nucleus tractus solitarius (NTS) of the medulla. There, information is received from arterial baro-, chemo-, and mechanoreceptors, and higher brain centers. The NTS has projections to the ventrolateral medulla (VLM), the vasomotor center where vagal and sympathetic projections to the heart reflexively control heart function. Heart rate variability (HRV) is an indication of the normal balance between the PNS and SNS in modulating heart rate [16].

Modulators of Cardiac Activity The Ca^{2+} current through L-type Ca^{2+} channels regulates the inotropic effect of increased heart rate and contributes to the inotropic effect of α and β-adrenergic stimulation [8, 11, 17–25]. Figure 9.5 is a cartoon of the signal transduction systems for β-adrenergic receptor and muscarinic receptor stimulations in a cardiac myocyte [18]. Figure 9.6 shows a depiction of the pathways from stimulation by adrenergic hormones to cardiovascular effects [17]. Figures 8.6 [26, 27] and 8.7 [21] demonstrate, respectively, how perfusion of positive and negative inotropic drugs alters the rate and strength of Ca^{2+} and LVP transients and the

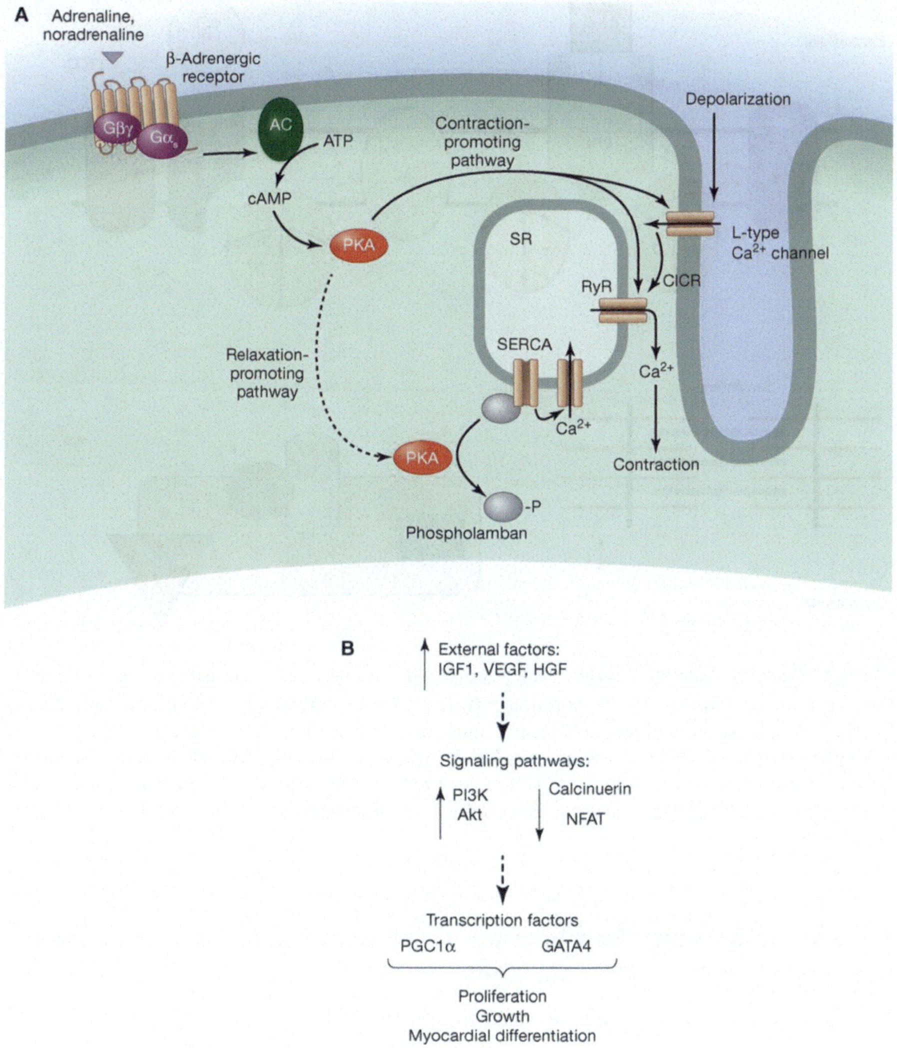

Fig. 9.4 Cardiac muscle contraction and changes with exercise: (**a**) Cardiac muscle contraction can occur as a result of Ca^{2+} entry through L-type Ca^{2+} channels, which activate ryanodine receptor (RyR) channels in the SR. Alternatively, β-adrenergic receptors on the cell membrane trigger activation of adenylyl cyclase (AC), which stimulates PKA. This can promote contraction by phosphorylating RyR and L-type Ca^{2+} channels or relaxation by phosphorylating the SERCA pump inhibitor phospholamban. (**b**) Changes with exercise lead to an activation of the PI3K/Akt pathway, and a down-regulation of NFAT and calcinurin. (Cartoon with scheme (Fig. 4) and modified legend resourced with permission from Cold Spring Harbor Laboratory Press, copyright owner of Kuo and Ehrlich [14] (perspective))

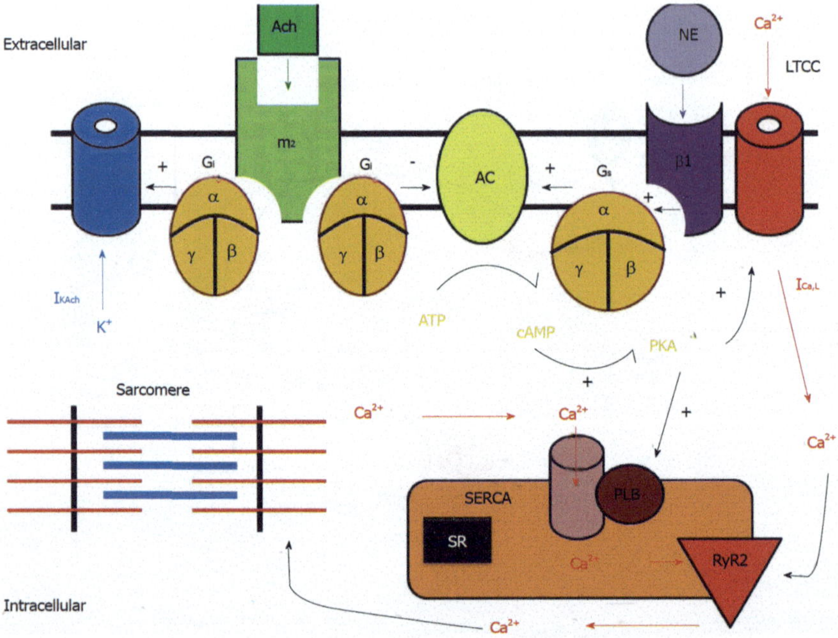

Fig. 9.5 Signal transduction systems for β-adrenergic receptor and muscarinic-receptor stimulations in a cardiac myocyte: *NE* Norepinephrine, *β1* Beta1-adrenergic receptor, G_s Stimulatory G-protein, *Ach* Acetylcholine, *m2* Type-2 muscarinic receptors, G_i Inhibitory G-protein, *AC* Adenylate cyclase, *PKA* Protein kinase A, $I_{Ca}L$ L-type Ca^{2+} channel, *RyR2* Ryanodine receptor 2, *SERCA* Sarcoplasmic reticulum Ca^{2+}-ATPase2a, *PLB* Phospholamban. (Scheme (Fig. 2) and legend resourced with license (https://creativecommons.org/licenses/BY-NC/4.0/) from Gordan et al. [18] (review))

Ca^{2+}-LVP relationship (loop orientation and area) before and after ischemia-reperfusion in isolated guinea pig hearts.

In addition to epinephrine and norepinephrine, there are many other neurotransmitters and neuromodulators that mediate the functions of the ANS and thereby affect cardiac function. These include substance P, neurotensin, acetylcholine, arginine vasopressin (AVP), enkephalins, dopamine, ATP, serotonin (derived from 5-hydroxytrytophan, 5-HT), and others [10, 18, 28, 29]. One grouping of these modulators is the renin-angiotensin-aldosterone system (RAAS). which if chronically activated, results in hemodynamic stress that can lead to exacerbation of heart failure [30]. These compounds exert hormonal control mostly via *G*-protein coupled receptors (GPCR) (Fig. 9.6) [17].

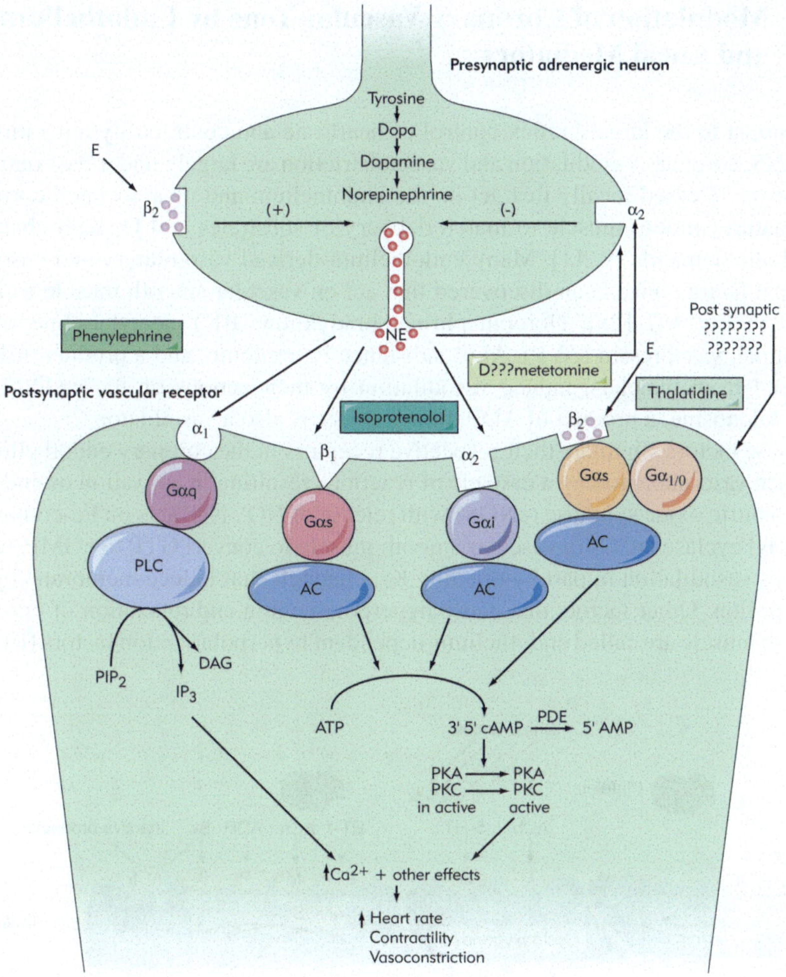

Fig. 9.6 Simplified schema of mechanism of adrenergic hormones to produce cardiovascular effects: The cardiac and vascular effects of noradrenaline (norepinephrine, NE) and adrenalin (epinephrine, E) depend on the pre- and postsynaptic receptors they occupy. Specific guanosine triphosphate-binding proteins (G subtypes) are attached to specific receptor subtypes (α1, β1) to activate or inactivate the effectors phospholipase C (PLC), adenylate cyclase (AC) and others. (+) and (−) refer to relative activation or inhibition. Downstream pathways diverge and converge, so effector pathways are not always well understood. Receptor occupation may activate more than one class of G protein, which in turn activate many kinase pathways (such as protein kinase A (PKA), with different effects. Representative drugs (in blue) block at these receptor sites. *PDE* phosphodiesterase, *PIP_2* phosphatidyl inositol 4,5 bisphosphate, *DAG* diacylglycerol, *IP3* inositol trisphosphate. (Cartoon (Fig. 41.1) and legend resourced with permission from copyright owner of Stowe [17] (book chapter))

9.3 Modulation of Coronary Vascular Tone by Endothelium and Local Mediators

In contrast to the largely reflex control of heart rate and contractility/relaxation by the ANS, coronary vasodilation and vasoconstriction are largely under the control of hormones released locally that act on the endothelium and its associated coronary vasculature smooth muscle to match delivery of substrates and O_2 with changing metabolic demands [9, 11]. Many endothelium-derived vasodilatory and vasoconstrictive factors have been discovered that act on vascular smooth muscle as illustrated in Fig. 9.7 [28]. Norepinephrine, bradykinin (BK), acetylcholine (Ach), histamine, vasopressin (AVP), ADP, substance P, serotonin, and a product of PGI_2, released from platelets, induce vasodilation by their actions on the endothelium [11]. Adenosine, a product of ADP during stress, is also a vasodilator.

These factors stimulate their respective receptors in the coronary endothelium to produce vasorelaxation by a cascade of reactions resulting in activation of endothelial (e) nitric oxide synthase (eNOS) with release of NO$^\bullet$. NO$^\bullet$ acts on Fe-containing guanylyl cyclase (sCG) in vascular smooth muscle to convert GTP to cGMP, which induces vasodilation in part by opening K_{Ca} channels that induce membrane hyperpolarization. Other factors that cause hyperpolarization and relaxation of coronary smooth muscle are called endothelium-dependent hyperpolarization factors (EDHF)

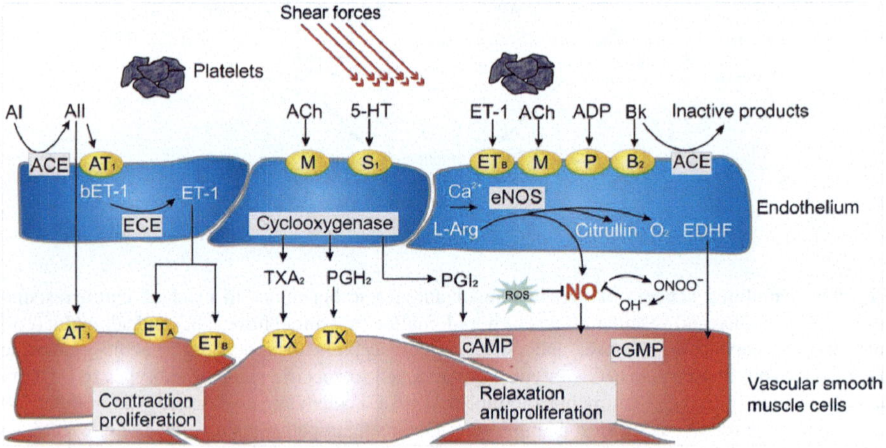

Fig. 9.7 Endothelium-derived vasoactive substances: See text for description of figure. Abbreviations: *ACE* angiotensin-converting enzyme, *Ach* acetylcholine, *AI* angiotensin I, *AII* angiotensin II, *AT1* angiotensin 1 receptor, *Bk* bradykinin, *COX* cyclooxygenase, *ECE* ET-converting enzyme, *EDHF* endothelium derived hyperpolarizing factor, *ETA and ETB* endothelin A and B receptors, *ET-1* endothelin-1, *l-Arg* l-arginine, *M* muscarinic acetylcholine receptor, PGH_2 prostaglandin H_2, *ROS* reactive oxygen species, *S1* serotoninergic receptor, *TX* thromboxane receptor, TXA_2 thromboxane, *5-HT* a.k.a. serotonin. (Original scheme from Gutierrez et al. (article Ref. [415]) with permission. Cartoon (Fig. 18) and legend resourced with permission from copyright owner of Goodwill et al. [28] (review))

and include H_2O_2 [31] and fatty acid epoxides [32]. Nitric oxide and H_2O_2 are discussed in more detail in Sects. 2.4, 6.1, and 8.4.

Coronary vasoconstriction receptors include those for the potent and long-lasting endothelins like ET-1, as well as thromboxane A_2, and 5-HT (serotonin); however, these antagonists of vasodilation may be activated only during pathological situations like endothelial damage. Acetylcholine is a direct vasoconstrictor of vascular smooth muscle, but if the overlying vascular endothelium is intact, the acetylcholine causes the endothelial release of a relaxing factor (EDRF) [33], the major one of which is now known to be $NO^{\bullet}$ as described in Sect. 8.4.

It is clear that the direct vasoconstrictive effect of 5-HT and other direct smooth muscle constrictors are diminished by the presence of endothelium [9, 11]. The endothelial cells also release local hormones that induce vasoconstriction; these include endothelin-1, the most potent [34, 35], angiotension II, and endothelium-dependent constriction factors (EDCFs) that are triggered for release by hypoxia, thrombin, and vascular stretch. Therefore, vasoconstrictor effects are largely designed to shut down blood flow during vascular or tissue damage. The endothelium of cardiac chambers may also exert contractile/relaxant effects as demonstrated by a modulation of peak contractile performance and relaxation of the underlying myocardium [36].

Figure 9.8 illustrates [14] several cascade of events in vascular smooth muscle with several receptor activators noted above that mediate competing vasodilation and vasoconstriction. One is the activation of gene protein-coupled receptors (GPCRs) utilizing Gq -induced activation, where one of the pathways, phospholipase C-β (PLCβ)-PIP_2 (phosphatidylinositol 4,5-bisphosphate), activates inositol 1,4,5-trisphophate (IP_3) receptor-induced release of Ca^{2+} from the endoplasmic reticulum (ER) Ca^{2+} storage site to promote vasoconstriction. The PIP_2 pathway also stimulates DAG (diacylglycerol) to stimulate protein kinase C (PKC), which leads to vasorelaxation. Another pathway is the GC-mediated increase in cGMP leading to PKG (and with PKC) via Gq-induced activation of myosin light chain (MLC) protein that promotes relaxation, or through MLC phosphatase (MLCP) that via Ca^{2+} promotes vasoconstriction.

9.4 O_2 and ATP Consumption in Cardiac Metabolism

All mammals and birds are dependent on the four chambered heart, two lungs, and two separate circulatory system to deliver foodstuffs, hormones, immune factors, and O_2 to mitochondria, and to remove CO_2, toxins and metabolic byproducts. In the mammalian heart, it is estimated that approximately 20–30% of the O_2 delivered during breathing is used to maintain $\Delta\Psi_m$ and attenuate waste of chemical energy as heat, 60–70% is utilized to supply ATP for operation of ATPase pumps that maintain Na^+, K^+, and Ca^{2+} gradients, and for protein synthesis, and 5% is consumed for resting actomyocin ATPase [37]. As ATP-activated pumps, Na^+/K^+-ATPase and Ca^{2+}-ATPase consume about 5% and 30%, respectively, of the ATP generated in the heart

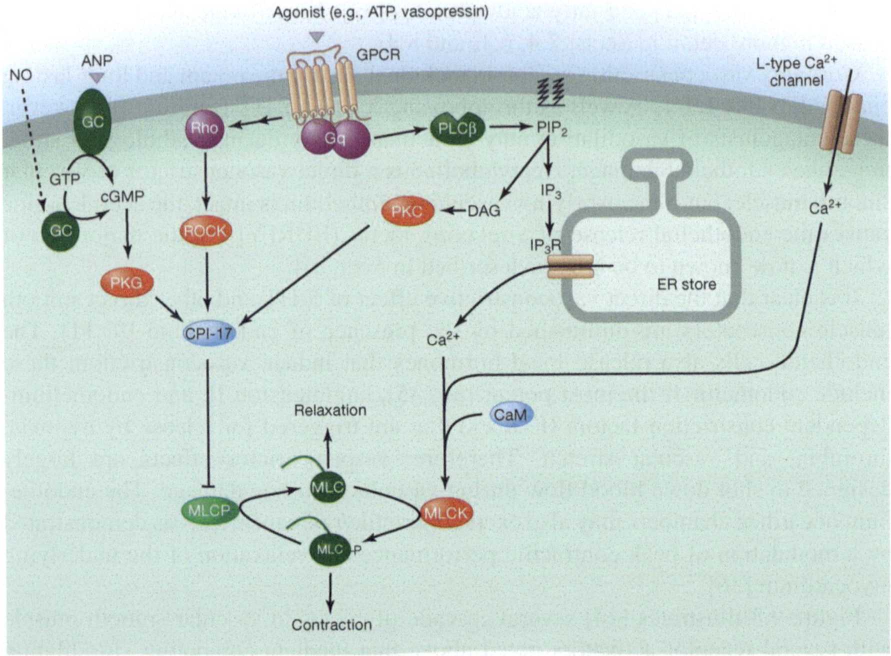

Fig. 9.8 Mechanism of smooth muscle contraction and relaxation: Calcium released by L-type calcium channels or IP_3Rs downstream from G*q* coupled cell-surface receptors causes smooth muscle contraction. It binds to calmodulin (CaM) and the resulting complex stimulates myosin light-chain (MLC) kinase (MLCK). This phosphorylates MLC to promote contraction. A RhoA/ROCK pathway and a diacylglycerol (DAG) pathway contribute to Ca^{2+} sensitization by altering the phosphorylation status of myosin light-chain phosphatase (MLCP). Relaxation is mediated through the cGMP/PKG pathway downstream from nitric oxide (NO•) and agonists such as atrial natriuretic peptide (ANP). (Cartoon with scheme (Fig. 5) and modified legend resourced with permission of Cold Spring Harbor Laboratory Press, copyright owner of Kuo and Ehrlich [14] (perspective))

[38]. These percentages are similar for total cardiac ATP consumption (P/O ratio) except that substrate level phosphorylation also contributes to ATP formation outside of OxPhos.

Cardiac Mitochondrial Power In expressing mitochondrial bioelectricity as physical electricity (V = IR and P = VI (Sects. 4.3 and 6.2), mitochondrial voltage is the *pmf*, as it incorporates the $\Delta\Psi_m$ gradient plus the ΔpH_m gradients, and so represents the total cation potential gradient (battery) across the IMM. Mitochondrial current (I, amperes) can be expressed potentially as electrons (*e*-) transferred/s, moles of H^+ pumped/s, moles of ATP produced/s, or moles of O_2 consumed/s. The latter, O_2 consumption, is the most practical method although it does not account for H^+ and *e*- leaks where the energy is lost as heat in partially uncoupled mitochondria. The number of calculated *e*- transferred/s from foodstuffs directly captured by moles of O_2 consumed/s would represent the true energy cost.

Mitochondrial power (P = VI) can be estimated as *pmf* • O_2 consumption. A hypothetical example, as could be measured in concentrated, rat cardiac isolated mitochondria [39], is −220 mv • 5 nmoles O_2 • mg^{-1} • s^{-1}, or a power of 1100 units • mg^{-1} • s^{-1} at state 2 (no ADP), and −160 mv • 25 nmoles O_2 • mg^{-1} • s^{-1}, or a power of 4000 units • mg^{-1} • s^{-1} during state 3 (ATP synthesis). In the whole heart, cardiac work is 95% dependent on OxPhos and the heart is the largest consumer of O_2 (and producer of CO_2) of all organs per weight. In humans, non-exercising myocardial O_2 consumption (coronary flow • arterial-venous O_2 content difference) is approximately 8–14 L O_2 • kg^{-1} • min^{-1} [20]. This is equivalent to 6–11 nmoles O_2 • mg^{-1} • s^{-1}. This estimate reflects the total mass of the heart rather than concentrated mitochondria and so is about half the value for isolated mitochondria during state 3 respiration.

Cardiac Fluid Power Because fluid power is equivalent to pressure times flow, a more common and practical estimate of assessing heart power than using O_2 consumption is cardiac power output (CPO); CPO is estimated as ((mean arterial pressure [MAP]—right atrial pressure [RAP]) • cardiac output)/451 [40]. Heart chamber pressures can be measured with pressure transducers and cardiac volume and output with echocardiography. Using values of 93.3 mmHg MAP, 3 mmHg RAP, and 5 L/min (CO), this gives an average, normal power of 1 W or 1 joule/s. Peak CPO increases about four to five times during peak exercise [41]. A power calculation of <0.6 W is considered a hemodynamic correlate of mortality in cardiogenic shock [42]. Power is equal to work done per unit time and cardiac work can be estimated as ventricular stroke work (VSW) • heart rate, where VSW is estimated from the LVP • volume loop area [40].

Cardiac O_2 Supply for O_2 Demand The many determinants of O_2 supply to demand are diagramed in Fig. 9.9 [28]. These include determinants of coronary flow and O_2 carrying capacity, as well as heart rate, contractility, left ventricular blood ejection fraction, and work demands. The supply of O_2 to match O_2 needs [19, 28] is dependent on adequate hemoglobin (4 O_2 molecules per hemoglobin molecule) in the blood but it is met largely by adjusting coronary blood flow rather than blood O_2 content. However, the differential roles of perfusion pressure, tissue pressure, myogenic, endothelial, neurohormonal, and metabolic factors during changing cardiac demand and stress are complex and not completely measurable or understood [18–20, 28]. The heart preferentially relies on FFA (70–90%) for transduction to bioenergy but also utilizes glucose, pyruvate, lactate [43], ketone bodies, amino acids, and other substrates depending on availability and energetic needs as diagramed in Fig. 9.10 [44], and in diseases, like diabetes [45].

As cardiac mitochondria adeptly respond to changing needs (Sects. 7.3 and 7.4), the heart overall is capable of remodeling the metabolic pathways in chronic pathophysiological conditions, which results in modulations of myocardial energetics and contractile function [46]. For example in the fed vs. fasting state, the utilization of glucose increases and FA oxidation decreases [47]. Another example is that exposure of the heart to moderate chronic hypoxia triggers O_2-sensitive transcriptional

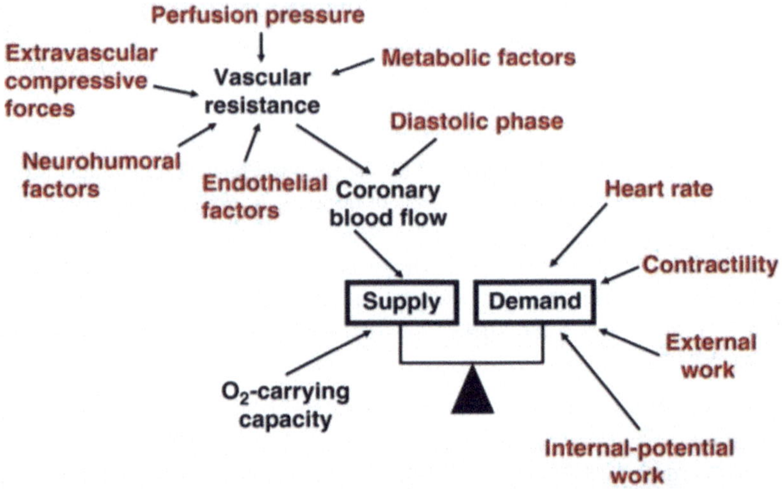

Fig. 9.9 Various determinants of myocardial oxygen supply and demand. (Adapted with permission, from Ardehali and Ports (see Ref. [16]) and reported by Tune (text Ref. [918]). Scheme (figure) and legend resourced with permission from copyright owner of Goodwill et al. [28] (review))

modulators that induce a switch to increased carbohydrate metabolism and to enhanced mitochondrial respiratory capacity in a bid to increase the efficiency of mitochondrial function [48].

Cardiomyocyte Hydrolysis of ATP ATP hydrolysis (Figure 1.17) [49] provides the energy for myosin to go through actin-myosin cycling [50, 51]. The released P_i causes release of actin, changes its conformation, and induces its contraction by myosin; this moves actin along myosin about 10 nm with each cycle. For the cycling to continue, ATP is indirectly required during each relaxation cycle to break the bond between the myosin heads and the actin filament. This is because Ca^{2+} within the myofilaments must be removed by Ca^{2+}-ATPase; without ATP the heart would remain in contracture. Thus, ATP is necessary for both contraction and relaxation. The energy pool for cardiomyocytes consists of ATP and creatine phosphate (CrP) as a source of energy and utilizes the CrP shuttle that acts as a circuit between the mitochondrial matrix and sarcoplasm.

Creatine Pool During decreased energy demand, ATP can donate its terminal P_i group to the creatine pool to make CrP, and with increased demand, creatine kinase can transfer its P_i group to ADP to make ATP much faster than via substrate level and oxidative phosphorylation [52]. Creatine kinase may also modulate respiration [53]. With the high rate of ATP utilization, but low energy stores, ATP must be continuously produced [54]. The entire ATP content of the heart must be hydrolyzed and resynthesized several times per minute [55]. For the human adult heart alone, this amounts to approximately a 30 kg ATP turnover every 24 h [56]. As noted

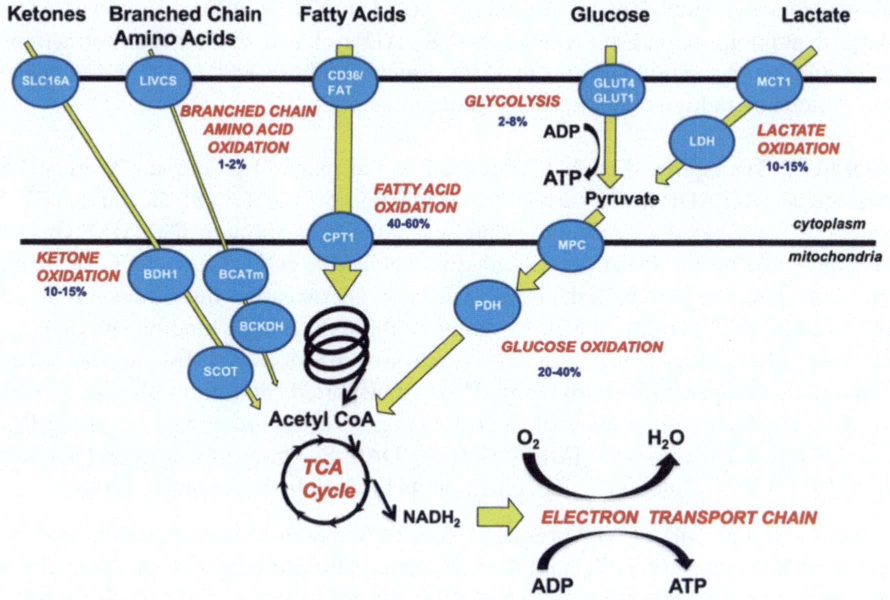

Fig. 9.10 Overview of different substrates contributing to energy metabolism in the normal heart: Glucose enters the cell via GLUT1 or GLUT4 (glucose transporter 1,4); it then undergoes glycolysis to produce pyruvate. Lactate is taken up by cardiomyocytes via MCT (monocarboxylic acid transporter) and is converted to pyruvate by LDH (lactate dehydrogenase). The pyruvate from glucose and lactate is transported into mitochondria via the mitochondrial pyruvate carrier (MPC) and is converted to acetyl CoA by PDH (pyruvate dehydrogenase). Fatty acids are transported into the cardiomyocyte, partly via CD36 and FATP-1 (FA transport protein-1), where they are esterified to fatty acyl-CoA. The acyl group is transferred to carnitine by CPT-1 (carnitine palmitoyl transferase) and transported into mitochondria where CPT-2 converts it back to fatty acyl-CoA, which can then undergo β-oxidation producing acetyl CoA. Ketones (i.e., β hydroxybutyrate) are transported into the cell via SLC16A1 (monocarboxylate transporter 1) where BDH1 (β-hydroxybutyrate dehydrogenase 1) catalyzes the oxidation of β-hydroxybutyrate (βOHB) to acetoacetate (AcAc). AcAc is then activated by SCOT (succinyl-CoA:3 oxoacid-CoA transferase) to acetoacetyl-CoA (AcAc-CoA), which undergoes a thiolysis reaction producing acetyl CoA. Branched chain amino acids (BCAAs) are transported into the cell by the branched chain amino acid cation symporter family (LIVCS). In mitochondria, BCAAs are converted to ketoacids by mitochondrial branched chain aminotransferase (BCATm). Acetyl CoA and succinyl CoA are subsequently formed from BCKDH (branched chain alpha-keto acid dehydrogenase). Acetyl CoA generated by fatty acid β-oxidation, glucose oxidation, ketone oxidation, and BCAA oxidation enter the TCA cycle, generating $FADH_2$ and NADH, which then enter the ETS, consuming O_2 to generate ATP. Numbers in blue represent the contribution of the individual pathways to overall ATP production. (Abridged scheme (Fig. 1) and modified legend resourced with permission from copyright owner of Lopaschuk et al. [44] (review))

above, approximately 40% of the energy stored in ATP is utilized to power ATP-dependent pumps (e.g., Ca^{2+}-ATPase, Na^+/K^+-ATPase), and ATP for the contraction/relaxation mechanism; the rest is used for protein synthesis and repair, and to maintain efficiency (a high coupling/uncoupling ratio).

ADP/ATP Transport The ATP produced in the matrix by complex V must be exchanged with ADP in the sarcoplasm (Figs. 6.9, 6.11 and 6.18). Because ADP^{3-} and ATP^{4-} are negatively charged and so cannot pass through the IMM, this is accomplished by the IMM protein adenine nucleotide ANT, a.k.a., ACC [57–61], the most abundant protein in the IMM. This monomeric carrier must transport 30 kg of free ADP/ATP per day in a 70 kg mammal through a single binding site on each enzyme, alternating from one side (matrix-open) of the IMM to the other (sarcoplasmic-open) [58, 56]. The P_i is transported first through the OMM voltage-dependent anion channel (VDAC) (Fig. 6.18) and then with H^+ at the P_iC carrier (Fig. 6.18) across the IMM [60, 62]. VDAC also transports ions and nucleotides [63]. VDAC function is dependent on its binding by hexokinase II [64].

For each molecule of ATP produced, four to five protons must enter the matrix; this tends to reduce the $\Delta\Psi_m$; and thus also Δp_m. This stimulates respiration which enhances pumping of protons to restore $\Delta\Psi_m$ [65, 66]. Since the 1:1 transfer of ADP to ATP is electrogenic, the added negative charge will tend to decrease $\Delta\Psi_m$ [67]. An increase in cardiac ATP consumption, e.g., due to enhanced cardiac rate (major factor) and contractility/relaxation [4], causes the ATP/ADP ratio to decreases. This will draw on more matrix ATP via ANT for the sarcoplasm and supply more ADP to the matrix. This is energy consuming and results in ATP consumption that is not directly utilized for cell metabolism. During maximal rest with lower cardiac O_2 and ATP consumption [37], $\Delta\Psi_m$ becomes higher, extra-matrix ATP/ADP higher, and matrix ATP/ADP lower.

Metabolic Reserve It is evident that by 200–150 Mya, the mammalian heart and the avian heart and their mitochondria had evolved with selective adaptions that allowed the heart to be quite capable of responding rapidly to changing bioenergetic demands; this was especially useful in animals that require high bursts of metabolism for survival. The evolutionary breakthroughs of mitochondrial endosymbiosis, and the compartmentalization of specific reactions in other nascent organelles, spurred development of multicellular lifeforms with templates for specialized cells like skeletal muscle and cardiac muscle cells with supporting circulatory systems and lungs. These cells arranged in cardiac and muscle tissues are capable of rapidly increasing and deceasing O_2 consumption, ATP production, contraction/relaxation, and cardiac output over five to six-fold from rest to maximal exercise. This is best exhibited in the activities of mammals and birds today with their high and quite efficient bioenergetic capabilities. Indeed, most birds have a much higher energy reserve and endurance than mammals [68].

References

1. Stephenson A, Adams JW, Vaccarezza M. The vertebrate heart: an evolutionary perspective. J Anat. 2017;231(6):787–97. Epub 20170914. https://doi.org/10.1111/joa.12687. PubMed PMID: 28905992; PMCID: PMC5696137.
2. Hetz SK, Bradley TJ. Insects breathe discontinuously to avoid oxygen toxicity. Nature. 2005;433(7025):516–9. https://doi.org/10.1038/nature03106. PubMed PMID: 15690040.
3. Harrison J, Kaiser A, VandenBrookes JM. Atmospheric oxygen level and the evolution of insect body size. Proc R Soc. 2010;277:1937–46. https://doi.org/10.1098/rspb.2010.0001.
4. Monahan-Earley R, Dvorak AM, Aird WC. Evolutionary origins of the blood vascular system and endothelium. J Thromb Haemost. 2013;11 Suppl 1(Suppl 1):46–66. https://doi.org/10.1111/jth.12253. PubMed PMID: 23809110; PMCID: PMC5378490.
5. Martynov A, Lundin K, Korshunova T. Ontogeny, phylotypic periods, paedomorphosis and ontogenetic systemics. Front Ecol Evol. 2020;10 https://doi.org/10.3389/fevo.2022.806414.
6. Medioni C, Senatore S, Salmand PA, Lalevee N, Perrin L, Semeriva M. The fabulous destiny of the Drosophila heart. Curr Opin Genet Dev. 2009;19(5):518–25. Epub 20090828. https://doi.org/10.1016/j.gde.2009.07.004. PubMed PMID: 19717296.
7. Olson EN. Gene regulatory networks in the evolution and development of the heart. Science. 2006;313(5795):1922–7. https://doi.org/10.1126/science.1132292. PubMed PMID: 17008524; PMCID: PMC4459601.
8. Katz AM, Lorell BH. Regulation of cardiac contraction and relaxation. Circulation. 2000;102(20 Suppl 4):IV69–74. https://doi.org/10.1161/01.cir.102.suppl_4.iv-69. PubMed PMID: 11080134.
9. Ebert T, Stowe DF. Peripheral circulation: recent insights into autonomic nervous control and endothelial factors relevant to cardiovascular disease and anesthesia. Curr Opin Anesthesiol. 1991;4:3–11. https://doi.org/10.1097/00001503-199102000-00002.
10. Ebert T, Stowe DF. Neural and endothelial control of the peripheral circulation—implications for anesthesia: Part I. Neural control of the peripheral vasculature. J Cardiothorac Vasc Anesth. 1996;10(1):147–58. https://doi.org/10.1016/s1053-0770(96)80190-x. PubMed PMID: 8634380.
11. Stowe DF, Ebert TJ. Neural and endothelial control of the peripheral circulation—implications for anesthesia: Part II, Endothelium-mediated effects in the normal and diseased circulation. J Cardiothorac Vasc Anesth. 1996;10(1):159–71. https://doi.org/10.1016/s1053-0770(96)80191-1. PubMed PMID: 8634382.
12. Stowe DF, Dujic Z, Bosnjak ZJ, Kalbfleisch JH, Kampine JP. Volatile anesthetics attenuate sympathomimetic actions on the guinea pig SA node. Anesthesiology. 1988;68(6):887–94. https://doi.org/10.1097/00000542-198806000-00009. PubMed PMID: 2897808.
13. MacDonald EA, Madl J, Greiner J, Ramadan AF, Wells SM, Torrente AG, Kohl P, Rog-Zielinska EA, Quinn TA. Sinoatrial node structure, mechanics, electrophysiology and the chronotropic response to stretch in rabbit and mouse. Front Physiol. 2020;11:809. Epub 20200722. https://doi.org/10.3389/fphys.2020.00809. PubMed PMID: 32774307; PMCID: PMC7388775.
14. Kuo IY, Ehrlich BE. Signaling in muscle contraction. Cold Spring Harb Perspect Biol. 2015;7(2):a006023. Epub 20150202. https://doi.org/10.1101/cshperspect.a006023. PubMed PMID: 25646377; PMCID: PMC4315934.
15. La Rovere MT, Porta A, Schwartz PJ. Autonomic control of the heart and its clinical impact. A personal perspective. Front Physiol. 2020;11:582. Epub 20200612. https://doi.org/10.3389/fphys.2020.00582. PubMed PMID: 32670079; PMCID: PMC7328903.
16. Shaffer F, Ginsberg JP. An overview of heart rate variability metrics and norms. Front Public Health. 2017;5:258. Epub 20170928. https://doi.org/10.3389/fpubh.2017.00258. PubMed PMID: 29034226; PMCID: PMC5624990.
17. Stowe DF. Cardiovascular pharmacology (Chapter 41, pp 499–509). In: Hemmings HJ, Hopkins PM, editor. Foundations of anaesthesia: basic sciences for clinical practice. 2nd ed. Elsevier Mosby Limited; 2006. p. 1–903. ISBN 0 3230 3707 0.

18. Gordan R, Gwathmey JK, Xie LH. Autonomic and endocrine control of cardiovascular function. World J Cardiol. 2015;7(4):204–14. https://doi.org/10.4330/wjc.v7.i4.204. PubMed PMID: 25914789; PMCID: PMC4404375.
19. Hoffman JI, Buckberg GD. The myocardial oxygen supply:demand index revisited. J Am Heart Assoc. 2014;3(1):e000285. Epub 20140121. https://doi.org/10.1161/JAHA.113.000285. PubMed PMID: 24449802; PMCID: PMC3959699.
20. Pagel P, Stowe DF. Cardiovascular anatomy and physiology. In: Powlovich L, Nemergut EC, Collins SR, editors. Barash's clinical anesthesia. Philadelphia: Lippincott Williams & Wilkins; 2017. p. 277–300. ISBN 13. 978-1496337009. https://doi.org/10.1213/ANE.0000000000003548.
21. Camara AK, Chen Q, Rhodes SS, Riess ML, Stowe DF. Negative inotropic drugs alter indexes of cytosolic [Ca^{2+}]-left ventricular pressure relationships after ischemia. Am J Physiol Heart Circ Physiol. 2004;287(2):H667–80. Epub 20040401. https://doi.org/10.1152/ajpheart.01142.2003. PubMed PMID: 15059780.
22. Chen Q, Camara AK, Rhodes SS, Riess ML, Novalija E, Stowe DF. Cardiotonic drugs differentially alter cytosolic [Ca^{2+}] to left ventricular relationships before and after ischemia in isolated guinea pig hearts. Cardiovasc Res. 2003;59(4):912–25. https://doi.org/10.1016/s0008-6363(03)00524-8. PubMed PMID: 14553831.
23. Gash M, Kandle PF, Murray IV, Varacallo M. Physiology, muscle contraction. Treasure Island: StatPearls Publishing; 2023. Available from: https://www.ncbi.nlm.nih.gov/books/NBK537140/.
24. Griffiths EJ. Mitochondrial calcium transport in the heart: physiological and pathological roles. J Mol Cell Cardiol. 2009;46(6):789–803. Epub 20090312. https://doi.org/10.1016/j.yjmcc.2009.03.001. PubMed PMID: 19285504.
25. Rhodes SS, Camara AK, Aldakkak M, Heisner JS, Stowe DF. Stretch-induced increase in cardiac contractility is independent of myocyte Ca^{2+} while block of stretch channels by streptomycin improves contractility after ischemic stunning. Physiol Rep. 2015;3(8) https://doi.org/10.14814/phy2.12486. PubMed PMID: 26290532; PMCID: PMC4562572.
26. Rhodes SS, Camara AK, Ropella KM, Audi SH, Riess ML, Pagel PS, Stowe DF. Ischemia reperfusion dysfunction changes model-estimated kinetics of myofilament interaction due to inotropic drugs in isolated hearts. Biomed Eng Online. 2006;5:16. Epub 20060302. https://doi.org/10.1186/1475-925X-5-16. PubMed PMID: 16512898; PMCID: PMC1431537.
27. Rhodes SS, Ropella KM, Camara AK, Chen Q, Riess ML, Pagel PS, Stowe DF. Ischemia-reperfusion injury changes the dynamics of Ca^{2+}-contraction coupling due to inotropic drugs in isolated hearts. J Appl Physiol (1985). 2006;100(3):940–50. Epub 20051110. https://doi.org/10.1152/japplphysiol.00285.2005. PubMed PMID: 16282437.
28. Goodwill AG, Dick GM, Kiel AM, Tune JD. Regulation of coronary blood flow. Compr Physiol. 2017;7(2):321–82. Epub 20170316. https://doi.org/10.1002/cphy.c160016. PubMed PMID: 28333376; PMCID: PMC5966026.
29. Stowe DF, Ebert TJ. Sympathomimetic and sympatholytic drugs. In: Evers A, Maze M, Kharasch E, editors. Anesthetic pharmacology: basic principles and clinical practice. Cambridge: Cambridge University Press; 2011. p. 648–65. https://doi.org/10.1017/CBO9780511781933.041.
30. Hartupee J, Mann DL. Neurohormonal activation in heart failure with reduced ejection fraction. Nat Rev Cardiol. 2017;14(1):30–8. Epub 20161006. https://doi.org/10.1038/nrcardio.2016.163. PubMed PMID: 27708278; PMCID: PMC5286912.
31. Zhang DX, Borbouse L, Gebremedhin D, Mendoza SA, Zinkevich NS, Li R, Gutterman DD. H_2O_2-induced dilation in human coronary arterioles: role of protein kinase G dimerization and large-conductance Ca^{2+}-activated K^+ channel activation. Circ Res. 2012;110(3):471–80. Epub 20111208. https://doi.org/10.1161/CIRCRESAHA.111.258871. PubMed PMID: 22158710; PMCID: PMC3272100.

32. Imig JD. Epoxides and soluble epoxide hydrolase in cardiovascular physiology. Physiol Rev. 2012;92(1):101–30. https://doi.org/10.1152/physrev.00021.2011. PubMed PMID: 22298653; PMCID: PMC3613253.
33. Furchgott RF, Zawadzki JV. The obligatory role of endothelial cells in the relaxation of arterial smooth muscle by acetylcholine. Nature. 1980;288(5789):373–6. https://doi.org/10.1038/288373a0. PubMed PMID: 6253831.
34. Stowe DF, O'Brien WC, Chang D, Knop CS, Kampine JP. Reversal of endothelin-induced vasoconstriction by endothelium-dependent and -independent vasodilators in isolated hearts and vascular rings. J Cardiovasc Pharmacol. 1997;29(6):747–54. https://doi.org/10.1097/00005344-199706000-00007. PubMed PMID: 9234655.
35. Haryono A, Ramadhiani R, Ryanto GRT, Emoto N. Endothelin and the cardiovascular system: the long journey and where we are going. Biology (Basel). 2022;11(5). Epub 20220516. https://doi.org/10.3390/biology11050759. PubMed PMID: 35625487; PMCID: PMC9138590.
36. Brutsaert DL, Meulemans AL, Sipido KR, Sys SU. Effects of damaging the endocardial surface on the mechanical performance of isolated cardiac muscle. Circ Res. 1988;62(2):358–66. https://doi.org/10.1161/01.res.62.2.358. PubMed PMID: 3338120.
37. Gibbs CL, Loiselle DS. Cardiac basal metabolism. Jpn J Physiol. 2001;51(4):399–426. https://doi.org/10.2170/jjphysiol.51.399. PubMed PMID: 11564278.
38. Clausen T, Van Hardeveld C, Everts ME. Significance of cation transport in control of energy metabolism and thermogenesis. Physiol Rev. 1991;71(3):733–74. https://doi.org/10.1152/physrev.1991.71.3.733. PubMed PMID: 2057526.
39. Agarwal B, Dash RK, Stowe DF, Bosnjak ZJ, Camara AK. Isoflurane modulates cardiac mitochondrial bioenergetics by selectively attenuating respiratory complexes. Biochim Biophys Acta. 2014;1837(3):354–65. https://doi.org/10.1016/j.bbabio.2013.11.006. Epub 20131217. PubMed PMID: 24355434; PMCID: PMC4084852.
40. Lim HS. Cardiac power output revisited. Circ Heart Fail. 2020;13(10):e007393. https://doi.org/10.1161/CIRCHEARTFAILURE.120.007393. Epub 20200930. PubMed PMID: 32993372.
41. Bromley PD, Hodges LD, Brodie DA. Physiological range of peak cardiac power output in healthy adults. Clin Physiol Funct Imaging. 2006;26(4):240–6. https://doi.org/10.1111/j.1475-097X.2006.00678.x. PubMed PMID: 16836698.
42. Fincke R, Hochman JS, Lowe AM, Menon V, Slater JN, Webb JG, LeJemtel TH, Cotter G, Investigators S. Cardiac power is the strongest hemodynamic correlate of mortality in cardiogenic shock: a report from the SHOCK trial registry. J Am Coll Cardiol. 2004;44(2):340–8. https://doi.org/10.1016/j.jacc.2004.03.060. PubMed PMID: 15261929.
43. Dong S, Qian L, Cheng Z, Chen C, Wang K, Hu S, Zhang X, Wu T. Lactate and myocardiac energy metabolism. Front Physiol. 2021;12:715081. https://doi.org/10.3389/fphys.2021.715081. Epub 20210817. PubMed PMID: 34483967; PMCID: PMC8415870
44. Lopaschuk GD, Karwi QG, Tian R, Wende AR, Abel ED. Cardiac energy metabolism in heart failure. Circ Res. 2021;128(10):1487–513. https://doi.org/10.1161/CIRCRESAHA.121.318241. Epub 20210513. PubMed PMID: 33983836; PMCID: PMC8136750.
45. Aon MA, Tocchetti CG, Bhatt N, Paolocci N, Cortassa S. Protective mechanisms of mitochondria and heart function in diabetes. Antioxid Redox Signal. 2015;22(17):1563–86. https://doi.org/10.1089/ars.2014.6123. Epub 20150331. PubMed PMID: 25674814; PMCID: PMC4449630.
46. Kolwicz SC Jr, Purohit S, Tian R. Cardiac metabolism and its interactions with contraction, growth, and survival of cardiomyocytes. Circ Res. 2013;113(5):603–16. https://doi.org/10.1161/CIRCRESAHA.113.302095. PubMed PMID: 23948585; PMCID: PMC3845521.
47. Iso T, Kurabayashi M. Fatty acid uptake by the heart during fasting. In: Preedy VP, editor. Handbook of famine, starvation, and nutrient deprivation. Cham: Springer; 2019. p. 1179–97. https://doi.org/10.1007/978-3-319-55387-0_13.

48. Essop MF. Cardiac metabolic adaptations in response to chronic hypoxia. J Physiol. 2007;584(Pt 3):715–26. https://doi.org/10.1113/jphysiol.2007.143511. Epub 20070830. PubMed PMID: 17761770; PMCID: PMC2276994.
49. Kolodiazhnyi OI. Phosphorus compounds of natural origin: prebiotic, stereochemistry, application. Symmetry. 2021;13:889. https://doi.org/10.3390/sym13050889.
50. Jafri MS, Dudycha SJ, O'Rourke B. Cardiac energy metabolism: models of cellular respiration. Annu Rev Biomed Eng. 2001;3:57–81. https://doi.org/10.1146/annurev.bioeng.3.1.57. PubMed PMID: 11447057.
51. Krans J. The sliding filament theory of muscle contraction. In: Cote G, De Tullio M, editors. Cell origins and metabolism: nature education; 2010. p. 66.
52. Guimaraes-Ferreira L. Role of the phosphocreatine system on energetic homeostasis in skeletal and cardiac muscles. Einstein (São Paulo). 2014;12(1):126–31. https://doi.org/10.1590/s1679-45082014rb2741. PubMed PMID: 24728259; PMCID: PMC4898252.
53. Kay L, Nicolay K, Wieringa B, Saks V, Wallimann T. Direct evidence for the control of mitochondrial respiration by mitochondrial creatine kinase in oxidative muscle cells in situ. J Biol Chem. 2000;275(10):6937–44. https://doi.org/10.1074/jbc.275.10.6937. PubMed PMID: 10702255.
54. Doenst T, Nguyen TD, Abel ED. Cardiac metabolism in heart failure: implications beyond ATP production. Circ Res. 2013;113(6):709–24. https://doi.org/10.1161/CIRCRESAHA.113.300376. PubMed PMID: 23989714; PMCID: PMC3896379.
55. Wu F, Zhang EY, Zhang J, Bache RJ, Beard DA. Phosphate metabolite concentrations and ATP hydrolysis potential in normal and ischaemic hearts. J Physiol. 2008;586(17):4193–208. https://doi.org/10.1113/jphysiol.2008.154732. Epub 20080710. PubMed PMID: 18617566; PMCID: PMC2652194.
56. Rosca MG, Tandler B, Hoppel CL. Mitochondria in cardiac hypertrophy and heart failure. J Mol Cell Cardiol. 2013;55:31–41. https://doi.org/10.1016/j.yjmcc.2012.09.002. Epub 20120913. PubMed PMID: 22982369; PMCID: PMC3805050.
57. Bertholet AM, Chouchani ET, Kazak L, Angelin A, Fedorenko A, Long JZ, Vidoni S, Garrity R, Cho J, Terada N, Wallace DC, Spiegelman BM, Kirichok Y. H^+ transport is an integral function of the mitochondrial ADP/ATP carrier. Nature. 2019;571(7766):515–20. https://doi.org/10.1038/s41586-019-1400-3. Epub 20190724. PubMed PMID: 31341297; PMCID: PMC6662629.
58. Kunji ERS, Ruprecht JJ. The mitochondrial ADP/ATP carrier exists and functions as a monomer. Biochem Soc Trans. 2020;48(4):1419–32. https://doi.org/10.1042/BST20190933. PubMed PMID: 32725219; PMCID: PMC7458400.
59. Klingenberg M. The ADP and ATP transport in mitochondria and its carrier. Biochim Biophys Acta. 2008;1778(10):1978–2021. https://doi.org/10.1016/j.bbamem.2008.04.011. Epub 20080502. PubMed PMID: 18510943.
60. Bround MJ, Bers DM, Molkentin JD. A 20/20 view of ANT function in mitochondrial biology and necrotic cell death. J Mol Cell Cardiol. 2020;144:A3–A13. https://doi.org/10.1016/j.yjmcc.2020.05.012. Epub 20200523. PubMed PMID: 32454061; PMCID: PMC7483807.
61. Bonora M, Patergnani S, Rimessi A, De Marchi E, Suski JM, Bononi A, Giorgi C, Marchi S, Missiroli S, Poletti F, Wieckowski MR, Pinton P. ATP synthesis and storage. Purinergic Signal. 2012;8(3):343–57. https://doi.org/10.1007/s11302-012-9305-8. Epub 20120412. PubMed PMID: 22528680; PMCID: PMC3360099.
62. Rochette L, Meloux A, Zeller M, Malka G, Cottin Y, Vergely C. Mitochondrial SLC25 carriers: novel targets for cancer therapy. Molecules. 2020;25(10):2417. https://doi.org/10.3390/molecules25102417. Epub 20200522. PubMed PMID: 32455902; PMCID: PMC7288124.
63. O'Rourke B. Mitochondrial ion channels. Annu Rev Physiol. 2007;69:19–49. https://doi.org/10.1146/annurev.physiol.69.031905.163804. PubMed PMID: 17059356; PMCID: PMC2712118.
64. Haloi N, Wen PC, Cheng Q, Yang M, Natarajan G, Camara AKS, Kwok WM, Tajkhorshid E. Structural basis of complex formation between mitochondrial anion channel VDAC1 and

Hexokinase-II. Commun Biol. 2021;4(1):667. https://doi.org/10.1038/s42003-021-02205-y. Epub 20210603. PubMed PMID: 34083717; PMCID: PMC8175357.
65. Berry BJ, Trewin AJ, Amitrano AM, Kim M, Wojtovich AP. Use the protonmotive force: mitochondrial uncoupling and reactive oxygen species. J Mol Biol. 2018;430(21):3873–91. https://doi.org/10.1016/j.jmb.2018.03.025. Epub 20180404. PubMed PMID: 29626541; PMCID: PMC6170739.
66. Malas KM, Lambert DS, Heisner JS, Camara AKS, Stowe DF. Time and charge/pH-dependent activation of K^+ channel-mediated K^+ influx and K^+/H^+ exchange in guinea pig heart isolated mitochondria; role in bioenergetic stability. Biochim Biophys Acta Bioenerg. 2022;1863(8):148908. https://doi.org/10.1016/j.bbabio.2022.148908. Epub 20220809. PubMed PMID: 35961396.
67. Villiers C, Michejda JW, Block M, Lauquin GJ, Vignais PV. The electrogenic nature of ADP/ATP transport in inside-out submitochondrial particles. Biochim Biophys Acta. 1979;546(1):157–70. https://doi.org/10.1016/0005-2728(79)90177-4. PubMed PMID: 36139.
68. Haeckel, E. (1866). Generelle Morphologie der Organismen. Bd. 1–2. Berlin: G. Reimer
69. Garstang, W. (1922). The theory of recapitulation: a critical restatement of the biogenetic law. Proc. Linn. Soc. Zool. 35, 81–101. https://doi.org/10.1111/j.1096-3642.1922.tb00464.x

Epilogue

All life on Earth is a response to the energy from our Sun, protection from cosmic radiation, the ideal temperature and pressure of our land and seas, the volume and chemical composition of the Earth's dry and wet surfaces, and our changing atmosphere. Through an evolutionary biochemistry approach, we observe that although the natural elements are spread throughout the universe, albeit unevenly, it is their bonding capabilities in our environment that underlies the developmental diversity and specificity of molecules and compounds that first permitted life to form, or if it did not derive only from Earth's chemicals, at least to propagate and evolve.

Although most of the lighter elements participate in the make-up of lifeforms, the first two key atoms are H and O, as they exist as H^+ (ion) and liquid H_2O that allows other ions to dissociate and ions to move around as currents in a circuit. O_2 at first was toxic to new lifeforms, but life later developed a use for it. C, N, P, and S are the second group of most essential elements as they form covalent bonds to construct the chemical "backbone" structures of lifeforms.

Ionic Na, K, Mg, Ca, and Cl are essential for ionic homeostasis and cell regulation; ionic Fe, Cu, and Mn, are required for *e*- transfer, ROS scavenging, catalysis, and carriage of O_2. These metals and other light elements are essential co-factors in biochemical reactions, whereas many heavier elements are toxic to lifeforms. In all normal animal cells, mitochondria are present to metabolize foodstuffs into high-energy reducing equivalents, which then step-wise release their energy and finally oxidize carbohydrates, FFAs, protein and other substrates to CO_2 and other metabolites when there is sufficient O_2 to pick up the 4 *e*- of 2 H^+ to reduce O_2 to $2H_2O$.

It is notable to remember that the evolution of bioenergetics that was essential for energy to support life started long before there were mitochondria. For prokaryotes the likely initial essential ingredients were a semipermeable membrane with selectivity to ions (chemiosmosis), sources of energy, and substrates that could be reduced and oxidized (like CO_2 and H_2) with processing machinery like multivalent ions and *e*- transfer. It is likely that life could not have evolved substantially beyond prokaryotes without endosymbiosis, and particularly for the mitochondrion to

D. F. Stowe, *Evolution of Bioenergetics from Elements to Life*,
https://doi.org/10.1007/978-3-031-76668-8

process an abundance of substrates into an abundance of energy to drive the natural selection of complex,eukaryotic, multicellular organisms.

The mammalian heart is but a four-valve, two-part muscular pump that delivers blood containing nutrients from digestion, and O_2 inhaled from the lungs to organs and tissues in a parallel manner, while simultaneously delivering CO_2 to be exhaled by the lungs and delivery of waste products of metabolism for excretion by the body. The heart and circulatory system of mammals (and birds) allow for high blood flows at a high pressure to the organs that have the highest energy needs, like the heart, brain, and skeletal muscles during exercise; thus, it was especially important for mammals and birds to have two separate circulations. The circulatory system also delivers hormones and immune factors to locations all over the body for regulating activity and protecting against invasion by toxins and foreign antigens.

In the cells of the heart, the abundance of the tiny powerplants called mitochondria furnish the abundant energy required for the heart to pump nutrients via vessels to all tissues and organs, and importantly to high energy consuming organs. These adaptions may help animals to survive, prosper, and reproduce in an ever-changing environment.

But, as natural selection drives evolution, we can expect more modifications to our chemical structures, anatomy, and biological functions, including bioenergetic mechanisms, as our environment changes, albeit hopefully not too fast. Over the coming eons, higher and lower animals may remain relatively unchanged, morph into beings we would no longer recognize, or if we can't adapt to our changing environment, cease to exist.

Index

D. F. Stowe, *Evolution of Bioenergetics from Elements to Life*,
https://doi.org/10.1007/978-3-031-76668-8

The manufacturer's authorised representative in the EU is Springer Nature Customer Service Centre GmbH, Europaplatz 3, 69115 Heidelberg, Germany. If you have any concerns regarding our products, please contact ProductSafety@springernature.com

Printed and bound by CPI Group (UK) Ltd, Croydon, CR0 4YY

30/09/2026

02225592-0015